Selected Titles in This Series

80 **Lindsay N. Childs,** Taming wild extensions: Hopf algebras and local Galois module theory, 2000

79 **Joseph A. Cima and William T. Ross,** The backward shift on the Hardy space, 2000

78 **Boris A. Kupershmidt,** KP or mKP: Noncommutative mathematics of Lagrangian, Hamiltonian, and integrable systems, 2000

77 **Fumio Hiai and Dénes Petz,** The semicircle law, free random variables and entropy, 2000

76 **Frederick P. Gardiner and Nikola Lakic,** Quasiconformal Teichmüller theory, 2000

75 **Greg Hjorth,** Classification and orbit equivalence relations, 2000

74 **Daniel W. Stroock,** An introduction to the analysis of paths on a Riemannian manifold, 2000

73 **John Locker,** Spectral theory of non-self-adjoint two-point differential operators, 2000

72 **Gerald Teschl,** Jacobi operators and completely integrable nonlinear lattices, 1999

71 **Lajos Pukánszky,** Characters of connected Lie groups, 1999

70 **Carmen Chicone and Yuri Latushkin,** Evolution semigroups in dynamical systems and differential equations, 1999

69 **C. T. C. Wall (A. A. Ranicki, Editor),** Surgery on compact manifolds, second edition, 1999

68 **David A. Cox and Sheldon Katz,** Mirror symmetry and algebraic geometry, 1999

67 **A. Borel and N. Wallach,** Continuous cohomology, discrete subgroups, and representations of reductive groups, second edition, 2000

66 **Yu. Ilyashenko and Weigu Li,** Nonlocal bifurcations, 1999

65 **Carl Faith,** Rings and things and a fine array of twentieth century associative algebra, 1999

64 **Rene A. Carmona and Boris Rozovskii, Editors,** Stochastic partial differential equations: Six perspectives, 1999

63 **Mark Hovey,** Model categories, 1999

62 **Vladimir I. Bogachev,** Gaussian measures, 1998

61 **W. Norrie Everitt and Lawrence Markus,** Boundary value problems and symplectic algebra for ordinary differential and quasi-differential operators, 1999

60 **Iain Raeburn and Dana P. Williams,** Morita equivalence and continuous-trace C^*-algebras, 1998

59 **Paul Howard and Jean E. Rubin,** Consequences of the axiom of choice, 1998

58 **Pavel I. Etingof, Igor B. Frenkel, and Alexander A. Kirillov, Jr.,** Lectures on representation theory and Knizhnik-Zamolodchikov equations, 1998

57 **Marc Levine,** Mixed motives, 1998

56 **Leonid I. Korogodski and Yan S. Soibelman,** Algebras of functions on quantum groups: Part I, 1998

55 **J. Scott Carter and Masahico Saito,** Knotted surfaces and their diagrams, 1998

54 **Casper Goffman, Togo Nishiura, and Daniel Waterman,** Homeomorphisms in analysis, 1997

53 **Andreas Kriegl and Peter W. Michor,** The convenient setting of global analysis, 1997

52 **V. A. Kozlov, V. G. Maz′ya, and J. Rossmann,** Elliptic boundary value problems in domains with point singularities, 1997

51 **Jan Malý and William P. Ziemer,** Fine regularity of solutions of elliptic partial differential equations, 1997

For a complete list of titles in this series, visit the AMS Bookstore at **www.ams.org/bookstore/**.

Taming Wild Extensions: Hopf Algebras and Local Galois Module Theory

Mathematical
Surveys
and
Monographs

Volume 80

Taming Wild Extensions: Hopf Algebras and Local Galois Module Theory

Lindsay N. Childs

American Mathematical Society

Work on this book was supported in part by National Security Agency grant #MDA9049410114

2000 *Mathematics Subject Classification.* Primary 11R33;
Secondary 11S15, 11S31, 12F10, 14L05, 14L15, 16W30.

ABSTRACT. This book studies generalizations of the Normal Basis Theorem for Galois extensions of local fields. Noether's theorem in local Galois module theory states that a finite Galois extension of local fields has a normal integral basis if and only if the extension is tamely ramified. Noether's theorem extends to wildly ramified Galois extensions of local fields, and more generally to Hopf Galois extensions of local fields, if the associated order is a Hopf order. This book presents results related to this extension of Noether's theorem. Topics include a review of Hopf algebras and Galois and tame extensions for Hopf algebras, a classification of Hopf Galois structures on separable field extensions, construction of several classes of Hopf orders over valuation rings of local fields, criteria for the associated order of a wildly ramified extension of local fields to be a Hopf order, and the application of Kummer theory of Lubin-Tate formal groups to construct some instructive examples. Most of the results presented here have been obtained since 1984 and have not previously appeared in book form.

Library of Congress Cataloging-in-Publication Data Childs, Lindsay.

Taming wild extensions : Hopf algebras and local Galois module theory / Lindsay N. Childs.
p. cm. — (Mathematical surveys and monographs, ISSN 0076-5376 ; v. 80)
Includes bibliographical references and index.
ISBN 0-8218-2131-8 (alk. paper)
1. Hopf algebras. 2. Galois modules(Algebra) 3. Field extensions(Mathematics) I. Title.
II. Mathematical surveys and monographs ; no. 80.
QA613.8 .C48 2000
512′.55—dc21 00-029301

∞ The paper used in this book is acid-free and falls within the guidelines
established to ensure permanence and durability.
Visit the AMS home page at URL: `http://www.ams.org/`

10 9 8 7 6 5 4 3 2 1 05 04 03 02 01 00

Contents

CHAPTER 0

Introduction

Galois module theory is the branch of algebraic number theory which studies rings of integers of Galois extensions of number fields as modules over the integral group ring of the Galois group.

The title of this book comes from [**Ch87**], which suggested that Hopf algebras could fruitfully broaden the domain of Galois module theory. That suggestion has led to a body of research towards understanding how Hopf algebras relate to wildly ramified extensions of local fields. The purpose of this book is to survey this work.

We begin by reviewing the classical theory.

Let L/K be a Galois extension of algebraic number fields with Galois group G. The normal basis theorem, dating back to the 19th century, is that L has a K-basis consisting of $\{\sigma(s)|\sigma \in G\}$, the conjugates of some element s of L–a normal basis. Thus as a module over the group ring KG, L is a free module of rank one with basis s.

Let R be the ring of integers of K (or the valuation ring of K, if K is a local field), and let S be the integral closure of R in L. Then the Galois group G acts on S, and one can ask if S has a normal basis as a free R-module (alternate terminology: L/K has a normal integral basis). Of course the first prerequisite is that S be a free R-module. In the local case, i.e. if K is a local field, this holds, but in the global case S need not be free over R.

The two classic theorems in Galois module theory are the Hilbert-Speiser Theorem, Satz 132 of Hilbert's Zahlbericht [**Hi97**], and Noether's Theorem [**No32**]. Noether's theorem is that in the local case, S has a normal basis over R if and only if L/K is tamely ramified (or, for short, "tame"): if $e_{L/K}$ is the ramification index of the extension L/K, then L/K is tamely ramified if $e_{L/K}$ is relatively prime to the characteristic of the residue field of R. Two equivalent characterizations of tameness are:

1) the trace map from S to R, $tr(s) = \sum_{\sigma\in G} \sigma(s)$, is onto,

and

2) the natural action of the group ring RG on S makes S into a projective RG-module.

The difficult part of Noether's theorem is that if R is the valuation ring of a local field K, and M is a finitely generated, projective RG-module so that $K \otimes_R M$ is KG-free, then M is RG-free. Thus the first really satisfactory proof of Noether's theorem was given only in 1960 by R. G. Swan [**Sw60**].

The Hilbert-Speiser Theorem asserts that if $K = \mathbb{Q}$ and L/K is any tamely ramified abelian extension, then the ring of integers S of L has a normal basis as a free $\mathbb{Z}$-module, or equivalently, S is a free $\mathbb{Z}G$-module of rank one. (It was recently proven [**GRRS 99**] that $\mathbb{Q}$ is the only base field K over which every tame abelian extension has a normal integral basis.)

The case of wild(ly ramified) extensions remained untouched by these results. Since wild extensions include all ramified Galois extensions of a local field K containing $\mathbb{Q}_p$ where the Galois group is a p-group, this was a substantial omission.

The first new approach to wild extensions was by Leopoldt [**Le59**], who proposed replacing the group ring RG by a larger order. More precisely, let L/K be a Galois extension of local or global number fields, with rings of integers (or valuation rings) S/R and Galois group G. Define the associated order of S in KG to be

$$\mathfrak{A} = \{\alpha \in KG | \alpha(s) \in S \text{ for all } s \in S\}.$$

Since G acts on S (elements of G send elements integral over R to elements integral over R), $RG \subset \mathfrak{A}$, but if L/K is wildly ramified, then $\mathfrak{A}$ will be larger than RG, but still an order over R in KG: that is, a subring of KG which is a finitely generated R-module.

If L/K is tame, then $\mathfrak{A} = RG$.

Leopoldt obtained a very satisfying generalization of the Hilbert-Speiser theorem, namely, if $L/\mathbb{Q}$ is any abelian extension, then the ring of integers S of L is a free $\mathfrak{A}$-module of rank one. This implies the corresponding result for abelian extensions of $\mathbb{Q}_p$.

However, it was soon found that if one varies the hypotheses: replace $\mathbb{Q}$ or $\mathbb{Q}_p$ by a more general base field, or allow G to be non-abelian, then even the local result is not valid. (However for extensions L/K with $L/\mathbb{Q}_p$ abelian, the local result remains true [**Lt98**].)

The observation of [**Ch87**] for abelian extensions and [**CM94**] in general, was that for wild Galois extensions, if the associated order $\mathfrak{A}$ is a Hopf order in KG, then S is free of rank one over $\mathfrak{A}$. Noether's theorem is the case where $\mathfrak{A} = RG$. This generalized Noether's theorem was perhaps the first general integral normal basis theorem for wildly ramified Galois extensions of local fields. (The converse is false: see (12.8).)

At about the same time, Greither and Pareigis [**GP87**] showed how ubiquitous Hopf Galois structures were on Galois, and even separable but non-Galois, extensions of local or global fields. Their observation expands the horizon greatly. For suppose L/K is a H-Hopf Galois extension of number fields, for some cocommutative K-Hopf algebra H (the terminology means that L is an H-module algebra). If S is the ring of integers of L, one can define the associated order of S in H in the same way as for $H = KG$: namely,

$$\mathfrak{A} = \{\alpha \in H | \alpha(s) \in S \text{ for all } s \in S\}.$$

Then the same result holds: if $\mathfrak{A}$ is a Hopf order in H, then S is free of rank one over $\mathfrak{A}$.

If the Galois group G of a Galois extension L/K is not cyclic of square-free order, then L/K has Hopf Galois structures other that given by the Galois group KG. For each Hopf algebra action $H \otimes L \to L$ which makes L/K into a Hopf Galois extension, there is an associated order $\mathfrak{A}_H$. Which is the "best "order? It need not be $\mathfrak{A}_{KG}$. N. Byott [**By99b**] has given a class of examples of Galois extensions of local fields for which the ring of integers S of L is not free over $\mathfrak{A}_{KG}$ but $\mathfrak{A}_H$ is a Hopf order for some other K-Hopf algebra H, and hence S is free over $\mathfrak{A}_H$: his examples, presented in section 40, conclude this work.

Chapter Summary. The main purpose of this book is to present mathematics related to the generalized Noether theorem cited above. Here is a brief description of the twelve chapters of this work.

Chapter 1 is an introduction to the basic concepts of Hopf algebras and Hopf Galois extensions. It includes an extended discussion of exact sequences of Hopf algebras, adapted from [**Sw69**] and [**Mt93**], and some results on Hopf orders and their integrals in section 5.

Chapter 2 surveys the Greither-Pareigis [**GP87**] classification of Hopf Galois structures on separable field extensions. Their classification transforms the problem of classifying Hopf Galois structures on a Galois extension into a purely group-theoretic problem involving the Galois group G, which, however, is often unmanagable. A reformulation by Byott [**By96a**] breaks up the problem into a possibly unmanageable collection of manageable problems. We then give Byott's classification of Galois extensions for which the classical Galois structure is the unique Hopf Galois structure, and survey results on the number of Hopf Galois structures on Galois extensions with Galois group G for various G, including cyclic p-groups [**Ko98**], and symmetric, alternating and simple groups [**CC99**].

Associated orders and the concept of tame H-extension, from [**CH86**], are considered in Chapter 3. Once one proves that an extension S/R is H-tame iff S is H-projective, then Noether's theorem comes down to showing that if A is a Hopf order over the valuation ring R of a local field K in some cocommutative K-Hopf algebra H, and P and Q are finitely generated projective left A-modules such that $K \otimes_R P \cong K \otimes_R Q$, then $P \cong Q$. This result, found for $H = KG$ in [**Sw60**] and also in [**CR81**], and proved in general by Schneider [**Sch77**], requires a very substantial amount of representation theory, so the proof is omitted. We include a proof, due to Waterhouse [**Wa92**] that if H is local and S/R is H-tame, then S/R is H- Galois.

In attempting to apply the generalized Noether theorem it has been helpful to understand the possible Hopf algebras over the valuation ring of a local field K which can be associated orders of valuation rings of extensions of K. Chapter 4, together with chapters 5 and 9, describe some families of Hopf orders over valuation rings of local fields. Chapter 4 presents the classification of Hopf algebras of rank p over valuation rings of local fields, due to Tate and Oort [**TO70**]. Mazur [**Mz70**] noted that describing all the possible abelian group schemes over the valuation ring of a local field "is a delicate matter", citing [**TO70**] as evidence. His remark remains valid 30 years later.

Chapter 5 presents Larson's construction [**La76**] of Hopf algebra orders in group rings which are defined by group valuations. Larson's construction remains the only general strategy for constructing Hopf orders in group rings of non-abelian groups. Section 20 obtains a class of examples of K-Hopf algebras which contain no Hopf orders at all.

Chapter 6 considers cyclic Galois extensions L/K of degree p, prime, of local number fields containing $\mathbb{Q}_p$ [**Ch87**], [**Gr92**]. In this case, a congruence condition on the ramification number, or break number, characterizes when the associated order of the valuation ring of L is a Hopf order.

Chapter 7, from [**Ch88**], considers L/K an extension of number fields, and briefly discusses orders over R in L, other than the maximal order, which may have associated orders which are Hopf.

Chapter 8 gives two theorems of Byott: one, from [**By95b**], giving conditions on when tame implies Galois, the other, from [**By97c**], giving necessary congruence conditions on the break numbers, or ramification numbers, of a totally and wildly ramified Galois extension of local fields for the associated order to be Hopf. Our treatment is slightly different than that in the literature.

In [**Gr92**], Greither classified (under suitable hypotheses) the R-Hopf orders in KG, where R is the valuation ring of a local field K containing $\mathbb{Q}_p$ and G has order p^2. In Chapter 9 we give a direct construction of these Greither algebras, following [**GC98**].

Chapter 10, from [**Gr92**] and [**Ch95**], examines cyclic Galois extensions of local fields L/K of degree p^2 with suitable ramification numbers, and determines when the associated order is Hopf. The results show that the necessary congruence condition on ramification numbers for the associated order to be Hopf found in Chapter 8 is far from sufficient. Also in Chapter 10 is a classification of Greither orders which are realizable, that is, can be the associated order of the valuation ring of some cyclic Galois extension L/K of order p^2.

The results in Chapter 10, as also those in Chapter 6, depend on a description of the Hopf algebra orders over R. Thus much effort in recent years has gone into describing Hopf algebras over valuation rings of local fields. In chapter 11 we present one general strategy for constructing Hopf algebras, namely, via formal groups. The Oort Embedding Theorem states that any local abelian R-Hopf algebra arises from an isogeny of formal groups. Thus Chapter 11 introduces formal groups, reviews Lubin-Tate and other examples of formal groups, and discusses this result.

Chapter 12 then presents the Kummer theory of formal groups, from [**CM94**], [**Mo94**], [**Mo96**], which shows that if a Hopf algebra H is constructed as in Chapter 11, then the principal homogeneous spaces for H, or equivalently, the Galois extensions for H^*, are easily and explicitly described. The results of chapters 8 and 12 are then combined to produce Byott's examples from [**By99b**], cited above.

We have not included other results on constructing Hopf algebras over valuation rings: among recent works are results of Underwood [**Un96**] constructing Hopf orders in KC_{p^3}, and a memoir by the author with Greither, Moss, Sauerberg and Zimmermann [**CGMSZ98**], and a thesis of H. Smith [**Sm97**], constructing Hopf orders in group rings of elementary abelian p-groups.

It will be evident to a reader of this work that while much progress has been made in the past 15 years, there are many open questions which need to be answered before this subject is well-understood. Perhaps this exposition will help act as a resource and motivation for further work in the field.

Beyond a general background in algebra at the level of [**Lg84**], the main prerequisite for this book is an acquaintance with some algebraic number theory. Serre's classic book, *Local Fields* (cited as [**CL**]), is a useful reference for local number theory.

This book focuses on results over local fields and their valuation rings. For a nice survey of some global results, including connections to L-functions, see [**TB92**].

The author wishes to thank his students and collaborators over the years for their inspiration, motivation and mathematical insights, particularly Susan Hurley, Steve Tesser, David Weinraub, Rob Underwood, David Moss, Maureen Cox, Alan Koch, Tim Kohl, Hal Smith, ManYiu Tse, and Scott Carnahan, Cornelius Greither, Jim Sauerberg and Karl Zimmermann. Thanks to Ellen Fisher for her assistance in typing parts of the manuscript. Special thanks go to Nigel Byott for his careful reading of the manuscript, and for his numerous contributions to the subject: his influence on the mathematics presented in this book has already been noted, and will become clearer in subsequent chapters.

Albany, NY
February, 2000

CHAPTER 1

Hopf Algebras and Galois Extensions

§**1. Basic definitions and notation.** Let R be a commutative ring with unity. We are interested in R-Hopf algebras H and R-algebras on which they act. Unadorned tensors are over R.

An *R-algebra* A is a ring with unity together with a ring homomorphism $\iota : R \to A$ whose image is contained in the center of A (= the set of elements of A which commute with every element of A). Then A becomes an R-module via the scalar multiplication $r \cdot a = \iota(r)a$, or which is the same, since ι maps to the center of A, via $a \cdot r = a\iota(r)$.

An alternative way of thinking of A is as an R-module together with a multiplication map $\mu : A \otimes_R A \to A$ and a unit map $\iota : R \to A$ so that μ is associative:

$$\begin{array}{ccc} A \otimes A \otimes A & \xrightarrow{\mu \otimes 1} & A \otimes A \\ {\scriptstyle 1\otimes\mu}\downarrow & & \downarrow{\scriptstyle \mu} \\ A \otimes A & \xrightarrow[1\otimes\mu]{} & A \end{array}$$

commutes, and the map $\mu \circ (1 \otimes \iota) : A \otimes_R R \to A \otimes_R A \to A$ is the same as the R-module multiplication map $A \otimes_R R \to A$, and also $\mu \circ (\iota \otimes 1)$ is the same as scalar multiplication: $R \otimes_R A \to A$:

$$(1.1) \qquad \begin{array}{ccc} A \otimes_R R & \xrightarrow{1\otimes\iota} & A \otimes_R A \\ \| & & \downarrow{\scriptstyle \mu} \\ A \otimes_R R & \xrightarrow{\textit{scalar mult.}} & A \end{array}$$

and

$$\begin{array}{ccc} R \otimes_R A & \xrightarrow{\iota\otimes 1} & A \otimes_R A \\ \| & & \downarrow{\scriptstyle \mu} \\ R \otimes_R A & \xrightarrow{\textit{scalar mult.}} & A \end{array}$$

commute.

An *R-bialgebra* H is an R-algebra with the following additional structure:

$$\Delta : H \to H \otimes H \quad \text{(comultiplication)}$$
$$\varepsilon : H \to R \quad \text{(counit)}$$

Here Δ and ε are R-algebra homomorphisms, where $H \otimes H$ has "componentwise" multiplication

$$(h \otimes j)(h' \otimes j') = hh' \otimes jj'.$$

These maps satisfy the following properties:
Coassociativity: The diagram

$$\begin{array}{ccc} H & \xrightarrow{\Delta} & H \otimes H \\ {\scriptstyle\Delta}\downarrow & & \downarrow{\scriptstyle\Delta\otimes 1} \\ H \otimes H & \xrightarrow[1\otimes\Delta]{} & H \otimes H \otimes H \end{array}$$

commutes: this is the "dual" diagram to that for associativity of the multiplication map μ;
Counitary: the diagrams

$$\begin{array}{ccc} H & \xrightarrow{\Delta} & H \otimes H \\ \| & & \downarrow{\scriptstyle 1\otimes\varepsilon} \\ H & \xleftarrow[\mu]{} & H \otimes R \end{array}$$

and

$$\begin{array}{ccc} H & \xrightarrow{\Delta} & H \otimes H \\ \| & & \downarrow{\scriptstyle \varepsilon\otimes 1} \\ H & \xleftarrow[\mu]{} & R \otimes H \end{array}$$

commute, where μ is the multiplication map: these are the dual properties to the properties (1.1) satisfied by μ and ι, described above.

An *R-coalgebra* is an R-module together with R-module homomorphisms Δ and ε satisfying the coassociativity and counitary properties.

Define $\tau : H \otimes H \to H \otimes H$ to be the *switch* map, $\tau(h_1 \otimes h_2) = h_2 \otimes h_1$.

An R-bialgebra is an *R-Hopf algebra* if there is an R-module homomorphism

$$\lambda : H \to H \text{ (antipode)}$$

which is both an R-algebra and an R-coalgebra antihomomorphism:

$$\lambda(h \otimes h') = \lambda(h') \otimes \lambda(h) \text{ and}$$
$$\Delta\lambda(h) = (\lambda \otimes \lambda)\tau\Delta,$$

and satisfies:

(1.2) ANTIPODE PROPERTY. $\mu(1 \otimes \lambda)\Delta = \iota\varepsilon$ and $\mu(\lambda \otimes 1)\Delta = \iota\varepsilon$.

An R-Hopf algebra, H is *cocommutative*, if $\tau\Delta = \Delta$, and *commutative* if H is commutative as an algebra. A Hopf algebra H is *abelian* if H is both commutative and cocommutative.

We adopt Sweedler's [**Sw69**] notation

$$\Delta(h) = \sum_{(h)} h_{(1)} \otimes h_{(2)} \in H \otimes H \ .$$

By coassociativity, $(\Delta \otimes 1)\Delta(h) = (1 \otimes \Delta)\Delta(h)$ so we denote either by

$$\sum_{(h)} h_{(1)} \otimes h_{(2)} \otimes h_{(3)},$$

etc. Cocommutativity becomes the condition:

$$\sum_{(h)} h_{(1)} \otimes h_{(2)} = \sum_{(h)} h_{(2)} \otimes h_{(1)}.$$

Sweedler notation is somewhat mysterious, but is remarkably efficient at deriving properties of Hopf algebras by looking at what happens to elements.

EXAMPLES. (1.3): THE GROUP RING. (1.3). The classical example of an R-Hopf algebra is $H = RG$, the group ring of a finite group G. Since Δ, ε and λ are R-linear homomorphisms, they are uniquely determined by their values on elements of G, which are:

$$\begin{aligned}\Delta(\sigma) &= \sigma \otimes \sigma \\ \varepsilon(\sigma) &= 1 \\ \lambda(\sigma) &= \sigma^{-1}\end{aligned}$$

for σ in G. The group ring RG is evidently cocommutative.

(1.4): THE DUAL OF A HOPF ALGEBRA. If H is an R-Hopf algebra which is a finitely generated and projective R-module, then the linear dual $H^* = Hom_R(H, R)$ is an R-Hopf algebra, where the structure maps on H^* are induced from H as follows:

$$\begin{aligned}\Delta(f) &= f \circ \mu \text{ that is, } \Delta(f)(h \otimes j) = f(hj), \\ \varepsilon(f) &= f(1), \\ \mu(f \otimes f')(h) &= (f \otimes f') \circ \Delta_H(h) \\ &= \sum_{(h)} f(h_{(1)}) f'(h_{(2)}); \\ i(r) &= r\varepsilon_H \ ; \\ \lambda(f)(h) &= f(\lambda_H(h)) \ .\end{aligned}$$

Then H^* is an R-Hopf algebra because the diagrams which describe the coassociativity and counitary properties of H translate into the associativity and unitary properties for H^* and vice versa.

Evidently H is cocommutative, resp. commutative, iff H^* is commutative, resp. cocommutative.

If H is a finitely generated projective R-module, then H^{**} is naturally isomorphic to H. In that case, for $f \in H^*, h \in H$, we may view f as a function on H, or

h as a function on H^* via the map from H to H^{**}. To avoid choosing a point of view, we will often use the map

$$\langle\ ,\ \rangle : H^* \otimes H \to R$$

by $\langle f, h\rangle = f(h)$.

For $H = RG$, G a finite group, $H^* = \mathrm{Hom}_R(RG, R)$. Let $\{e_\sigma : \sigma \in G\}$ be the dual basis for $\{\sigma \in G\}$: $\langle e, \sigma\rangle = e_\sigma(\tau) = \delta_{\sigma,\tau}$ for σ, τ in G. Then for any ρ in G, $(e_\sigma \cdot e_\tau)(\rho) = (e_\sigma \otimes e_\tau)(\rho \otimes \rho) = \delta_{\sigma\rho}\delta_{\tau\rho}$, so $e_\sigma e_\tau = \delta_{\sigma\tau} e_\sigma$ and the (e_σ) are pairwise orthogonal idempotents; also the identity of H^* is $1 = \sum_{\sigma\in G} e_\sigma$. Then $\Delta(e_\sigma)(\tau \otimes \rho) = e_\sigma(\tau\rho) = \delta_{\sigma,\tau\rho}$, from which one can verify easily that

$$\Delta(e_\sigma) = \sum_{\tau\rho=\sigma} e_\tau \otimes e_\rho\ ;$$

also,

$$\varepsilon(e_\sigma) = \delta_{\sigma,1} \text{ and } \lambda(e_\sigma) = e_{\sigma^{-1}}\ .$$

Note that H^* is always commutative, but is cocommutative iff G is abelian.

Here are all the rank 2 examples:

(1.5) PROPOSITION ([**Kr82**], [**Wa79, p. 19**]). *Let R be a domain. Let H be an R-Hopf algebra which is free of rank 2 over R. Then $H \cong R[x]/(x^2 - bx)$ with $\Delta(x) = x \otimes 1 + 1 \otimes x + a(x \otimes x)$, $\varepsilon(x) = 0$ and $\lambda(x) = x$, where $ab = -2$.*

This result is a special case of the Tate-Oort classification of Hopf algebras of prime rank p [**TO70**], which we will present in Chapter 4. However, the case $p = 2$ is easy enough to prove independently.

PROOF. Let $M = ker\ \varepsilon$. Then $H = R \cdot 1 \oplus M$ as R-modules, and since H is R-free, the second exterior power of the R-module H, $\Lambda^2 H$, is R-free. But $\Lambda^2 H \cong M$, hence M is free, $M = Rx$ for some x with $\varepsilon(x) = 0$, and H is a free R-module with basis $\{1, x\}$. Since ε is an algebra homomorphism, $\varepsilon(x^2) = \varepsilon(x)^2 = 0$, so x^2 is in M. Hence $x^2 = bx$ for some $b \in R$.

Now $\varepsilon \circ \lambda = \varepsilon$, and so $\lambda(x) \in M$, $\lambda(x) = cx$ for some $c \in R$. Also, if

$$\Delta(x) = p(1 \otimes 1) + q(x \otimes 1) + r(1 \otimes x) + a(x \otimes x)$$

for some p, q, r, $a \in R$, then

$$\begin{aligned} x &= (1 \otimes \varepsilon)\Delta(x) = p + qx \\ &= (\varepsilon \otimes 1)\Delta(x) = p + rx\ . \end{aligned}$$

Since $\{1, x\}$ is an R-basis of H, $p = 0$ and $q = r = 1$; hence

$$\Delta(x) = x \otimes 1 + 1 \otimes x + a(x \otimes x)$$

for some $a \in R$.

Now $\mu(1 \otimes \lambda)\Delta = \iota\varepsilon$, so

$$0 = \iota\varepsilon(x) = x + cx + acx^2 = (1 + c)x + acbx$$

which implies that $1 + c + acb = 0$.

Also, Δ is an algebra homomorphism, so

$$\begin{aligned} b\Delta(x) &= \Delta(x^2) = \Delta(x)^2, \text{ hence} \\ b(x \otimes 1 + 1 \otimes x &+ a(x \otimes x)) \\ &= (x \otimes 1 + 1 \otimes x + a(x \otimes x))^2 \\ &= b(x \otimes 1) + b(1 \otimes x) + a^2b^2(x \otimes x) \\ &+ 2(x \otimes x) + 2ab(x \otimes x) + 2ab(x \otimes x)\ , \end{aligned}$$

which implies that $ab = a^2b^2 + 2 + 4ab$, or $a^2b^2 + 3ab + 2 = 0$, hence $ab = -2$ or $ab = -1$. Since $abc = -c - 1$, $ab \neq -1$; hence $ab = -2$, then $c = 1$, so $\lambda(x) = x$. $\square$

We shall see many more examples of Hopf algebras later.

In these notes R-Hopf algebras are usually finitely generated and projective as R-modules, in which case we say that H is a *finite* R-Hopf algebra.

We observed (Example (1.3)) that if $H = KG$, G a group, then for all σ in G, $\Delta(\sigma) = \sigma \otimes \sigma$. This comultiplication on group elements of a group ring generalizes to the following definition:

A non-zero element h in H is *grouplike* if $\Delta(h) = h \otimes h$.

(1.6) PROPOSITION. *If R has no idempotents but 0 and 1 and h is a grouplike element of H, then h satisfies $\varepsilon(h) = 1$. The set $G(H)$ of grouplike elements of H is a subgroup of the multiplicative group of units of H, with identity element 1.*

PROOF. Let h be a grouplike element of H, then by counitarity,

$$h = \mu(1 \otimes \varepsilon)\Delta(h) = h\varepsilon(h)$$

hence $\varepsilon(h) = \varepsilon(h)\varepsilon(h)$. Thus $\varepsilon(h)$ is an idempotent of R, hence $\varepsilon(h) = 0$ or 1. But $h \neq 0$ and $h = h\varepsilon(h)$, so $\varepsilon(h) = 1$.

Since Δ is an algebra homomorphism, if $h, j \in G(H)$ then

$$\Delta(hj) = \Delta(h)\Delta(j) = (h \otimes h)(j \otimes j) = hj \otimes hj\ ,$$

so $G(H)$ is closed under multiplication. Also, $G(H)$ is closed under the antipode λ. For if $h \in G(H)$, since $\Delta \cdot \lambda = (\lambda \otimes \lambda) \cdot \tau \cdot \Delta$, then $\Delta(\lambda(h)) = \lambda(h) \otimes \lambda(h)$. Finally, for $h \in G(H)$, $\lambda(h) = h^{-1}$, since $1 = \iota\varepsilon(h) = \mu(1 \otimes \lambda)\Delta(h) = h\lambda(h)$, and so the elements of $G(H)$ are units of H. $\square$

REMARK. An element f in $H^* = Hom_R(H, R)$ is grouplike iff f is an R-algebra homomorphism from H to R. For

$$\Delta(f)(x \otimes y) = f(xy).$$

Thus $\Delta(f) = f \otimes f$ iff $f(xy) = f(x)f(y)$ for all $x, y \in H$.

(1.7) PROPOSITION. *If R is a field, then distinct non-zero grouplike elements of H are linearly independent over R.*

PROOF. Let $h_1, \dots, h_n$ be linearly independent grouplike elements, and suppose h is a grouplike element which is a linear combination of $h_1, \dots, h_n$. Then

$$h = \sum r_i h_i$$

for some $r_1, \dots, r_n$ in R. Applying Δ gives

$$h \otimes h = \sum_i r_i (h_i \otimes h_i),$$

while substituting for h in $h \otimes h$ gives

$$h \otimes h = \sum_{i,j} r_i r_j (h_i \otimes h_j).$$

So

$$\sum_i r_i (h_i \otimes h_i) = \sum_{i,j} r_i r_j (h_i \otimes h_j).$$

Since $h_1, \dots, h_n$ are linearly independent in H, $\{h_i \otimes h_j\}$ are linearly independent in $H \otimes H$. So

$$r_i r_j = 0 \text{ for } i \neq j$$

and

$$r_i^2 = r_i \text{ for } i = 1, \dots, n.$$

Then $r_i = 1$ for at most one i, and so $h = 0$ or $h = h_i$ for some i. □

Convolution. The antipode can be better understood in terms of the convolution product.

(1.8) DEFINITION. Let C be an R-coalgebra, A an R-algebra. Then $Hom_R(C, A)$ has an operation, convolution, defined by

$$f * g = \mu(f \otimes g)\Delta$$

for $f, g \in Hom_R(C, A)$, or, for $c \in C$,

$$(f * g)(c) = \sum_{(c)} f(c_{(1)}) g(c_{(2)}).$$

Associativity of $*$ follows easily from the associativity of A and the coassociativity of C. Also, $\iota\varepsilon$ is the identity under $*$, where ε is the counit of C and $\iota : R \to A$ the unit map.

Let $C = A = H$, an R-Hopf algebra. Then $\lambda \in End_R(H)$ may be identified as the inverse of the identity map under convolution, for the identities

$$I * \lambda = \mu(I \otimes \lambda)\Delta = \iota\varepsilon$$

and

$$\lambda * I = \mu(\lambda \otimes I)\Delta = \iota\varepsilon$$

are restatements of the antipode property (1.2).

(1.9) DEFINITION. An *R-Hopf algebra homomorphism* $f : H \to H'$ is an R-module homomorphism which respects $\iota, \mu, \varepsilon, \Delta$ and λ: for example, $\Delta(f(x)) = (f \otimes f)\Delta(x)$.

For abelian (i.e. commutative and cocommutative) Hopf algebras the convolution yields some useful structure on the set of Hopf algebra endomorphisms of H.

(1.10) PROPOSITION. *Let H be an abelian R-Hopf algebra. Then $End_{R-Hopf}(H)$, the set of R-Hopf algebra endomorphisms of H, is an abelian group under convolution, and a ring under convolution and composition.*

PROOF. If $f, g \in End_{R-Hopf}(H)$ it is routine to verify that $f*g$ is an R-algebra homomorphism and that for $h \in H$,

$$\Delta(f * g)(h) = ((f * g) \otimes (f * g))\Delta(h),$$

$$\varepsilon(f * g)(h) = \varepsilon(h),$$

and

$$(f * g)\lambda(h) = \lambda(f * g)(h) :$$

commutativity and cocommutativity of H are needed, especially in the last identity. Thus $End_{R-Hopf}(H)$ is closed under $*$.

To check that $End_{R-Hopf}(H)$ under $*$ is an abelian group, we note that since H is abelian, $*$ is commutative, while associativity was noted above. The identity $\iota\varepsilon$ is a Hopf algebra morphism since ι and ε are; and the convolution inverse of $f \in End_{R-Hopf}(H)$ is $f\lambda = \lambda f$:

$$\begin{aligned}(f * f\lambda)(h) &= \sum f(h_{(1)})f(\lambda(h_{(2)})) \\ &= \sum f(h_{(1)}\lambda(h_{(2)})) \\ &= f(\varepsilon(h)) \\ &= \varepsilon(h),\end{aligned}$$

so $End_{R-Hopf}(H)$ is an abelian group.

Now $End_{R-Hopf}(H)$ is closed under composition and the identity element under composition is I, the identity map from H to itself. To show that $End_{R-Hopf}(H)$ is a ring under $*$ and composition, we need to check the two distributive laws,

$$f \circ (g' * g'') = (f \circ g') * (f \circ g''),$$

which follow easily since f is an algebra homomorphism, and

$$(f' * f'') \circ g = (f' \circ g) * (f'' \circ g),$$

which follows easily since g is a coalgebra homomorphism. □

We will revisit this result in Section 16.

Finally, we remark:

(1.11) PROPOSITION. *If H is commutative or cocommutative, then $\lambda \circ \lambda = I$.*

PROOF. Since $\lambda(r) = r$ for $r \in R$, we have for $h \in H$,

$$\iota\varepsilon(h) = \lambda\iota\varepsilon(h) = \lambda(\mu(\lambda \otimes 1)\Delta(h)) = \sum \lambda^2(h_{(2)})\lambda(h_{(1)}).$$

If H is commutative, this is $(\lambda * \lambda^2)(h)$; If H is cocommutative, this is $(\lambda^2 * \lambda)(h)$. If $\iota\varepsilon = \lambda * \lambda^2$, then

$$I = I * \iota\varepsilon = I * \lambda * \lambda^2 = \iota\varepsilon * \lambda^2 = \lambda^2.$$

If $\iota\varepsilon = \lambda^2 * \lambda$, then

$$I = \iota\varepsilon * I = \lambda^2 * \lambda * I = \lambda^2 * \iota\varepsilon = \lambda^2.$$

□

If H is neither commutative nor cocommutative, then λ^2 need not be the identity: see [**Mt93, 1.5.6**] for an example.

§**2. Hopf algebras and R-algebras.** Let H be an R-Hopf algebra and S an R-algebra which is an H-module. S is an H-module algebra if $h(st) = \sum h_{(1)}(s)h_{(2)}(t)$ and $h(1) = \varepsilon(h)1$ for all $h \in H$, $s, t \in S$.

If we view $S \otimes S$ as an H-module via Δ, that is,

$$h(s \otimes t) = \sum_{(h)} h_{(1)}(s) \otimes h_{(2)}(t),$$

and R as an H-module via ε, then S is an H-module algebra iff the multiplication $\mu_S : S \otimes S \to S$ and the unit map $\iota_S : R \to S$ are H-module homomorphisms.

Let H be a Hopf algebra, S an R-module. S is a right H-comodule if there is an R module homomorphism

$$\alpha : S \to S \otimes_R H$$

such that $(\alpha \otimes 1)\alpha = (1 \otimes \Delta)\alpha$ (coassociativity) and $(1 \otimes \varepsilon)\alpha = id : S \to S \otimes H \to S \otimes_R R \cong S$ (counitary).

The Sweedler notation is adapted to comodule actions as

$$\alpha(s) = \sum_{(s)} s_{(0)} \otimes s_{(1)}, \ s_{(0)} \in S, s_{(1)} \in H.$$

Then coassociativity becomes

$$(\alpha \otimes 1)\alpha(s) = (1 \otimes \Delta)\alpha(s) = \sum_{(s)} s_{(0)} \otimes s_{(1)} \otimes s_{(2)}.$$

Call S an H-comodule algebra if S is an R-algebra and μ_S and ι_S are H-comodule homomorphisms. Here we view R as a right H-comodule via $\iota : R \to H$:

$$\alpha(r) = r \otimes \iota(1_R) = r \otimes 1_H$$

and $S \otimes S$ as a right H-comodule via the multiplication on H: for $s, t \in S$,

$$\alpha(s \otimes t) = \sum_{(s),(t)} s_{(0)} \otimes t_{(0)} \otimes \mu_H(s_{(1)} \otimes t_{(1)}).$$

Thus μ is an H-comodule morphism iff $\alpha(st) = \sum s_{(0)}t_{(0)} \otimes s_{(1)}t_{(1)} = \alpha(s)\alpha(t)$: that is, iff α is an R-algebra homomorphism.

If S is a right H-comodule, then S becomes a left H^*-module via

$$f \cdot s = \sum_{(s)} s_{(0)} \langle f, s_{(1)} \rangle \tag{2.1}$$

for $f \in H^*$; if S is a right H-comodule algebra, then S becomes a left H^*-module algebra:

$$\begin{aligned} f \cdot (st) &= \sum_{(s)(t)} s_{(0)}t_{(0)} \langle f, s_{(1)}t_{(1)} \rangle \ , \\ &= \sum_{(f)} \sum_{(s)(t)} s_{(0)}t_{(0)} \langle f_{(1)}, s_{(1)} \rangle \langle f_{(2)}, t_{(1)} \rangle \ , \\ &= \sum_{(f)(s)(t)} s_{(0)} \langle f_{(1)}, s_{(1)} \rangle t_{(0)} \langle f_{(2)}, t_{1)} \rangle = \sum_{(f)} (f_{(1)} \cdot s)(f_{(2)} \cdot t) \ . \end{aligned}$$

(2.2). If S is a left H-module and H is a finite R-Hopf algebra then S is a right H^*-comodule as follows. Let $\{f_i, h_i\}_{i=1}^n$ be a projective coordinate system, so that for all $h \in H$,

$$h = \sum_{i=1}^{n} \langle f_i, h \rangle h_i.$$

Then for all $s \in S$,

$$\alpha(s) = \sum_{i=1}^{n} h_i s \otimes f_i.$$

Various properties should be checked. For example, the coassociativity property,

$$(\alpha \otimes 1)\alpha(s) = (1 \otimes \Delta)\alpha(s) \text{ for all } s \in S,$$

follows from the associativity of the H-action on S, as follows:

$$\alpha(h_i s) = \sum_{j} h_j(h_i s) \otimes f_j,$$

so

$$(\alpha \otimes 1)\alpha(s) = \sum_{i,j} h_j(h_i s) \otimes f_j \otimes f_i,$$

while since

$$\Delta(f_i) = \sum_{(f_i)} f_{i(1)} \otimes f_{i(2)},$$

we have

$$(1 \otimes \Delta)\alpha(s) = \sum_{i,(f_i)} h_i s \otimes f_{i(1)} \otimes f_{i(2)}$$

both in $S \otimes H^* \otimes H^*$. For $a, b \in H$, we compute $((\alpha \otimes 1)\alpha(s))(a \otimes b)$ and $((1 \otimes \Delta)\alpha(s))(a \otimes b)$:

$$\begin{aligned}((\alpha \otimes 1)\alpha(s))(a \otimes b) &= \sum_{i,j} h_j(h_i s) \otimes \langle f_j \otimes f_i, a \otimes b\rangle \\ &= \sum h_j(h_i(s))\langle f_j, a\rangle\langle f_i, b\rangle \\ &= \sum h_j(h_i(s)\langle f_i, b\rangle\langle f_j, a\rangle \\ &= (\sum\langle f_j, a\rangle h_j(\sum\langle f_i, b\rangle h_i(s))) \\ &= a(bs)\end{aligned}$$

since $\{f_i, h_i\}$ is a projective coordinate system, while

$$\begin{aligned}(1 \otimes \Delta)(\alpha(s))(a \otimes b) &= (\sum h_i s \otimes \langle f_{i(1)} \otimes f_{i(2)}, a \otimes b\rangle \\ &= \sum(h_i s)\langle f_i, ab\rangle \\ &= (ab)s \quad .\end{aligned}$$

It is worthwhile to check that the processes of going from the H-module action to the H^*-comodule action and the opposite are inverse operations: given an H-module action on S, the induced coaction α is given by $\alpha(s) = \sum_{i=1}^n h_i(s) \otimes f_i$ for some projective coordinate system $\{h_i, f_i\}$ for H (and for H^*). Then for any $h \in H, s \in S$, the induced H-module action on S is given by

$$\begin{aligned}h(s) &= \sum_i h_i(s)\langle f_i, h\rangle \\ &= \sum_i \langle f_i, h\rangle h_i s = hs,\end{aligned}$$

so the induced H-module action is the original H-module action. Also, if $s \mapsto \sum s_{(0)} \otimes s_{(1)}$ is the original H^*-comodule structure on S, then the induced H-module action on S is

$$h(s) = \sum s_{(0)}\langle s_{(1)}, h\rangle,$$

so the induced H^*-comodule structure is

$$\begin{aligned}\alpha(s) &= \sum_i h_i(s) \otimes f_i \\ &= \sum_{i,(s)} s_{(0)}\langle s_{(1)}, h_i\rangle \otimes f_i \\ &= \sum_{(s)} s_{(0)} \sum_i \langle s_{(1)}, h_i\rangle \otimes f_i \\ &= \sum_{(s)} s_{(0)} \otimes s_{(1)} \ .\end{aligned}$$

We leave as an exercise the verification that if the H- action is an H-module algebra action on S then S becomes an H^*-comodule algebra, and conversely.

(2.3) EXAMPLE. Let H be a finite R-Hopf algebra with dual H^*, then H^* is a right H^*-comodule via the comultiplication map

$$\Delta : H^* \to H^* \otimes H^*.$$

So H^* is a left H-module via

$$h \cdot f = \sum f_{(1)} \langle h, f_{(2)} \rangle.$$

Then for k in H,

$$\langle k, h \cdot f \rangle = \sum \langle k, f_{(1)} \rangle \langle h, f_{(2)} \rangle = \langle kh, f \rangle.$$

Since Δ is an R-algebra homomorphism, Δ makes H^* into a right H^*-comodule algebra, hence H^* becomes a left H-module algebra. We note for future reference that

$$\varepsilon(h \cdot f) = \sum \varepsilon(f_{(1)} \langle h, f_{(2)} \rangle = \langle h, f \rangle.$$

(2.4). The notions of H-module algebra and H-comodule algebra generalize the concepts of actions and gradings of algebras by groups.

If S is a finite commutative R-algebra and G is a subgroup of the group of R-algebra automorphisms of S, then for any $\sigma \in G$, $\sigma(st) = \sigma(s)\sigma(t)$. Defining an RG-module action on S by

$$(\sum r_\sigma \sigma)(s) = \sum r_\sigma \sigma(s)$$

makes S into an RG-module algebra:

$$\begin{aligned} \Delta(\sum r_\sigma \sigma)(s \otimes t) &= \sum r_\sigma (\sigma \otimes \sigma)(s \otimes t) \\ &= \sum r_\sigma \sigma(s)\sigma(t) \\ &= \sum r_\sigma \sigma(st) \\ &= (\sum r_\sigma \sigma)(st) . \end{aligned}$$

DEFINITION. For G an abelian group, S is a *G-graded R-algebra* if

$$S = \bigoplus_{\sigma \in G} S_\sigma$$

with $R \subset S_1$, such that

$$S_\sigma \cdot S_\tau \subseteq S_{\sigma\tau} .$$

A G-grading on S induces an R-linear map

$$\alpha : S \to S \otimes_R RG$$

defined by $\alpha(s) = s \otimes \sigma$ for $s \in S_\sigma$. The property $S_\sigma \cdot S_\tau \subseteq S_{\sigma\tau}$ insures that α is an R-algebra homomorphism, and so the G-grading yields an RG-comodule algebra structure α on S.

These ideas are explored at greater length by Bergman [**Bg85**], c.f [**Da82**].

Galois extensions.

(2.5) DEFINITION. Let R be a commutative ring, S a finite commutative R-algebra (that is, a commutative R-algebra which is a finitely generated projective R-module). Let G be a finite group of R-algebra automorphisms of S. Then S is a *Galois extension of R with group G* if any of the following three equivalent conditions (labeling from Theorem 1.3 of [**CHR65**]) hold.
(c) the R-module map j from the crossed product

$$D(S,G) = \{\sum_{\sigma \in G} s_\sigma \sigma | s_\sigma \text{ in } S\}$$

to $End_R(S)$ given by $j(s_\sigma \sigma)(t) = s_\sigma \sigma(t)$ for all t in S, is bijective;
(e) the R-module map

$$h : S \otimes_R S \to S \otimes_R Hom_R(RG, R) \cong Hom_S(SG, S)$$

given by $h(s \otimes t)(\sigma) = s\sigma(t)$ is an isomorphism;
(f) for any maximal ideal $\mathfrak{m}$ of S and any $\sigma \neq 1$ in G, there is some s in S so that $\sigma(s) - s \notin \mathfrak{m}$.

(2.6) EXAMPLE. If L/K is a finite Galois extension of fields, and G is the Galois group of L/K, then L is a Galois extension of K with group G in the sense of (2.5). For the theorem on linear independence of characters asserts that the elements of G are L-linearly independent as endomorphisms of L. This means that the map j is 1-1. Since $|G| = \dim_K L$, $D(L,G)$ and $End_K(L)$ have the same dimension as vector spaces over K, and so j is then onto.

Now suppose S and R are the rings of integers of a Galois extension $L \supseteq K$ of local or global algebraic number fields with Galois group G. Then G acts as a group of R-algebra automorphisms of S. Condition (f) says that for any non-zero prime ideal P of S, the inertia group $I_P = \{\sigma \text{ in } G | \sigma(s) \equiv s \pmod{P}\}$ is trivial, which is equivalent to P being unramified over R. Thus S/R is a Galois extension with group G iff every prime ideal P of R is unramified in S.

It follows that the notion of Galois extension of rings has broader applicability in local number theory than in global number theory. For the discriminant $disc_R(S)$ (c.f. Section 22, below) is divisible by exactly those primes of R which ramify in S, and so $disc_R(S)$ measures the failure of S/R to be a Galois extension. If S, R are rings of integers of global number fields, it is seldom the case that the discriminant is trivial and S is a Galois extension of R with group G (indeed, if $R = \mathbb{Z}$, never, because for any $L/\mathbb{Q}$ with ring of integers S, the discriminant $disc_{\mathbb{Z}}(S)$ of $S/\mathbb{Z}$ is always non-trivial); on the other hand, if $S \supseteq R$ are rings of integers of a Galois extension $L \supseteq K$ of algebraic number fields, then except for finitely many primes $\mathfrak{p}$ of R (i.e. those dividing $disc_R(S)$), the localization $S_\mathfrak{p}$ of S at $\mathfrak{p}$ will be a Galois extension of $R_\mathfrak{p}$ with group G.

The extension of the notion of Galois extension from group actions to Hopf algebra actions, due to Chase and Sweedler [**CS69**], is based on properties (c) and (e).

(2.7) DEFINITION. Let H be a finite cocommutative R-Hopf algebra. A finite commutative R-algebra S is an *H-Galois extension*, or for short, *"H-Galois"*, if S

is a left H-module algebra, and the R module homomorphism

$$j : S \otimes_R H \to End_R(S)$$
$$j(s \otimes h)(t) = sh(t)$$

$s, t \in S$, $h \in H$, is an isomorphism.

The map j becomes an isomorphism of R-algebras if we define a suitable multiplication on $S \otimes H$:

(2.8) DEFINITION. Let S be a left H-module algebra. Then the *smash product* $S\#H$ is the R-algebra which as R-module is $S \otimes_R H$ and with multiplication

$$(s\#x)(t\#y) = \sum_{(x)} sx_{(1)}(t)\#x_{(2)}y$$

for $s, t \in S, x, y \in H$.

One checks easily that with this multiplication, j becomes a homomorphism of rings from $S\#H$ to $End_R(S)$.

If j is an isomorphism, then the "fixed ring"

$$S^H = \{s \in S | h \cdot s = \varepsilon(h)s \text{ for all } h \in H\}$$

is equal to R. For evidently $R \subset S^H$. On the other hand, if $s \in S^H$, then in $S\#H$, $s\#1$ commutes with $t\#h$ for all $t \in S, h \in H$ (a short computation), so $j(s\#1)$ commutes with $j(t\#h)$ for all $s, t \in S, h \in H$, so $j(s\#1)$ is in R, the center of $End_R(S)$. Hence $s \in R$.

For H a finite cocommutative R-Hopf algebra, let

$$H^* = Hom_R(H, R),$$

then H^* is a finite commutative R-Hopf algebra.

If S is a right H^*-comodule algebra via $\alpha : S \to S \otimes H^*$ then S is an *H^*-Galois object*, or *H^*-principal homogeneous space*, if the map

$$\gamma : S \otimes S \to S \otimes H^*$$

by $\gamma(s \otimes t) = (s \otimes 1)\alpha(t)$ is an isomorphism. (Surjectivity of γ suffices, as [**KT81**] observed.)

If $Y = Spec(S)$, $X = Spec(R)$ and $G = Spec(H^*)$ (which makes sense if H is cocommutative), then the condition (e) that γ is an isomorphism is a translation of the statement that $Y \to X$ is a principal homogeneous space for the finite group scheme G. (See [**Wa79, (18.3)**].)

The following result shows that the H-Galois extension and H^*-Galois object properties are equivalent.

(2.9) PROPOSITION. *Let H be a finite R-Hopf algebra and S a finite commutative R-algebra which is a left H-module algebra. Then*

$$S \otimes H \xrightarrow{j} End_R(S),$$

$$j(s \otimes h)(t) = sh(t)$$

is an isomorphism iff

$$\gamma : S \otimes S \to S \otimes H^*,$$
$$\gamma(s \otimes t) = (s \otimes 1)\alpha(t)$$

is an isomorphism.

PROOF. Consider the diagram

$$\begin{array}{ccc} S \otimes H & \xrightarrow{j} & Hom_R(S,S) \\ \downarrow \eta & & \downarrow \beta \\ Hom_S(S \otimes H^*, S) & \xrightarrow{\gamma^*} & Hom_S(S \otimes S, S) \end{array},$$

where $\eta(s \otimes h)(t \otimes f) = st\langle h, f\rangle$, and $\beta(f)(s \otimes t) = sf(t)$ are isomorphisms, (recall that $\langle\ ,\ \rangle : H \otimes H^* \to R$ is evaluation) and

$$\gamma^*(f)(s \otimes t) = f(\gamma(s \otimes t)) = f((s \otimes 1)\alpha(t)) = f(\sum_{(t)} st_{(0)} \otimes t_{(1)}).$$

The diagram commutes: given $s \otimes h \in S \otimes H$, $j(s \otimes h)(t) = sh(t)$, so for t, u in S,

$$\beta j(s \otimes h)(t \otimes u) = tj(s \otimes h)(u) = tsh(u).$$

On the other hand, $\eta(s \otimes h)(t \otimes f) = st\langle h, f\rangle$, so

$$\begin{aligned} \gamma^*\eta(s \otimes h)(t \otimes u) &= \eta(s \otimes h)(\gamma(t \otimes u)) \\ &= \eta(s \otimes h)(\sum_{(u)} tu_{(0)} \otimes u_{(1)}) \\ &= st \sum_{(u)} u_{(0)}\langle h, u_{(1)}\rangle \\ &= sth(u)\ . \end{aligned}$$

Since γ is an isomorphism iff γ^* is an isomorphism, it follows that j is an isomorphism iff γ is an isomorphism. □

(2.10) COROLLARY. *Let H be a finite cocommutative R-Hopf algebra and S a commutative H-module algebra. Then S is H-Galois iff S is an H^*-principal homogeneous space.*

(2.10A) EXAMPLE. H^* is an H^*-principal homogeneous space, hence an H-Galois extension. For the map

$$\gamma : H^* \otimes H^* \to H^* \otimes H^*$$

by

$$\gamma(f \otimes g) = \sum_{(g)} fg_{(1)} \otimes g_{(2)}$$

has an inverse δ defined by

$$\delta(f \otimes g) = \sum_{(g)} f\lambda(g_{(1)}) \otimes g_{(2)}.$$

H is called the *trivial* H-Galois extension.

(2.11) Base change. If B is a faithfully flat R-algebra and S is an H-Galois extension of R, then $B \otimes_R S$ is an $B \otimes_R H$-Galois extension of B, since if $j : S \otimes_R H \to End_R(S)$ is an isomorphism, then $B \otimes j : (B \otimes_R S) \otimes_B (B \otimes_R H) \to B \otimes_R End_R(S) \cong End_B(B \otimes_R S)$ is an isomorphism; conversely, if S is an H-module algebra and $B \otimes_R S$ is an $B \otimes_R H$-Galois extension of B with action of $B \otimes_R H$ on $B \otimes_R S$ induced from that of H on S, then S is an H-Galois extension of R, since if $B \otimes j$ is an isomorphism, then by faithful flatness, j is an isomorphism.

This applies in two important special cases: B is a finite R-algebra (recall: B is an R-algebra which is a finitely generated and projective R-module), and $B = \prod_p R_p$ where the sum runs through all prime ideals of R, and R_p denotes localization at the prime ideal p. This latter case implies that being H-Galois is a local property: an H-module S is H-Galois over R iff $R_p \otimes_R S$ is an $R_p \otimes_R H$-Galois extension of R_p for all prime ideals p of R.

(2.12) Galois descent. Condition (c) is the key to Galois descent.

Let S be a Galois extension of R with Galois group G. Then S is a finite R-algebra, so if A is an R-module (or R-algebra, or R-Hopf algebra), then $S \otimes_R A$ is an S-module (or S-algebra, or S-Hopf algebra). Similarly, if $f : A \to B$ is a homomorphism of R-modules, then $S \otimes f : S \otimes_R A \to S \otimes_R B$, given by $(S \otimes f)(s \otimes a) = s \otimes f(a)$, is a homomorphism of S-modules.

Descent theory asks, given an S-module A, under what conditions is $A \cong S \otimes_R A_0$ as S-modules for some R-module A_0? Also, if

$$f : S \otimes_R A_0 \to S \otimes_R B_0$$

is an S-module homomorphism, under what conditions is $f = S \otimes f_0$ for some R-module homomorphism $f_0 : A_0 \to B_0$?

If S is a finite R-algebra, then by Morita theory (see [**CR81**]), there is an equivalence of categories between the category of left R-modules and the category of left $End_R(S)$-modules, given by the base change functor $S \otimes_R -$. An S-module M therefore descends, that is, is isomorphic to a module of the form $S \otimes_R M_0$ for some R-module M_0, iff the S-action on M extends to an action by $End_R(S)$.

Now if S/R is a Galois extension with Galois group G, then $j : D(S, G) \to End_R(S)$ is an isomorphism, so a left $End_R(S)$-module is simply a left S-module with a compatible action by G (i.e. so that for $m \in M$, $\sigma \in G$, and $e \in S$, $\sigma(em) = \sigma(e)\sigma(m)$. In that case the inverse to the base change functor is the fixed module functor

$$M \mapsto M^G = \{x \in M | \sigma(x) = x \text{ for all } x \in G\}.$$

More generally:

(2.13) LEMMA. *Let S be an R-algebra which is a finitely generated faithful projective R-module and an H-module algebra so that $j : S \# H \to E := End_R(S)$ is an isomorphism. Then for any left E-module M, $M \cong S \otimes_R M^H$ via the map $s \otimes m \mapsto sm$.*

SKETCH OF PROOF. Since S is a faithful finitely generated projective R-module, Morita theory [**CR81**] yields an equivalence from the category of left R-modules to the category of left E-modules given by $N \mapsto S \otimes N$, N an R-module, with inverse $M \mapsto Hom_E(S, E) \otimes_E M$ for M a left E-module. Thus for M a left E-module,

$$M \cong S \otimes_R Hom_E(S, E) \otimes_E M.$$

Now $Hom_E(S, E) \otimes_E M \cong Hom_E(S, M)$ since S is a projective left E-module (the isomorphism is clear if $S = E$, hence for S a free E-module, hence for S a direct summand of a free E-module). Finally, if $E \cong S\#H$, then

$$Hom_E(S, M) \cong Hom_{S\#H}(S, M).$$

We claim:

$$Hom_{S\#H}(S, M) \cong M^H$$

via the homomorphism $\phi \mapsto \phi(1)$ for ϕ in $Hom_{S\#H}(S, M)$. For $\phi(s) = \phi((s\#1)1) = (s\#1)\phi(1)$, so ϕ is uniquely determined by $\phi(1)$; moreover, for $h \in H$,

$$\begin{aligned} h\phi(1) &= (1\#h)\phi(1) \\ &= \phi((1\#h)(1)) \\ &= \phi(\varepsilon(h)1) \\ &= \varepsilon(h)\phi(1) \end{aligned}$$

so $\phi(1) \in M^H$. Conversely, if $m \in M^H$, then the map $\phi : S \to M$ defined by $\phi(s) = sm$ is easily seen to be in $Hom_{S\#H}(S, M)$. □

The functors apply to maps as well as to objects. Let $f : A \to B$ be an S-module homomorphism of $D(S, G)$-modules which is G-equivariant, that is:

$$f(\sigma(a)) = \sigma(f(a))$$

for all $a \in A$, $\sigma \in G$. Then f is a $D(S, G)$-module homomorphism, so $f = S \otimes f_0$ where $f_0 : A^G \to B^G$ is an R-module homomorphism.

Suppose given a commutative diagram of $D(S, G)$-module homomorphisms, such as

$$\begin{array}{ccc} A \otimes_S A \otimes_S A & \xrightarrow{\mu} & A \otimes_S A \\ {\scriptstyle\mu}\downarrow & & \downarrow{\scriptstyle\mu\otimes 1} \\ A \otimes_S A & \xrightarrow[1\otimes\mu]{} & A \end{array}$$

(which defines associativity of A). Since $(\ )^G$ is a functor, we get a commutative diagram of fixed submodules:

$$\begin{array}{ccc} (A \otimes_S A \otimes_S A)^G & \xrightarrow{\mu} & (A \otimes_S A)^G \\ {\scriptstyle\mu}\downarrow & & \downarrow{\scriptstyle\mu\otimes 1} \\ (A \otimes_S A)^G & \xrightarrow[1\otimes\mu]{} & A^G \end{array}.$$

(Since $A^G \otimes_R A^G \otimes_R A^G$ embeds in $(A \otimes_S A \otimes_S A)^G$, it follows that A^G is also associative.)

This implies that if A is an S-algebra, or an S-Hopf algebra, and G acts on A as algebra homomorphisms, or as Hopf algebra homomorphisms, then the S-module homomorphisms which define the structure on A are $D(S,G)$-module homomorphisms, and so the structure maps on A, and properties of those structure maps definable by commutative diagrams, induce structure maps on A^G with the same properties. Thus if A is an S-(Hopf) algebra and G acts as (Hopf) algebra homomorphisms on A, then A^G is an R-(Hopf) algebra.

Similarly, suppose H is an S-Hopf algebra, A an S-algebra, and

$$f : H \otimes_S A \to A$$

makes A into an H-module algebra. Recall that this means: for $h \in H, a, b \in A$,

$$h(ab) = \mu(\Delta(h)(a \otimes b))$$

and

$$h(1) = \varepsilon(h) \cdot 1.$$

Both properties may be expressed by commutative diagrams; the first is:

$$\begin{array}{ccc} H \otimes A \otimes A & \xrightarrow{\mu} & H \otimes H \otimes A \otimes A \\ {\scriptstyle 1\otimes\mu_A}\downarrow & & \downarrow{\scriptstyle \mu_A \circ f \otimes f \circ (1 \otimes \tau \otimes 1)} \\ H \otimes A & \xrightarrow[f]{} & A \end{array} \tag{2.14}$$

So if G acts on H and A so as to respect the structures on H and A, and f is G-equivariant, then f descends, diagrams such as (2.14) commute after the functor $(\ )^G$ is applied, and A^G is an H^G-module algebra.

For a general treatment of descent theory, see [**KO74**].

(2.15) Ramification. Condition (f) for Galois extensions with group G does not extend to Hopf Galois extensions. In fact, if S is a Hopf Galois extension of R with Hopf algebra H, where S, R are rings of integers of number fields, then the discriminant of $S/R =$ the discriminant of H^*/R. This observation of Greither [**Gr92**] (see (22.13), below), which follows from the isomorphism $\gamma : S \otimes S \to S \otimes H^*$, reduces to the unramified condition when S is a Galois extension with group G, for then $H^* = RG^* \cong R \oplus R \oplus \cdots \oplus R$ ($|G|$ copies of R) as R-algebra, hence the discriminant of $H^* = R =$ the discriminant of S over R, so no prime of R ramifies in S.

Since ramification is possible for Hopf Galois extensions of number rings, Hopf Galois theory has interesting potential applications in Galois module theory.

The starting point for classical Galois module theory is Noether's theorem, that if L/K is a tamely ramified Galois extension of number fields with Galois group G, and S, R are the rings of integers of L and K, respectively, then S is a locally free rank one RG-module. A highly special case of Noether's theorem is the case where L/K is unramified, i.e. S is a Galois extension of R with group G. The next result generalizes that special case to Hopf Galois extensions.

(2.16) THEOREM. *If S/R is H-Galois, then S is a locally free rank one H-module.*

PROOF. If S/R is H-Galois then for any prime ideal p of R, S_p/R_p is H_p-Galois, where R_p is the localization of R at p and $H_p = R_p \otimes H$ and $S_p = R_p \otimes_R S$. So we may assume R is local. Now S is a free R-module of rank n, and the isomorphism $\gamma : S \otimes_R S \cong S \otimes_R H^*$ by $\gamma(s \otimes t) = \sum st_{(0)} \otimes t_{(1)}$ is an H-module homomorphism, H acting on the right factors, since

$$\begin{aligned}\gamma(s \otimes ht) &= \gamma(s \otimes \sum t_{(0)}\langle h, t_{(1)}\rangle) \\ &= \sum st_{(0)} \otimes t_{(1)}\langle h, t_{(2)}\rangle) \\ &= \sum st_{(0)} \otimes h \cdot t_{(1)} \\ &= h\gamma(s \otimes t).\end{aligned}$$

Hence, as H-modules

$$S \oplus \cdots \oplus S \cong H^* \oplus \cdots \oplus H^* \ .$$

Now H is a finite algebra over a local ring, so by the Krull-Schmidt theorem (see [**CR81**]), both sides decompose uniquely into indecomposable H-modules, from which it follows that $S \cong H^*$ as H-modules. Since $H^* \cong H$ as H-modules (see (3.3) below), we obtain $S \cong H$ as H-modules, as we wished to show. □

(2.17) COROLLARY. *If L/K is a Galois extension of fields with Galois group G, then L has a normal basis as a K-module.*

PROOF. By (2.16), $L \cong (KG)^*$ as KG-modules. If $\{e_\sigma | \sigma \in G\}$ is the dual basis in KG^* to G in KG, then $\tau(e_\sigma) = e_{\sigma\tau^{-1}}$, so e_1 generates KG^* as a KG-module. Let α correspond to e_1 under the KG-isomorphism $L \cong KG^*$, then $\{\sigma(\alpha) | \sigma \in G\}$ is a K-basis of L, that is, a normal basis for L. □

There is a partial converse which we will examine in Chapter 3.

(2.18) Non-commutative Galois extensions. Let H be a finite K-Hopf algebra, K a field. Let A be a K- algebra. Then A is a right H-comodule algebra if A is a right H-comodule via $\alpha : A \to A \otimes_K H$ which is a K-algebra homomorphism (i. e. $\alpha(xy) = \sum x_{(0)}y_{(0)} \otimes x_{(1)}y_{(1)}$ for $x, y \in A$, and $\alpha(1) = 1 \otimes 1$.

Let A be a right H-comodule algebra. Then the subalgebra $B = A^{co(H)}$ of H-coinvariants is defined by

$$A^{co(H)} = \{a \in A | \sum_{(a)} a_{(0)} \otimes a_{(1)} = a \otimes 1\}.$$

A/B is called an H-extension, and A/B is called a (right) H-Galois extension if A is injective as right H-comodule, and

$$\gamma : A \otimes A \to A \otimes H,$$

$\gamma(x \otimes y) = \sum xy_{(0)} \otimes y_{(1)}$, is bijective. This definition extends the notion of H-Galois object to arbitrary finite Hopf algebra comodule algebra actions on non-commutative algebras. See [**Sch90**], [**Sch94**] for this theory.

Note that Schneider uses "H-extension" for an H-comodule action, while we use "H-extension" for an H-module action on A. There is no consistency of terminology in the literature on Galois theory of rings.

(2.19). Suppose R is a commutative ring, H is a finite R-Hopf algebra, and S is a finite commutative R-algebra and an H^*-comodule algebra so that

$$\gamma : S \otimes S \to S \otimes H^*$$

is an isomorphism. Then H^* is commutative. For since S is commutative, γ is an S-algebra homomorphism:

$$\begin{aligned}\gamma(xx' \otimes yy') &= \sum xx' y_{(0)} y'_{(0)} \otimes y_{(1)} y'_{(1)} \\ &= \gamma(x \otimes y)\gamma(x' \otimes y').\end{aligned}$$

Thus $S \otimes_R S \cong S \otimes_R H^*$ as S-algebras. Since S is finitely generated and projective as R-module, it follows that H^* is commutative.

Thus in (2.18), if one considers commutative K-algebras A which are H-Hopf Galois extensions for some K-Hopf algebras H in the sense of (2.18), the only Hopf algebras H which can arise are commutative. This explains the cocommutativity assumption on the Hopf algebras in (2.7).

§3. Integrals.

(3.1) Definition. Let H be an R-Hopf algebra. An element $\theta \in H$ is a *left integral* if for all $x \in H$, $x\theta = \varepsilon(x)\theta$. An element $\theta \in H$ is a *right integral* if for all $x \in H$, $\theta x = \varepsilon(x)\theta$.

Recall that if M is a left H-module, then the submodule of invariants of M,

$$M^H = \{m \in M | xm = \varepsilon(x)m \text{ for all } x \in H\}.$$

If we view H as a left H-module via multiplication, the the set of left integrals of H is H^H. One sees easily that H^H is not only an R-submodule of H but is a two-sided ideal of H: if θ is a left integral of H then for all $x, y \in H$,

$$y(x\theta) = (yx)\theta = \varepsilon(yx)\theta = \varepsilon(y)\varepsilon(x)\theta = \varepsilon(y)x\theta,$$

and

$$y(\theta x) = (y\theta)x = \varepsilon(y)\theta x,$$

hence $x\theta$ and θx are left integrals of H.

Example. If $H = RG$, G a finite group, then

$$\theta = \sum_{\sigma \in G} \sigma$$

is a left and a right integral, and every left or right integral is an R-multiple of θ. If $H = RG^* = \sum Re_\sigma$ with $e_\sigma(\tau) = \delta_{\sigma,\tau}$, then e_1 is a left and right integral, for $\varepsilon(e_\sigma) = \delta_{1,\sigma}$ and so $e_\sigma e_1 = \delta_{1,\sigma} e_1 = \varepsilon(e_\sigma)e_1$ for all $\sigma \in G$, hence for any $f \in RG^*$, $fe_1 = \varepsilon(f)e_1$.

(3.2) DEFINITION. A Hopf algebra H is *unimodular* if the module of left integrals of H = the module of right integrals of H.

Clearly any commutative Hopf algebra is unimodular, as is the group ring of any finite group.

Integrals help illuminate the structure of a Hopf algebra as a module over its dual. To illustrate this idea, recall that if H is a finite Hopf algebra with dual H^*, then H acts on H^* by

$$x \cdot f = \sum_{(f)} f_{(1)} \langle x, f_{(2)} \rangle .$$

In particular, for G a finite group, RG acts on RG^* by

$$\sigma \cdot e_\tau = \sum_\rho e_\rho \langle \sigma, e_{\rho^{-1}\tau} \rangle = e_{\tau\sigma^{-1}}$$

Thus

$$RG^* = RGe_1,$$

the free RG-module with basis e_1, a left integral of RG^*. Similarly, RG^* acts on RG by $e_\sigma \cdot \tau = \delta_{\sigma,\tau}\tau$, so

$$RG = RG^*(\sum \sigma),$$

the free RG^*-module with basis $\sum_{\sigma \in G} \sigma$, a left integral of RG.

This description of a finite Hopf algebra as a module over its dual holds in general.

(3.3) THEOREM (LARSON-SWEEDLER). *For H any finite R-Hopf algebra, the action of H on H^* defines an isomorphism $H^* \cong H \otimes_R J$ where J is the module of left integrals of H^*.*

Proofs of this standard result may be found in [**LS69, Proposition 1**]; [**Sw69, Theorem 5.1.3**]; [**Pa71**];[**Ra74, Lemma 6**]; [**Ja87, 8.12**]; and [**PS80**].

(3.4) COROLLARY. *If H is a finite R-Hopf algebra, then the module J of left integrals of H^* is a rank one projective R-module.*

PROOF. Recall that "finite" means finitely generated and projective as R-module. We may assume R is connected (has no idempotents but 0 and 1), so that H has rank n as a projective R-module for some n. Then H^* also has rank n. Since

$$H^* \cong H \otimes_R J$$

as R-modules and R is an R-direct summand of H (the counit ε splits the unit map i), J is a projective R-module, hence projective of rank one. □

In particular, if R is a principal ideal domain, then J is free of rank one over R. (We saw that for RG and RG^* for any R.) If $J = R\theta$, then $H^* = H\theta$ is a free H-module on the left integral θ: the map

$$H \to H^*$$

by $x \mapsto x \cdot \theta$ is a left H-module isomorphism.

(3.5) COROLLARY [**Pa71**]. *Let R be a principal ideal domain and H a finite R-Hopf algebra. Then H is a Frobenius algebra: in particular, if θ generates the module of left integrals of H^*, then*

$$\beta : H \otimes H \to R$$

by

$$\beta(h,k) = \langle hk, \theta \rangle$$

is a non-degenerate, associative bilinear form on H.

PROOF. β is clearly bilinear; also $\beta(hg,k) = \beta(h,gk)$ for $h,g,k \in H$, so β is associative.

To show non-degeneracy, let $\{h_1, \dots, h_n\}$ be an R- basis of H, and $\{\hat{h}_1, \dots, \hat{h}_n\}$ be the dual basis in H^*. Then $\hat{h}_i = k_i \cdot \theta$ for some $k_i \in H$, and so $\{k_1, \dots, k_n\}$ is an R-basis of H and (c.f. (2.3))

$$\delta_{i,j} = \langle h_i, k_j \cdot \theta \rangle = \varepsilon_{H^*}(h_i \cdot (k_j \cdot \theta)) = \varepsilon_{H^*}((h_i k_j) \cdot \theta) = \langle h_i k_j, \theta \rangle = \beta(h_i, k_j).$$

Then for any $h, k \in H$, if $h = \sum r_i h_i$, $k = \sum s_j k_j$, then $\beta(h,k) = \sum r_i s_j$. Hence given k, if $\beta(h,k) = 0$ for all h, then $k = 0$, and given h, if $\beta(h,k) = 0$ for all k, then $h = 0$. □

(3.6) COROLLARY. *For any basis $\{h_1, \dots, h_n\}$ of H, the matrix $(\langle h_i h_j, \theta \rangle)$ is invertible.*

PROOF. Let $\{k_1, \dots, k_n\}$ be the preimage of the dual basis of $\{h_1, \dots, h_n\}$ as in the proof of (3.5). Write $h_i = \sum_\ell r_{i,\ell} k_\ell$, then the $n \times n$ matrix $(r_{i,\ell})$ is invertible and

$$\langle h_i h_j, \theta \rangle = \sum_\ell r_{j,\ell} \langle h_i k_\ell, \theta \rangle = \sum_\ell r_{j,\ell} \delta_{i,\ell} = r_{j,i}.$$

So $(\langle h_i h_j, \theta \rangle) = (r_{j,i})$. □

We conclude this section with the following useful property of integrals.

(3.7) LEMMA. *If θ is a left integral of H, then for all $x \in H$,*

$$(x \otimes 1)((1 \otimes \lambda)\Delta(\theta)) = ((1 \otimes \lambda)\Delta(\theta))(1 \otimes x).$$

A special case of this is when $H = RG$ and $\theta = \sum \sigma$:

$$\sum \tau\sigma \otimes \sigma^{-1} = \sum \sigma \otimes \sigma^{-1}\tau.$$

PROOF. (A classic example of the use of Sweedler notation, due to Sweedler [**Sw69**].)

$$\begin{aligned}(x \otimes 1)((1 \otimes \lambda)\Delta(\theta)) &= \sum x\theta_{(1)} \otimes \lambda(\theta_{(2)}) \\ &= \sum x_{(1)}\varepsilon(x_{(2)})\theta_{(1)} \otimes \lambda(\theta_{(2)}) \\ &= \sum x_{(1)}\theta_{(1)} \otimes \lambda(\theta_{(2)})\varepsilon(x_{(2)}) \\ &= \sum x_{(1)}\theta_{(1)} \otimes \lambda(\theta_{(2)})\lambda(x_{(2)})x_{(3)} \\ &= \sum x_{(1)}\theta_{(1)} \otimes \lambda(x_{(2)}\theta_{(2)})x_{(3)} \\ &= \sum (1 \otimes \lambda)\Delta(x_{(1)}\theta)(1 \otimes x_{(2)}) \\ &= \sum (1 \otimes \lambda)\Delta(\varepsilon(x_{(1)})\theta)(1 \otimes x_{(2)}) \\ &= \sum (1 \otimes \lambda)\Delta(\theta)(1 \otimes \varepsilon(x_{(1)})x_{(2)}) \\ &= (1 \otimes \lambda)\Delta(\theta)(1 \otimes x).\end{aligned}$$

□

(3.8) REMARK. Let K be a field and H a finite K-Hopf algebra with dual H^*. Then H^* is a left H-module via $\langle h \cdot \psi, g\rangle = \langle \psi, gh\rangle$ and a right H-module via $\langle \psi \cdot h, g\rangle = \langle \psi, hg\rangle$ for $\psi \in H^*, g, h \in H$. If θ is a non-zero left integral of H^*, then the H-module homomorphism $H \to H \otimes_K K\theta \to H^*$ by $h \mapsto h \cdot \theta$ is an isomorphism (3.3). Also the map $T : H \to H^*$, $T(g) = \theta \cdot g$ is injective by (3.5), hence bijective. Pareigis calls T the Fourier transform from H to H^*. Then T specializes to the Fourier transform of a finite group, converts multiplication in H to a kind of convolution in H^*, and yields the Plancherel formula (invariance of the duality pairing $H \otimes H^* \to R$ under the Fourier transform) if H is abelian. For details and connections of the integral to functional analysis, see [**Pa98**].

§**4. Short exact sequences of Hopf algebras.** In this section we translate into a Hopf algebra setting and generalize the following standard results:

Suppose $G, \overline{G}, N$ are finite groups.

If $f : N \to G$ is an injective group homomorphism such that $f(N)$ is a normal subgroup of G, then $G/f(N)$ is a group and N is isomorphic to the kernel of the canonical morphism $G \to G/f(N)$.

If $g : G \to \overline{G}$ is a surjective group homomorphism, then $N = ker(g)$ is a normal subgroup of G and $\overline{G}$ is isomorphic to the cokernel G/N of the inclusion map $N \to G$.

From either starting point, N or $\overline{G}$, we get a short exact sequence of groups

(4.1) $$1 \to N \xrightarrow{f} G \xrightarrow{g} \overline{G} \to 1$$

i.e., $N \cong ker(g)$, $\overline{G} \cong cok(f)$.

Now suppose R is a commutative ring. Then the group rings $H_1 = RN, H = RG$, and $\overline{H} = R\overline{G}$ are R-Hopf algebras. A group homomorphism between groups induces a R-Hopf algebra homomorphism between the corresponding group rings, and also an R-Hopf algebra homomorphism between the duals. Thus given the short exact sequence (4.1) of groups, we have the corresponding sequences

$$R \to RN \to RG \to R\overline{G} \to R \tag{4.2}$$

and

$$R \to (R\overline{G})^* \to (RG^*) \to (RN)^* \to R \tag{4.3}$$

which we wish to view as short exact sequences of Hopf algebras. (Note: $R = RE = (RE)^*$ for E the trivial group.)

The sequences are not exact as R-modules, since every map sends 1 to 1.

We will need to define what it means for a sequence of Hopf algebra homomorphisms

$$R \to H_1 \xrightarrow{f} H \xrightarrow{g} \overline{H} \to R$$

to be exact. After doing so, we will reexamine (4.2) and (4.3).

If H is a R-Hopf algebra, the kernel of the counit map $\varepsilon : H \to R$ is denoted by H^+.

We first introduce some concepts related to the morphism $f : H_1 \to H$.

(4.4) DEFINITION. Let $f : H_1 \to H$ be a Hopf algebra homomorphism. Then f is *normal* if $f(H_1^+)H = Hf(H_1^+)$.

Evidently if H is commutative, then f is normal.

A two-sided ideal I of H is a *Hopf ideal* if $\Delta(I) \subset H \otimes_R I + I \otimes_R H$, $\varepsilon(I) = 0$ and $\lambda(I) \subset I$.

(4.5) PROPOSITION. *H/I is a Hopf algebra iff I is a Hopf ideal.*

PROOF. We first show: if I is a two-sided ideal of H, then

$$H/I \otimes H/I \cong (H \otimes H)/(H \otimes I + I \otimes H).$$

The obvious map $g : H \otimes H \to H/I \otimes H/I$ has kernel containing $I \otimes H + H \otimes I$; on the other hand, if for $i = 1, 2$, the maps

$$j_i : H \to (H \otimes H)/(H \otimes I + I \otimes H)$$

are defined by $j_1(x) = x \otimes 1, j_2(x) = 1 \otimes x$, then j_i factors through H/I, so yields a map

$$j_1 \otimes j_2 : H/I \otimes H/I \to (H \otimes H)/(H \otimes I + I \otimes H)$$

which is the inverse of the map induced by g. Thus

$$H/I \otimes H/I \cong (H \otimes H)/(H \otimes I + I \otimes H).$$

Hence the comultiplication Δ on H yields a well-defined map

$$\overline{\Delta} : H/I \to H/I \otimes H/I$$

iff for all $x \in I, \Delta(x) \in H \otimes I + I \otimes H$. Similarly, ε and λ are well-defined on H/I iff $\varepsilon(I) = 0$ and $\lambda(I) \subset I$. $\square$

Of course, $H/H^+ \cong R$, which is a R-Hopf algebra, hence H^+ is a Hopf ideal. More generally, we have

(4.6) PROPOSITION. *If $f : H_1 \to H$ is a normal morphism of Hopf algebras, then $Hf(H_1^+)$ is a Hopf ideal of H.*

PROOF. Since f is normal, $Hf(H_1^+)$ is a two-sided ideal of H. We have

$$\varepsilon(f(H_1^+)H) \subset \varepsilon(f(H_1^+))H = f(\varepsilon(H_1^+)) = 0,$$

$$\lambda(f(H_1^+)H) \subset \lambda(H)\lambda(f(H_1^+)) = \lambda(H)f(\lambda(H_1^+)) \subset Hf(H_1^+),$$

and

$$\begin{aligned}\Delta(f(H_1^+)H) &\subset \Delta(f(H_1^+))(H \otimes H)\\ &\subset (f \otimes f)(\Delta(H_1^+))(H \otimes H)\\ &\subset (f(H_1^+) \otimes f(H_1) + f(H_1) \otimes f(H_1^+))(H \otimes H)\end{aligned}$$

since H_1^+ is a Hopf ideal of H_1;

$$\subset (f(H_1^+))H \otimes H + H \otimes (f(H_1^+)).$$

Thus $f(H_1^+)H$ is a Hopf ideal. □

(4.7) DEFINITION. If $f : H_1 \to H$ is a normal Hopf algebra morphism, set $H//f(H_1) = H/f(H_1^+)H$, the Hopf cokernel of f.

Now we introduce concepts related to $g : H \to \overline{H}$.

(4.8) DEFINITION. Let $g : H \to \overline{H}$ be a Hopf algebra homomorphism. The *algebra of right coinvariants*, $H^{co(g)}$, is the equalizer of the two maps $(1 \otimes g) \circ \Delta$ and $(1 \otimes i\varepsilon) \circ \Delta : H \to H \otimes \overline{H}$, where $i : R \to \overline{H}$ is the unit map.

The algebra of left coinvariants, ${}^{co(g)}H$, is the equalizer of the two maps $(g \otimes 1) \circ \Delta$ and $(i\varepsilon \otimes 1) \circ \Delta : H \to \overline{H} \otimes H$.

We note that for any $x \in H$,

$$\begin{aligned}(1 \otimes i\varepsilon)\Delta(x) &= \sum_{(x)} x_{(1)} \otimes \varepsilon(x_{(2)})\overline{1}\\ &= \sum_{(x)} x_{(1)}\varepsilon(x_{(2)}) \otimes \overline{1}\\ &= x \otimes \overline{1}\end{aligned}$$

in $H \otimes \overline{H}$, and similarly,

$$(i\varepsilon \otimes 1)\Delta(x) = \overline{1} \otimes x$$

in $\overline{H} \otimes H$.

(4.9) DEFINITION. The Hopf algebra morphism $g : H \to \overline{H}$ is *conormal* if ${}^{co(g)}H = H^{co(g)}$.

If $\overline{H}$ is cocommutative, then any Hopf algebra morphism $g : H \to \overline{H}$ is conormal, since if $x \in H^{co(g)}$, that is,

$$(1 \otimes g)\Delta(x) = x \otimes \overline{1},$$

then, letting τ be the switch map as in Section 1 (above (1.2)),

$$\begin{aligned}\overline{1}\otimes x &= \tau(x\otimes\overline{1})\\ &= \tau(1\otimes g)\Delta(x)\\ &= (g\otimes 1)\tau\Delta(x)\\ &= (g\otimes 1)\Delta(x)\\ &= \overline{1}\otimes x\end{aligned}$$

by cocommutativity. Thus $x \in^{co(g)} H$, and conversely.

(4.10) PROPOSITION. *If g is conormal and $H_1 = H^{co(g)}$ is an R-module direct summand of H, then H_1 is a sub-Hopf algebra of H.*

PROOF. $H_1 = H^{co(g)}$ is the difference kernel

$$\{x \in H | (1\otimes i\varepsilon)\Delta(x) = (1\otimes g)\Delta(x)\}$$

of the two algebra homomorphisms $(1\otimes i\varepsilon)\Delta$ and $(1\otimes g)\Delta(x) : H \to H\otimes\overline{H}$. The difference kernel of any two R-algebra homomorphisms is easily seen to be a R-algebra.

As for the remaining Hopf structure:

$\varepsilon : H_1 \to R$ works by restriction.

For λ: if $x \in H_1$, then

$$(1\otimes g)\Delta(\lambda(x)) = (1\otimes g)(\lambda\otimes\lambda)\tau\Delta(x)$$

where τ is the switch map $H\otimes H \to H\otimes H$;

$$\begin{aligned}&= (\lambda\otimes\lambda)(1\otimes g)\tau\Delta(x)\\ &= (\lambda\otimes\lambda)\tau(g\otimes 1)\Delta(x)\\ &= (\lambda\otimes\lambda)\tau(i\varepsilon\otimes 1)\Delta(x)\end{aligned}$$

since $H_1 =^{co(g)} H = H^{co(g)}$

$$\begin{aligned}&= (\lambda\otimes\lambda)\tau(\overline{1}\otimes x)\\ &= \lambda(x)\otimes\overline{1}\\ &= (1\otimes i\varepsilon)\Delta(\lambda(x)).\end{aligned}$$

Hence $\lambda(H_1) \subset H_1$.

Now we show that $\Delta(H_1) \subset H\otimes H_1$. Since

$$\begin{aligned}H_1 = H^{co(g)} &= \{x \in H | (1\otimes i\varepsilon)\Delta(x) = (1\otimes g)\Delta(x) \in H\otimes\overline{H}\}\\ &= ker\{(1\otimes(i\varepsilon - g))\Delta : H \to H\otimes\overline{H}\},\end{aligned}$$

we have

$$\begin{aligned}H\otimes H_1 &= H\otimes ker\{(1\otimes(i\varepsilon - g))\Delta\}\\ &= ker\{1\otimes(1\otimes(i\varepsilon - g))\Delta\}.\end{aligned}$$

So we must show that if $x \in H_1$, then

$$(1\otimes 1\otimes g)(1\otimes\Delta)\Delta(x) = (1\otimes 1\otimes i\varepsilon)(1\otimes\Delta)\Delta(x).$$

By coassociativity, this is the same as

$$(1 \otimes 1 \otimes g)(\Delta \otimes 1)\Delta(x) = (1 \otimes 1 \otimes i\varepsilon)(\Delta \otimes 1)\Delta(x)$$

or

$$(\Delta \otimes 1)(1 \otimes g)\Delta(x) = (\Delta \otimes 1)(1 \otimes i\varepsilon)\Delta(x)$$

which is true since for $x \in H_1$,

$$(1 \otimes g)\Delta(x) = (1 \otimes i\varepsilon)\Delta(x).$$

Similarly, since $H_1 =^{co(g)} H$, $\Delta(H_1) \subset H_1 \otimes H$. Since H_1 is a R-direct summand of H, we have that

$$(H_1 \otimes H) \cap (H \otimes H_1) = H_1 \otimes H_1$$

and so $\Delta(H_1) \subset H_1 \otimes H_1$. □

(4.11) DEFINITION. If $g : H \to \overline{H}$ is conormal and $H^{co(g)}$ is a Hopf algebra, then $H^{co(g)} = hker(g)$ is the *Hopf kernel* of g.

(4.12) DEFINITION. A *short exact sequence* of R-Hopf algebras is a sequence of Hopf algebra maps $f : H_1 \to H$, $g : H \to \overline{H}$:

$$H_1 \xrightarrow{f} H \xrightarrow{g} \overline{H}$$

such that f is injective and normal, g is surjective and conormal, $\overline{H} \cong H//H_1$ via g, and $H_1 \cong hker(g)$ via f.

Finite groups. Using this terminology, we reexamine the sequence of Hopf algebras arising from a short exact sequence of groups

$$1 \to N \xrightarrow{f} G \xrightarrow{g} \overline{G} \to 1, \tag{4.13}$$

i. e. when $f(N)$ is a normal subgroup of G and $G/f(N) \cong \overline{G}$. Let $H_1 = RN, H = RG, \overline{H} = R\overline{G}$, and denote the Hopf algebra homomorphisms induced from $f : N \to G$ and $g : G \to \overline{G}$ also by f and g.

(4.14) PROPOSITION. *$f : H_1 \to H$ is normal with cokernel $\cong \overline{H}$ and $g : H \to \overline{H}$ is conormal with kernel $\cong H_1$.*

PROOF. Identify N with its image $f(N)$ in G, and identify $\overline{G}$ with G/N, so that f becomes the inclusion map and g the canonical map.

We first show f is normal, i. e., (after the identification) $H_1^+ H = HH_1^+$.

Since H_1^+ is generated as an H_1-module by $\{\eta - 1 | \eta \in N\}$, the left ideal HH_1^+ of H is also generated as an H-ideal by $\{\eta - 1 | \eta \in N\}$. To show that HH_1^+ is normal, that is, a two-sided ideal of H, it suffices to observe that for $\sigma \in G, \eta \in N$,

$$(\eta - 1)\sigma = \sigma(\sigma^{-1}\eta\sigma - 1)$$

which is in HH_1^+ since N is normal in G.

Now we show that $H_1^+ H$ is the kernel of the canonical ring homomorphism $g : H \to \overline{H}$.

First, for $\eta \in N$, $g(\eta - 1) = \overline{\eta} - \overline{1} = \overline{0}$, so $HH_1^+ \subset ker(g)$. Conversely, let T be a transversal of N in G. Then any ξ in $H = KG$ may be written as

$$\xi = \sum_{\tau \in T} \sum_{\eta \in N} a_{\eta,\tau} \eta \tau$$

and

$$g(\xi) = \sum_{\tau \in T} \sum_{\eta \in N} a_{\eta,\tau} \overline{\tau}.$$

If $g(\xi) = \overline{0}$ in $\overline{H}$, then $\sum_{\eta \in N} a_{\eta,\tau} = 0$ for each $\tau \in T$, so

$$\xi = \sum_{\tau \in T} (\sum_{\eta \in N} a_{\eta,\tau}(\eta - 1))\tau \in H_1^+ H.$$

Thus $ker(g) = H_1^+ H$, and hence $\overline{H} = H//H_1$.

Now we show that $H_1 = KN$ is the set of left and right coinvariants of the canonical map $g : H \to \overline{H}$; that is, $H_1 = KN$ is the difference kernel, or equalizer, of the two maps $(g \otimes 1)\Delta$ and $(i\varepsilon \otimes 1)\Delta : H \to \overline{H} \otimes H$, and also of the two maps $(1 \otimes g)\Delta$ and $(1 \otimes i\varepsilon)\Delta : H \to H \otimes \overline{H}$. This will show that $H \to \overline{H}$ is conormal with kernel H_1.

We consider the first difference kernel ${}^{co(g)}H$:

$$(g \otimes 1)\Delta(\sum_\sigma a_\sigma \sigma) = \sum_\sigma a_\sigma \overline{\sigma} \otimes \sigma$$

while

$$(i\varepsilon \otimes 1)\Delta(\sum_\sigma a_\sigma \sigma) = \sum_\sigma a_\sigma \overline{1} \otimes \sigma.$$

Since $\{\overline{1} \otimes \sigma | \sigma \in G\}$ is an $\overline{H}$-basis of $\overline{H} \otimes H$,

$$\sum_\sigma a_\sigma \overline{\sigma} \otimes \sigma = \sum_\sigma a_\sigma \overline{1} \otimes \sigma$$

iff for all $\sigma \in G$, $a_\sigma \overline{\sigma} = a_\sigma \overline{1}$, iff $\overline{\sigma} = \overline{1}$ or $a_\sigma = 0$, iff $\sigma \in N$ or $a_\sigma = 0$.

So ${}^{co(g)}H$, the difference kernel of $(g \otimes 1)\Delta$ and $(\varepsilon \otimes 1)\Delta$, is RN. Since RG is cocommutative, ${}^{co(g)}H = H^{co(g)}$ and g is conormal. □

For the sequence (4.3) of duals arising from (4.13) we note that g^* is normal because RG^* is commutative. We also have

(4.15) PROPOSITION. *For $f : N \to G$ a homomorphism of groups, f^* is conormal iff $f(N)$ is a normal subgroup of G.*

PROOF. Let $f^* : RG^* \to RN^*$. Then

$$\begin{aligned} RG^{*co(f^*)} &= \{\phi \in RG^* | (1 \otimes f^*)\Delta(\phi) = \phi \otimes \overline{1} \in RG^* \otimes RN^*\} \\ &= \{\phi \in RG^* | \sum \phi_{(1)} \otimes (\phi_{(2)} f) = \phi \otimes \overline{1}\} \\ &= \{\phi \in RG^* | \sum \phi_{(1)}(\sigma)\phi_{(2)}(f(\eta)) = \phi(\sigma)\varepsilon(\eta) \text{ for all } \sigma \in G, \eta \in N\} \\ &= \{\phi \in RG^* | \phi(\sigma f(\eta)) = \phi(\sigma) \text{ for all } \sigma \in G, \eta \in N\}. \end{aligned}$$

So $RG^{*co(f^*)}$ consists of functions constant on left cosets $\sigma f(N)$ for $\sigma \in G$.

Similarly, ${}^{*co(f^*)}RG$ consists of functions constant on right cosets $f(N)\sigma$ for $\sigma \in G$. Thus f^* is conormal iff any function on G which is constant on cosets on one side is constant on cosets on the other side, hence, iff the right cosets of $f(N)$ in G and the left cosets of $f(N)$ in G are equal, i. e., iff $f(N)$ is a normal subgroup of G. □

We leave the results that $R(G/f(N))^* = hker(f^*)$ and $RN^* = RG^*//\overline{RG}^*$ as exercises.

Exactness for Hopf algebras. We wish to extend these results to more general finite Hopf algebras. We will show that if R is a field, then:

(4.16). If $f : H_1 \to H$ is an injective normal Hopf algebra morphism of finite R-Hopf algebras and $g : H \to H//f(H_1)$ is the canonical map, then g is conormal and $f(H_1) = {}^{co(g)}H$;

If $g : H \to \overline{H}$ is a surjective conormal Hopf algebra morphism and $f : {}^{co(g)}H \to H$ is the inclusion, then f is normal and $\overline{H} \cong H//{}^{co(g)}H$.

We begin with the first result.

(4.17) Proposition [**Mt93, 3.4.3**]. *Let $f : H_1 \to H$ be a normal injective Hopf algebra morphism, H finite. Suppose R is a field or H is a free $f(H_1)$- module. Let $\overline{H} = H/(f(H_1^+))H$ and $g : H \to \overline{H}$ the canonical map. Then $f(H_1) = {}^{co(g)}H = H^{co(g)}$.*

Proof. Identify H_1 with its image $f(H_1)$, then we can assume f is the inclusion map. If R is a field and H is a finite R-Hopf algebra, then H is a free H_1-module by the Nichols-Zoeller Theorem [**Mt93, Chapter 3**].

We show $H_1 = H^{co(g)}$.

Since H is free over H_1, H_1 is the difference kernel of the two maps

$$e_1, e_2 : H \to H \otimes_{H_1} H$$

given by $e_1(x) = x \otimes 1, e_2(x) = 1 \otimes x$: that is,

$$0 \to H_1 \to H \xrightarrow{e_1 - e_2} H \otimes_{H_1} H \tag{4.18}$$

is exact as R-modules. Similarly, by definition,

$$0 \to H^{co(g)} \to H \xrightarrow{(1\otimes g)\Delta - (1 \otimes i\varepsilon)\Delta} H \otimes_K \overline{H} \tag{4.19}$$

is exact as R-modules.

Define $\gamma_0 : H \otimes_R H \to H \otimes_R \overline{H}$ by

$$\gamma_0(x \otimes y) = \sum_{(y)} xy_{(1)} \otimes g(y_{(2)}).$$

Then γ_0 induces a well-defined homomorphism

$$\gamma : H \otimes_{H_1} H \to H \otimes_R \overline{H} :$$

to see this, observe that any $h \in H_1$ may be written as $h = \varepsilon(h) + h_1$ with $h_1 \in H_1^+$, hence $g(h) = \varepsilon(h)$; hence

$$\begin{aligned}\gamma_0(x \otimes hy) &= \sum xh_{(1)}y_{(1)} \otimes g(h_{(2)})g(y_{(2)}) \\ &= \sum xh_{(1)}\varepsilon(h_{(2)})y_{(1)} \otimes g(y_{(2)}) \\ &= \sum xhy_{(1)} \otimes g(y_{(2)}) \\ &= \gamma_0(xh \otimes y).\end{aligned}$$

Then γ has an inverse induced by $\beta : H \otimes_R H \to H \otimes_{H_1} H$ given by

$$\beta(x \otimes y) = \sum_{(y)} x\lambda(y_{(1)}) \otimes y_{(2)}.$$

It is easy to see that if $y \in H_1^+ H$ then $\beta(x \otimes y) = 0$, so β induces a well-defined map

$$\overline{\beta} : H \otimes_K \overline{H} \to H \otimes_{H_1} H$$

which is the inverse of γ.

Also, $H_1 \subset H^{co(g)}$, that is, for $x \in H_1$,

$$(1 \otimes g)\Delta(x) = (1 \otimes i\varepsilon)\Delta(x) :$$

to see this, we have

$$(1 \otimes g) \sum x_{(1)} \otimes x_{(2)} = \sum x_{(1)} \otimes g(x_{(2)});$$

now since $\overline{H} = H/H_1^+ H$, the image of H_1 in $\overline{H}$ is $H_1/H_1^+ \cong R$, where the isomorphism is induced by ε. Hence for $x \in H_1$,

$$\sum x_{(1)} \otimes g(x_{(2)}) = \sum x_{(1)} \otimes \varepsilon(x_{(2)}) = (1 \otimes i\varepsilon)\Delta(x).$$

Now the diagrams

$$\begin{array}{ccc} H & \xrightarrow{e_1} & H \otimes_{H_1} H \\ \| & & \downarrow \gamma \\ H & \xrightarrow{(1\otimes i\varepsilon)\Delta} & H \otimes_R \overline{H} \end{array}$$

and

$$\begin{array}{ccc} H & \xrightarrow{e_2} & H \otimes_{H_1} H \\ \| & & \downarrow \gamma \\ H & \xrightarrow{(1\otimes g)\Delta} & H \otimes_R \overline{H} \end{array}$$

commute. Hence we have a commutative diagram of R-modules involving the exact sequences (4.18) and (4.19):

$$\begin{array}{ccccccc} 0 & \longrightarrow & H_1 & \longrightarrow & H & \xrightarrow{e_1 - e_2} & H \otimes_{H_1} H \\ & & \downarrow i & & \| & & \downarrow \gamma \\ 0 & \longrightarrow & H^{co(g)} & \longrightarrow & H & \xrightarrow{(1\otimes g)\Delta - (1\otimes i\varepsilon)\Delta} & H \otimes_K \overline{H} \end{array}$$

where i is the inclusion map. So $H_1 = H^{co(g)}$. The argument that $H_1 = {}^{co(g)}H$ is similar. □

Before we prove the analogous conormal theorem, we need the following result.

(4.20) LEMMA. *Let $H, \overline{H}$ be finite R-Hopf algebras, R a commutative ring, and $\rho : H \to \overline{H}$ be a surjective Hopf algebra morphism. Then there is a right $\overline{H}$-comodule splitting $j : \overline{H} \to H$ with $\rho j = 1_{\overline{H}}$.*

PROOF. The right $\overline{H}$-comodule structure on $\overline{H}$ is via $\Delta_{\overline{H}}$; that on H is by $(1 \otimes \rho)\Delta : H \to H \otimes \overline{H}$. Since $\overline{H}$ is a right $\overline{H}$-comodule, it is a left $\overline{H}^*$-module, and is in fact a projective left $\overline{H}^*$-module of rank one via the multiplication map $\overline{H}^* \otimes I \to \overline{H}$, where I is the module of left integrals of $\overline{H}$ (3.3). Hence the map $\rho : H \to \overline{H}$ splits as a map of left $\overline{H}^*$-modules, hence as an $\overline{H}$-comodule map. □

(4.21) THEOREM. *Let H be a finite R-Hopf algebra, R a commutative ring. Let $g : H \to \overline{H}$ be a surjective, conormal Hopf algebra homomorphism, and let $H^{co(g)} = H_1$. Then $H_1^+ H = H H_1^+$ and $\overline{H} \cong H//H_1$.*

PROOF. We follow [**Sw69, 16.0.2**]. We need to show that $ker(g) = H_1^+ H = H H_1^+$. We will show that $ker(g) = H_1^+ H$: a symmetric argument will show that $ker(g) = H H_1^+$.

We first show $H_1^+ H \subset ker(g)$. Since $ker(g)$ is a two-sided ideal of H it suffices to show that $H_1^+ \subset ker(g)$.

If $h \in H^+$ and $(1 \otimes g)\Delta(h) = h \otimes 1$, then

$$\begin{aligned} g(h) &= \mu(\varepsilon \otimes 1)\Delta_{\overline{H}}(g(h)) \\ &= \mu(\varepsilon \otimes 1)(g \otimes g)\Delta(h) \\ &= \mu(\varepsilon g \otimes 1)(1 \otimes g)\Delta(h) \\ &= \mu(\varepsilon g \otimes 1)(h \otimes \overline{1}) \\ &= \mu(\varepsilon(g(h)) \otimes \overline{1}) \\ &= \varepsilon(g(h)) \\ &= \varepsilon(h) = 0 \end{aligned}$$

since $h \in H_1^+$.

For the opposite inclusion, we let j be a right $\overline{H}$-comodule splitting of g given by (4.20), and define $k : H \to H$ by

$$k = \mu_H(jg \otimes \lambda)\Delta_H = jg * \lambda : H \to H$$

where $*$ here is convolution (c.f. (1.8)). We show that $Im(k) \subset H_1$, after which we will show that $ker(g) \subset H_1^+ H$, which will complete the proof.

To show that $Im(k) \subset H_1 = H^{co(g)}$ we show: for all $h \in H$,

$$(1 \otimes g)\Delta(k(h)) = k(h) \otimes 1,$$

as follows:

$$\begin{aligned}(1\otimes g)\Delta(k(h)) &= (1\otimes g)(\Delta(\sum jg(h_{(1)})\lambda(h_{(2)})))\\ &= (1\otimes g)(\sum \Delta(jg(h_{(1)}))\Delta(\lambda(h_{(2)})))\\ &= (1\otimes g)(\sum \Delta(jg(h_{(1)}))(\lambda(h_{(3)})\otimes\lambda(h_{(2)})))\\ &= \sum(1\otimes g)(\Delta(jg(h_{(1)})))(\lambda(h_{(3)})\otimes g(\lambda(h_{(2)})))\end{aligned}$$

Now $(1\otimes g)\Delta j = (j\otimes 1)\Delta$, since $gj = 1$ and j is a right $\overline{H}$-comodule homomorphism. So the last expression

$$\begin{aligned}&= \sum(j\otimes 1)\Delta(g(h_{(1)})))(\lambda(h_{(3)})\otimes g(\lambda(h_{(2)})))\\ &= \sum jg(h_{(1)})\otimes g(h_{(2)}))(\lambda(h_{(4)})\otimes g(\lambda(h_{(3)})))\\ &= \sum jg(h_{(1)})\lambda(h_{(4)})\otimes g(h_{(2)})g\lambda(h_{(3)})\\ &= \sum jg(h_{(1)})\lambda(h_{(3)})\varepsilon(h_{(2)})\otimes 1\\ &= k(h)\otimes 1.\end{aligned}$$

Now we show that $ker(g)\subset H_1^+H$:

We have $ker(g)\subset Im(jg-1)$, for if $h\in ker(g)$, then $h=(jg-1)(-h)$. To show $Im(jg-1)\subset H_1^+H$, we show:

$$(jg-1) = (1-\varepsilon)k * I = \mu((1-\varepsilon)k\otimes 1)\Delta, \tag{4.22}$$

Since $k: H\to H$ maps into H_1, $(1-\varepsilon)k$ maps H into H_1^+. Thus the right hand side of (4.22) maps to H_1^+H, and so (4.22) implies $Im(jg-1)\subset H_1^+H$.

To show (4.22) we observe that $k = jg*\lambda$, so $k*I = jg*\lambda*I = jg$. Also, setting $\varepsilon=\varepsilon_H$, we have $\varepsilon k=\varepsilon$. For

$$\begin{aligned}\varepsilon(\sum jg(h_{(1)})\lambda(h_{(2)})) &= \sum \varepsilon(jg(h_{(1)}))\varepsilon(\lambda(h_{(2)}))\\ &= \sum \varepsilon(jg(h_{(1)})\varepsilon(h_{(2)}))\\ &= \varepsilon(jg(h))\\ &= \varepsilon_{\overline{H}}(g(jg(h)))\\ &= \varepsilon_{\overline{H}}(g(h))\\ &= \varepsilon(h)\end{aligned}$$

since for all $h\in H$, $\varepsilon_H(h)=\varepsilon_{\overline{H}}(g(h))$ and $gj=1$. Hence

$$\begin{aligned}(1-\varepsilon)k*I &= (k-\varepsilon)*I\\ &= (k*I)-(\varepsilon*I)\\ &= jg-I,\end{aligned}$$

proving (4.22). That completes the proof. □

These results show that for R a field and H a finite R-Hopf algebra, if $f : H_1 \to H$ is a normal injective Hopf algebra morphism, respectively $g : H \to \overline{H}$ is a conormal surjective Hopf algebra morphism, then f, resp. g, is part of a short exact sequence of R-Hopf algebras. Thus the analogue of the situation for finite groups described at the beginning of this section is valid.

For H not necessarily finite, the theory is more difficult. See [**Mt93, chapter 3**] for a discussion.

Now we consider duals of finite R-Hopf algebras and homomorphisms.

(4.23) PROPOSITION. *Let $f : H_1 \to H$ be a Hopf algebra homomorphism between finite R-Hopf algebras. Then f is normal iff the induced map $f^* : H^* \to H_1^*$ on duals is conormal.*

PROOF. f is normal iff $f(H_1)^+H = Hf(H_1)^+$; f^* is conormal iff $H^{*co(f^*)} = {}^{co(f)}H^*$, that is, iff

$$\{\alpha \in H^* | (1 \otimes f^*)\Delta(\alpha) = \alpha \otimes 1\} = \{\alpha \in H^* | (f^* \otimes 1)\Delta(\alpha) = 1 \otimes \alpha\}.$$

Now

$$\begin{aligned} H^{*co(f)} &= \{\alpha \in H^* | \alpha_{(1)} \otimes (\alpha_{(2)} f) = \alpha \otimes 1\} \\ &= \{\alpha \in H^* | \alpha_{(1)}(h)\alpha_{(2)}(f(k)) = \alpha(h)1(k)\} \text{ for all } h \in H, k \in H_1 \\ &= \{\alpha \in H^* | \alpha_{(1)}(h)\alpha_{(2)}(f(k)) = \alpha(h)\varepsilon(k)\} \\ &= \{\alpha \in H^* | \alpha(hf(k)) = \alpha(h)\varepsilon(k)\} \\ &= \{\alpha \in H^* | \alpha(h(f(k) - \varepsilon(k) + \varepsilon(k))) = \alpha(h)\varepsilon(k)\} \\ &= \{\alpha \in H^* | \alpha(h(f(k - \varepsilon(k)))) = 0\} \\ &= \{\alpha \in H^* | \alpha(h(f(\ell)) = 0 \text{ for all } h \in H, \ell \in H_1^+\} \end{aligned}$$

and similarly for ${}^{co(f^*)}H^*$. Thus if $Hf(H_1^+) = f(H_1^+)H$, then $\alpha \in \overline{H}^{*co(f^*)}$ iff $\alpha \in^{co(f^*)} H^*$. Thus if f is normal, then f^* is conormal.

Conversely, if f^* is conormal, then

$$\{\alpha \in H^* | \alpha(H(fH_1^+)) = 0\} = \{\alpha \in H^* | \alpha(f(H_1^+)H) = 0\}.$$

But if the orthogonal complements of two subspaces of H are equal, then the spaces must be equal: $H(f(H_1^+) = f(H_1^+)H$. Hence f is normal. □

(4.24) COROLLARY. *The sequence*

$$R \to H_1 \xrightarrow{f} H \xrightarrow{g} \overline{H} \to R$$

is exact iff

$$R \to \overline{H}^* \xrightarrow{g^*} H^* \xrightarrow{f^*} H_1^* \to R$$

is exact.

PROOF. We showed that f is normal iff f^* is conormal, and g is conormal iff g^{**} is conormal, iff g^* is normal. We need to show that $\overline{H}^* = H^{*co(f^*)}$ iff $\overline{H} = H//f(H_1)$. But we just showed that

$$\begin{aligned} H^{*co(g^*)} &= \{\alpha \in H^* | \alpha(H(fH_1^+)) = 0\} \\ &= (H/Hf(H_1^+))^* \\ &= (H//f(H_1))^*. \end{aligned}$$

By the same argument on duals, $H_1 = H^{co(g)}$ iff $H_1^* = H^*//g(\overline{H}^*)$. Thus

g is conormal with Hopf kernel H_1 and f is normal with Hopf cokernel $\overline{H}$, iff f^* is conormal with Hopf kernel $\overline{H}^*$ and g^* is normal with Hopf cokernel H_1^*.□

§**5. Hopf orders.** Now we consider Hopf orders and exact sequences.

(5.1) DEFINITION. Let R be a Dedekind ring with quotient field K of characteristic 0. Let A be a finite K-Hopf algebra. A Hopf order over R in A is a finitely generated projective R-submodule H of A such that $KH = A$ and H is an R-Hopf algebra with operations induced from those on A.

For this to make sense, note that the inclusion map from H to A induces a one-to-one homomorphism ν from $H \otimes_R H$ to $A \otimes_K A$ by $\nu(\sum a_i \otimes_R b_i) = \sum a_i \otimes_K b_i$. (To show that ν is 1-1, we can assume R is local and H is R-free, in which case, if $\{h_1, \ldots, h_n\}$ is an R-basis of H, then $\{h_i \otimes_R h_j | i, j = 1, \ldots, n\}$ is an R-basis of $H \otimes_R H$ and ν maps that basis to the K-basis $\{h_i \otimes_K h_j | i, j = 1, \ldots, n\}$ of $A \otimes_K A$. Hence ν is 1-1.) We can then identify $H \otimes_R H$ as a subset of $A \otimes_K A$.

Given that identification, H is a Hopf order over R in A if

$$\mu : A \otimes_K A \to A \text{ restricts to } \mu : H \otimes_R H \to H;$$

$$\iota : K \to A \text{ restricts to } \iota : R \to H;$$

$$\Delta : A \to A \otimes_K A \text{ restricts to } \Delta : H \to H \otimes_R H;$$

$$\varepsilon : A \to K \text{ restricts to } \varepsilon : H \to R;$$

$$\lambda : A \to A \text{ restricts to } \lambda : H \to H.$$

EXAMPLE. Let $A = KG$, G a finite group. Then $H = RG$ is a Hopf order in A. and, in fact, is minimal among all Hopf orders in $A = KG$:

(5.2) PROPOSITION. *If H is a Hopf order over R in KG, then $RG \subset H$.*

PROOF. Let $H^* = Hom_R(H, R)$. Since H is a finitely generated free R-module, H^* is a finite R-Hopf order in $(KG)^* = Hom_K(KG, K)$. Now $KG^* = \sum_{\sigma \in G} Ke_\sigma$ is commutative and has a basis consisting of the pairwise orthogonal idempotents which form the dual basis to the elements of G in KG. Hence KG^* has a unique maximal order, namely $RG^* = \sum_{\sigma \in G} Re_\sigma$, as is easily checked. Then $H^* \subset RG^*$, hence $RG \subset H$. □

An arbitrary finite cocommutative K-Hopf algebra A need not have any Hopf orders over R, as we shall show in (20.5).

Now we consider exact sequences of Hopf orders.

(5.3) PROPOSITION. *Let R be a domain with quotient field K. Let A_1 be the Hopf kernel of a conormal morphism $g : A \to \overline{A}$ of K-Hopf algebras, so that*

$$K \to A_1 \to A \xrightarrow{g} \overline{A} \to K$$

is a short exact sequence of finite K-Hopf algebras (c.f. (4.12)). Let H be a Hopf order over R in A. Let $H_1 = H \cap A_1$ and $\overline{H} = g(H) \subset \overline{A}$. Then

$$R \to H_1 \to H \xrightarrow{g} \overline{H} \to R \tag{5.4}$$

is a short exact sequence of R-Hopf orders.

PROOF. Evidently g is surjective. In order to show (5.4) is exact, we must show (4.12) that H_1 is a normal subHopf algebra of H, i. e. $H_1^+H = HH_1^+$; g is conormal, i. e. ${}^{co(g)}H = H^{co(g)}$; $\overline{H} \cong H/H_1^+H$ via g; and $H_1 = H^{co(g)}$. Clearly ${}^{co(g)}H = H \cap {}^{co(g)}A = H_1 = H \cap A_1 = H \cap A^{co(g)} = H^{co(g)}$, so $g|_H$ is conormal. We need to show that

$$H_1^+H = ker(g|_H) = (A_1^+A) \cap H.$$

But since $H_1 = H^{co(g)}$, the first equality follows from (4.21); the second is clear. Since the inclusion map $A_1 \to A$ is normal, it follows that $H_1^+H = A_1^+A \cap H = AA_1^+ \cap H = HH_1^+$ and so H_1 is normal in H. □

Now we consider integrals for orders.

(5.5) PROPOSITION. *Let R be a principal ideal domain, and H and H^* be finite unimodular (3.2) R-Hopf algebras. Let θ generate the module of integrals of H. Then there is an α in H^*, a generator of the module of integrals of H^*, such that $\alpha \cdot \theta = 1$ in H.*

PROOF. Recall that H^* acts on H by $f \cdot h = \sum \langle f, h_{(2)} \rangle h_{(1)}$. This action yields the scalar multiplication isomorphism m from $H^* \otimes R\theta$ onto H, where $m(f, r\theta) = f \cdot r\theta$ (c. f. (3.3)). So $1 = \alpha \cdot \theta$ for some unique $\alpha \in H^*$. Then for any $f \in H^*$, $f \cdot 1 = f\alpha \cdot \theta$. But

$$f \cdot 1 = \langle f, 1 \rangle 1 = \varepsilon(f)1 = \varepsilon(f)(\alpha \cdot \theta) = (\varepsilon(f)\alpha) \cdot \theta.$$

Since m is an H^*-module isomorphism, $\varepsilon(f)\alpha = f\alpha$ for all $f \in H^*$, hence α is an integral of H^*. If β is a generator of the integrals of H^* and $\alpha = s\beta$ for $s \in R$, then

$$1 = \alpha \cdot \theta = s(\beta \cdot \theta)$$

and $\beta \cdot \theta \in H$. Thus s is a unit of H, hence of R, and so α generates the integrals of H^*. □

(5.6) PROPOSITION. *Let*

$$K \to A_1 \xrightarrow{\iota} A \xrightarrow{\pi} \overline{A} \to K$$

be a short exact sequence of finite, unimodular K-Hopf algebras and H an R-Hopf order in A. Let $\overline{H} = \pi(H)$ and $H_1 = \{x \in A_1 | \iota(x) \in H\}$. Let $\overline{\theta}, \theta_1$ be integrals of $\overline{H}, H_1$, respectively. Let θ be a preimage in H of $\overline{\theta}$. Then $\iota(\theta_1)\theta$ is an integral of H.

PROOF. Let $x \in H$. Then

$$\pi(x\theta) = \pi(x)\overline{\theta} = \varepsilon(\pi(x))\overline{\theta} = \varepsilon(x)\overline{\theta} = \pi(\varepsilon(x)\theta)$$

since π is a Hopf algebra homomorphism. So

$$x\theta - \varepsilon(x)\theta = y \in H \cap A\iota(A_1^+) = ker\pi|_H.$$

Then

$$x\theta\iota(\theta_1) = \varepsilon(x)\theta\iota(\theta_1) + y\iota(\theta_1).$$

Since $y = \sum_i z_i\iota(w_i)$ for some $z_i \in A, w_i \in A_1^+$, we have

$$y\iota(\theta_1) = \sum z_i\iota(w_i\theta_1) = \sum z_i\iota(\varepsilon(w_i)\theta_1) = 0.$$

So $\theta\iota(\theta_1)$ is an integral of H. □

Suppose K has characteristic zero and A is cocommutative and finite of rank n over K, and let

$$K \to A_1 \to A \to \overline{A} \to K$$

be a short exact sequence of K-Hopf algebras. Then A^* is commutative and a separable K-algebra [**Wa79, (16.3)**]. Hence there is a Galois extension L of K with Galois group Ω so that $L \otimes A^*$ splits, that is, $L \otimes A^* \cong L \times \ldots \times L$ (n copies), and also $L \otimes A_1$ and $L \otimes \overline{A}$ split. Let $e_i : L \otimes A^* \to L$ be the ith coordinate function, then $\{e_1, \ldots e_n\}$ is an L-basis of pairwise orthogonal idempotents of $L \otimes A^*$. Let $\{f_1, \ldots f_n\}$ be the dual basis in $L \otimes A$. Then each f_i is grouplike (see remark above (1.7), or proof of (6.3) below), hence $\{f_1, \ldots f_n\}$ is a group G and $L \otimes A \cong LG$, a group ring. Similarly, $L \otimes A_1$ and $L \otimes \overline{A}$ are group rings LN and $L\overline{G}$, respectively, where $N \subset G$ and $\overline{G} = G/N$. Then Ω acts as group automorphisms on L, G, N and $\overline{G}$ and $A = (LG)^\Omega, A_1 = (LN)^\Omega, \overline{A} = (L\overline{G})^\Omega$. Similarly,

$$K \to \overline{A}^* \to A^* \to A_1^* \to K$$

is the sequence of Ω-fixed subHopf algebras of

$$L \to L\overline{G}^* \to LG^* \to LN^* \to L.$$

(5.7) LEMMA. *If A is a K-Hopf algebra and $A = (LG)^\Omega$ for some action of $\Omega = Gal(L/K)$ on G, then A is unimodular.*

PROOF. Let $\Sigma_G = \sum_{\sigma \in G} \sigma$, then $\Sigma_G \in LG^\Omega$, hence is a non-zero two-sided integral of A. The left (right) integrals of A are a one-dimensional K-space, so every left (right) integral of A is a multiple of Σ_G. □

Let $\Sigma_N = \sum_{\sigma \in N} \sigma, \Sigma_{\overline{G}} = \sum_{\overline{\sigma} \in \overline{G}} \overline{\sigma}$. Then Σ_N and $\Sigma_{\overline{G}}$ span the spaces of integrals of $A_1, \overline{A}$, respectively, since $N, \overline{G}$ are Ω-sets. Similarly, set

$$\begin{aligned} \rho_N(\sigma) &= 1 \text{ for } \sigma = 1 \\ &= 0 \text{ for } \sigma \neq 1, \sigma \in N \\ \rho_G(\sigma) &= 1 \text{ for } \sigma = 1 \\ &= 0 \text{ for } \sigma \neq 1, \sigma \in G \\ \rho_{\overline{G}}(\overline{\sigma}) &= 1 \text{ for } \overline{\sigma} = \overline{1} \\ &= 0 \text{ for } \overline{\sigma} \neq 1, \sigma \in \overline{G}. \end{aligned}$$

Then $\rho_N, \rho_G, \rho_{\overline{G}}$ are in $A_1^*, A^*, \overline{A}^*$, respectively, since for $\omega \in \Omega, f \in LG^*$,

$$(\omega f)(\sigma) = \omega(f(\omega^{-1}(\sigma)))$$

and $\omega(1) = 1$. Hence $\rho_N, \rho_G, \rho_{\overline{G}}$ span the spaces of integrals of $A_1^*, A^*, \overline{A}^*$, respectively. Also,

$$\langle \Sigma_G, \rho_G \rangle = \sum_{\sigma \in G} \rho_G(\sigma) = 1$$

and similarly,

$$\langle \sigma_N, \rho_N \rangle = 1 = \langle \Sigma_{\overline{G}}, \rho_{\overline{G}} \rangle.$$

Lemma (5.7) applies to any finite cocommutative K-Hopf algebra (when K has characteristic zero). Thus any integral of a Hopf order in a finite cocommutative K-Hopf algebra is a multiple of the integral Σ_G of a suitable group ring over an extension of K. Our next result describes the relationship among the integrals of an exact sequence of Hopf orders in terms of the integrals of the corresponding group rings.

(5.8) PROPOSITION. *Let R be the valuation ring of a local field K. Let*

$$K \to A_1 \to A \to \overline{A} \to K$$

be a short exact sequence of finite, cocommutative K-Hopf algebras, let H be a Hopf order over R in A, and let

$$R \to H_1 \to H \to \overline{H} \to R$$

be the corresponding short exact sequence of R-Hopf algebras (c.f. (5.3)). Let $\theta_1 = \frac{1}{n}\Sigma_N$ generate the module of integrals of H_1, let $\theta = \frac{1}{b}\Sigma_G$ generate the module of integrals of H, and $\overline{\theta} = \frac{1}{m}\Sigma_{\overline{G}}$ generate the module of integrals of $\overline{H}$, for some $m, n, b \in R$. Then $mnR = bR$.

(5.9) LEMMA. *If $\theta = \frac{1}{b}\Sigma_G$ generates the module of integrals of H, then $\phi = b\rho_G$ generates the module of integrals of H^*.*

PROOF OF (5.9). Given $\theta = \frac{1}{b}\Sigma_G$, let ϕ be an integral of H^* so that $\phi \cdot \theta = 1$ (c.f. (5.5)). The H^*-module action isomorphism

$$H^* \otimes_R R\theta \to H$$

is the restriction of the corresponding isomorphism

$$A^* \otimes_K K\theta \to A.$$

Now $\rho_G \cdot \Sigma_G = 1$ in A, so $b\rho_G \cdot \theta = 1 = \phi \cdot \theta$. Hence $\phi = b\rho_G$. □

PROOF OF (5.8). Let θ' be a preimage in H of $\overline{\theta} \in \overline{H}$. Then $\theta'\theta_1$ is an integral of H by (5.6), and so $\theta'\theta_1 = c\theta$ for some $c \in R$, or

$$\theta' \frac{1}{n} \Sigma_N = c(\frac{1}{b} \Sigma_G).$$

Apply π: we get

$$\frac{1}{n}[N:1] \cdot \frac{1}{m} \Sigma_{\overline{G}} = c \frac{1}{b} [N:1] \Sigma_{\overline{G}}.$$

Hence $nm = \frac{b}{c}$. Now we do the same thing with the duals. By Lemma (5.9), $n\rho_N, m\rho_{\overline{G}}$ and $b\rho_G$ generate the modules of integrals of $H_1^*, \overline{H}^*$ and H^*. Let ρ be a preimage in H^* of $n\rho_N$ in A_1^*, and let $\rho'_{\overline{G}}$ be the image of $\rho_{\overline{G}}$ in H^*, then $\rho'_{\overline{G}}(\sigma) = \rho_{\overline{G}}(\overline{\sigma})$. Then $m\rho'_{\overline{G}}\rho$ is an integral of H^* by (5.6), hence

$$m\rho'_{\overline{G}}\rho = db\rho_G$$

for some $d \in R$. Applying the two sides to $1 \in N$ yields

$$m\rho'_{\overline{G}}(1)\rho(1) = db\rho_G(1).$$

Since $\rho(1) = n\rho_N(1)$ and $1 = \rho_G(1) = \rho_N(1) = \rho'_{\overline{G}}(1)$, we get $mn = db$. Hence $cd = 1$ in R and $mnR = bR$. □

The next result, due to Byott, will yield information on tame extensions in section 28.

THEOREM (5.10). *Let K be a local field with valuation ring R. Let G be a finite group and N a normal subgroup of G, $\overline{G} = G/N$, and $\pi : KG \to K\overline{G}$ the Hopf algebra homomorphism induced by the canonical map from G to $\overline{G}$. Let H be a Hopf order over R in KG, $H_1 = KN \cap H$ and $\overline{H} = \pi(H)$, so that*

$$R \to H_1 \to H \to \overline{H} \to R$$

is an exact sequence of R-Hopf algebras. Let N act on KG by left translation. Let $I(H_1)$ denote the module of integrals of H_1. Then $H^N = I(H_1)H$.

Before beginning the proof, recall that if Σ_N, Σ_G and $\Sigma_{\overline{G}}$ are the sums of the elements of $N, G, \overline{G}$, respectively, and $\frac{1}{n}\Sigma_N$ and $\frac{1}{m}\Sigma_{\overline{G}}$ generate the integrals of H_1 and $\overline{H}$, respectively, for some m, n in R, then $\frac{1}{mn}\Sigma_G$ generates the integrals of H by (5.8).

PROOF. Let

$$e_N = \frac{\Sigma}{|N|} = \frac{n}{|N|}(\frac{\Sigma}{n}),$$

a scalar multiple of the generator $\frac{\Sigma}{n}$ of the module of integrals for H_1. Then e_N is an idempotent of KN and $KGe_N = KG^N$, as we will show shortly. The idea of the proof is to understand how He_N relates to H^N. Note that since N is a normal subgroup of G, e_N is in the center of KG.

First, observe that $H^N \subset He_N \subset KG^N$. For if $x = \sum_{\sigma \in G} r_\sigma \sigma \in H^N$, then $r_{\eta\sigma} = r_\sigma$ for all $\eta \in N$, so

$$x = \sum_{\tau \in T} r_\tau (\sum_{\eta \in N} \eta)\tau$$

where T is a transversal of N in G; hence

$$\begin{aligned} x &= \sum_{\tau \in T} r_\tau \Sigma_N \tau \\ &= \sum_{\tau \in T} r_\tau \Sigma_N e_N \tau \\ &= x e_N. \end{aligned}$$

Also, for $\eta \in N$,

$$\begin{aligned} \eta(xe_N) &= \sum_\sigma r_\sigma \eta\sigma e_N = \sum_\sigma r_\sigma \eta e_N \sigma \\ &= \sum_\sigma r_\sigma e_N \sigma = x e_N. \end{aligned}$$

We will show that

$$H^N = \frac{|N|}{n} He_N = (\frac{\Sigma}{n})H = I(H_1)H.$$

To see this, define $\theta : He_N \to KG^N$ by $\theta(y) = \frac{|N|}{n} y$. To show that $\theta(He_N) = H^N$, we factor θ through $(H^*)^N$.

Since $\frac{1}{b}\Sigma_G$ generates the module of integrals of H, the module of integrals of H^* is generated by $b\rho_G$, and the H-action on H^* defines an isomorphism

$$H \cong H \otimes_R Rb\rho_G \cong H^*$$

by $y \mapsto yb\rho_G$. Let θ_2 be the restriction to H^N of the H-action isomorphism $H \to H^*$, then θ_2 is an isomorphism from H^N onto $(H^*)^N$.

We seek an isomorphism

$$\theta_1 : He_N \to (H^*)^N$$

so that $\theta_2^{-1}\theta_1 = \theta$. We obtain θ_1 as a composite of three isomorphisms, as follows.

First, since $(KN)^+ e_N = 0$, the canonical map π yields by restriction a map

$$\pi_r : KGe_N \to K\overline{G}$$

by $\pi_r(xe_N) = \pi(xe_N) = \pi(x)\pi(e_N) = \pi(x)$. Then π_r is clearly onto. To show π_r is 1-1: if $\pi_r(xe_N) = 0$ then $\pi(x) = 0$, hence $x \in (KN)^+KG$; but since $(KN)^+e_N = 0$, therefore $xe_N = 0$. Thus π_r is an isomorphism. If we denote the restriction of π_r to He_N also by π_r, then since $\pi : H \to \overline{H}$ is onto by definition of $\overline{H}$, π_r maps He_N onto $\overline{H}$, and is 1-1 since it is the restriction of a 1-1 map over K. Hence $\pi_r : He_N \to \overline{H}$ is an isomorphism.

Now $\overline{H} \cong (\overline{H})^*$: the module of integrals of $\overline{H}^*$ is generated by $m\rho_{\overline{G}}$ where $\frac{1}{m}\Sigma_{\overline{G}}$ generates the module of integrals of $\overline{H}$. Then the $\overline{H}$-module structure on $\overline{H}^*$ yields an isomorphism

$$\overline{H} \cong \overline{H} \otimes m\rho_{\overline{G}}R \cong \overline{H}^*$$

given by $y \mapsto ym\rho_{\overline{G}}$.

Finally, we have an isomorphism $(\overline{H})^* \to (H^*)^N$, defined as follows. The 1-1 homomorphism

$$\pi^* : Hom_K(K\overline{G}, K) \to Hom_K(KG, K)$$

by $\pi^*(f) = f \cdot \pi$, clearly maps $K\overline{G}^*$ into

$$(KG^*)^N = Hom_K(KG, K)^N = \{f : KG \to K | f(\eta\sigma) = f(\sigma) \text{ for all } \eta \in N\}$$

(N acts trivially on K) since $f(\pi(\eta\sigma)) = f(\pi(\sigma))$ for all $\eta \in N, \sigma \in G$. But π^* maps onto $(KG^*)^N$: for if $f \in Hom_K(KG, K)$ and $f(\eta\sigma) = f(\sigma)$ then the map $f_0 \in K\overline{G}^*$ defined by $f_0(\overline{\sigma}) = f(\sigma)$ is well-defined. Then π^* yields an isomorphism

$$\overline{H}^* = H^* \cap K\overline{G}^* \to (H^*)^N$$

since the intersection $\overline{H}^* = H^* \cap K\overline{G}^*$ is the intersection in KG^* of $\pi^* K\overline{G}^* = (KG^*)^N$ with H^*. Thus the image in $(H^*)^N$ of $\rho_{\overline{G}}$ is $\pi^*\rho_{\overline{G}} = \hat{\rho}_N$, defined by $\hat{\rho}_N(\sigma) = 1$ if $\sigma \in N$, $= 0$ if $\sigma \notin N$; and for $y \in H$, $\pi^*(\pi(y)\rho_{\overline{G}}) = y\hat{\rho}_N$, since for $z \in H_1$,

$$\begin{aligned}\langle \pi^*(\pi(y)\rho_{\overline{G}}), z\rangle &= \langle \pi(y)\rho_{\overline{G}}, \pi(z)\rangle = \langle \rho_{\overline{G}}, \pi(zy)\rangle \\ &= \langle \hat{\rho}_N, zy\rangle = \langle y\hat{\rho}_N, z\rangle.\end{aligned}$$

Let θ_1 be the composite isomorphism:

$$He_N \to \overline{H} \to \overline{H}^* \to (H^*)^N.$$

Then for $xe_N \in He_N$,

$$xe_N \mapsto \pi(xe_N) \mapsto \pi(xe_N)m\rho_{\overline{G}} \mapsto xe_N m\hat{\rho}_N.$$

Now $\theta_2 : H^N \cong (H^*)^N$ is defined by $\theta_2(y) = yb\rho_G$. But if $y \in H^N$ then

$$\theta_2(y) = yb\rho_G = bye_N\rho_G = by\frac{1}{|N|}\hat{\rho}_N$$

since

$$e_N\rho_G = e_N\ell_1 = \sum_{\sigma \in G} \ell_\sigma \langle e_N, \ell_{\sigma^{-1}}\rangle = \sum_{\sigma \in N} \ell_\sigma \frac{1}{|N|} = \frac{1}{|N|}\hat{\rho}_N$$

where $\ell_\sigma : G \to K$ is defined by $\ell_\sigma(\tau) = \delta_{\sigma,\tau}$. Since $\theta : He_N \to KG^N$ is defined by $\theta(xe_N) = \frac{|N|}{n} xe_N$, $\theta_2^{-1}\theta_1 = \theta$, hence θ is an isomorphism from He_N onto H^N, as we needed to show. □

COROLLARY (5.11). *Let M be a free rank one left H-module. Then $M^N = I(H_1)M$.*

PROOF. Since $M = Hz$ is free of rank one over H, this is clear. □

CHAPTER 2

Hopf Galois Structures on Separable Field Extensions

Chase and Sweedler [**CS69**] developed the concept of Hopf Galois extension in the late 1960s with an intention of applying it to purely inseparable field extensions L/K. While the concept had some value for "height one "purely inseparable extensions, it turned out that a Hopf algebra of rank equal to the dimension of L over K was too small to encompass the full "automorphism "structure of a purely inseparable extension L/K of arbitrary height. (See Chase [**Cs76**] for a definitive treatment of such extensions.)

It came as something of a surprise, therefore, when Greither and Pareigis, in 1987, demonstrated that the concept of Hopf Galois extension could be applied extensively to separable field extensions, and especially to classical Galois (normal and separable) field extensions.

The purpose of this chapter is to describe the classification of Hopf Galois structures on a separable field extension due to Greither and Pareigis, and subsequent work of Byott, the author and others. The existence of non-classical Hopf Galois structures on Galois extensions of number fields suggests intriguing potential applications to Galois module theory.

§**6. Greither-Pareigis theory [GP87].** Recall (2.7) that if $L \supset K$ is a finite extension of fields, and H a finite cocommutative K-Hopf algebra, then L is an H- Galois extension of K ("L/K is H- Galois "), if L is an H-module algebra and the K-linear map $j : L \otimes_K H \to End_K(L)$, defined by $j(s \otimes h)(t) = s(ht)$ for $h \in H$, $s, t \in L$, is bijective.

Suppose L is a separable field extension of K with normal closure E so that $Gal(E/K) = G$. Suppose L/K is H-Galois. Then (2.11) $E \otimes_K L/E$ is $E \otimes_K H$-Galois, and one can recover the action of H on L by identifying H and L with the fixed rings $(E \otimes_K H)^G$ and $(E \otimes_K L)^G$, where G acts via its action on E.

This observation leads to Greither and Pareigis' classification of Hopf Galois structures on L/K: the strategy is to classify those Hopf Galois structures on $E \otimes_K L$ fixing E on which G acts, and then take the rings of invariants under the G-action.

This strategy is facilitated by the special form of $E \otimes_K L$, and hence $E \otimes_K H$, which permits a complete description of Galois structures of $E \otimes_K L$ over E.

We first see what an H-Galois extension L/K looks like after base change to E.

Let $G = Gal(E/K)$, $G' = Gal(E/L)$ and $X = G/G'$. Tensoring with E yields an $E \otimes_K H$-Galois extension $E \otimes_K L$ of E, arising from an $E \otimes_K H$-module algebra action

$$(E \otimes_K H) \otimes_E (E \otimes_K L) \to E \otimes_K L.$$

(6.1) PROPOSITION. *The map*

$$\overline{\gamma}' : E \otimes L \to Map(G/G', E)$$

defined by

$$\overline{\gamma}'(n \otimes \ell)(\overline{\sigma}) = n\sigma(\ell)$$

is an E-algebra, G-module isomorphism, where G acts on $Map(G/G, E)$ by

$$\tau(f)(\overline{\sigma}) = \tau(f(\overline{\tau^{-1}\sigma})),$$

where $\tau \in G$, $\overline{\sigma} \in G/G$ is the left coset of $\sigma \in G$, and $f \in Map(G/G, E)$.

PROOF. The map $\gamma' : E \otimes L \to Map(G, E)$, $\gamma'(n \otimes \ell) = n\sigma(\ell)$ is the restriction to $E \otimes L$ of the E-algebra, G-module homomorphism

$$\gamma : E \otimes E \to Map(G, E),$$

$\gamma(n \otimes m)(\sigma) = n\sigma(m)$. The map γ is a G-module, E-algebra homomorphism where G and E act on $E \otimes E$ via the left factor, and is an isomorphism since E/K is a Galois extension. One sees easily that γ' factors through the natural map $Map(G/G', E) \to Map(G, E)$, yielding the 1-1 map $\overline{\gamma}'$. Then $\overline{\gamma}'$ is bijective since both domain and range have the same dimension over K. □

(6.2) DEFINITION. Let $Perm(X)$ be the group of permutations of the set X. A subgroup $N \subset Perm(X)$ is *regular* if any two of the following are satisfied:

i) N and X have the same cardinality,
ii) N acts transitively on X (i.e. for all x and $y \in X$ there exists $\eta \in N$ such that $\eta x = y$)
iii) The stabilizer $Sta_N(x) = \{\eta \in N | \eta x = x\}$ is trivial (contains only the identity element of N) for all $x \in X$.

Thus $N \subset Perm(X)$ is regular iff for some $x \in X$ (hence for all x in X), the map from N to X by $\eta \mapsto \eta(x)$ is bijective.

Let X be a finite set, E a field, and for x in X let $u_x : X \to E$ be defined by $u_x(y) = \delta_{x,y}$ for all $y \in X$. Then $\{u_x | x \in X\}$ is an E-basis of the E-vector space $XE = Map(X, E)$ and is a set of primitive pairwise orthogonal idempotents of XE viewed as an E-algebra via pointwise operations.

The next result describes the Hopf Galois structures on such "split "algebras XE:

(6.3) THEOREM. *Let X be a finite set, E a field, and $XE = Map(X, E)$. If H is an E-Hopf algebra such that XE is an H-Hopf Galois extension of E, then H is a group ring EN for N some group of the same cardinality as X. N may be identified as a subgroup of $Perm(X)$ where the action of N on X is defined by*

$u_{\eta(x)} = \eta(u_x)$ for all $x \in X, \eta \in N$: N is then a regular subgroup of $Perm(X)$. Conversely, if N is a regular subgroup of $Perm(X)$, then XE is EN-Galois.

PROOF. Suppose XE is H-Galois over E. Then

$$\begin{aligned} E \times \ldots \times E \quad (n^2 \text{ copies}) &\cong Map(X \times X, E) \\ &\cong XE \otimes_E XE \\ &\cong XE \otimes_E H^* \\ &\cong H^* \times \ldots H^* \quad (n \text{ copies}) , \end{aligned}$$

where $n = |X|$, $H^* = Hom_E(H, E)$ is the dual E-Hopf algebra, and all isomorphisms are as E-algebras: the third isomorphism is given since XE is H-Galois (see (2.9)). By standard theory of semisimple rings,

$$H^* \cong E \times \ldots \times E \ (n \text{ copies})$$

as E-algebras.

Let $\eta : H^* \to E$ be the ith coordinate function. Then $N = \{\eta_i\}$ is a basis of $H^{**} \cong H$. For each i, η_i is an E-algebra homomorphism, hence is grouplike (1.6) in H. Since the η_i are a basis of H, N consists of all the grouplike elements of H, hence N is a group (1.7) and H is the group ring EN.

Now we show that N acts as a group of permutations of X.

Let $\{u_x | x \in X\}$ be a basis of orthogonal idempotents of XE, $u_x(y) = \delta_{x,y}$ for all x, y in X. Since XE is an $H = EN$-module algebra, for $\eta \in N$, $\eta(u_x) = \eta(u_x \cdot u_x) = \Delta(\eta)(u_x \otimes u_x) = \eta(u_x)\eta(u_x)$ for $x \in X$, and $0 = \eta(u_x \cdot u_y) = \eta(u_x)\eta(u_y)$ for $x \neq y$ in X. So $\eta \in N$ maps the primitive idempotents of XE to pairwise orthogonal idempotents of XE. We show η maps primitive idempotents to primitive idempotents, as follows.

The identity element of N is the identity 1 of EN. For all $\eta \in N$, $x \in X$, $1 \cdot u_x = \eta^{-1} \cdot \eta(u_x) = u_x$, hence $\eta(u_x) \neq 0$. Also, $\eta(1) = \varepsilon(\eta) = 1$ since η is grouplike, and $1 = \sum_{x \in X} u_x$ in XE, so

$$1 = \eta(1) = \eta(\sum_{x \in X} u_x) = \sum_{x \in X} \eta(u_x) .$$

Each $\eta(u_x)$ is a sum of primitive idempotents of XE. Since $1 = \sum_{x \in X} \eta(u_x)$ and $\eta(u_x)\eta(u_y) = 0$ for $x \neq y$, each primitive idempotent can occur as a summand of $\eta(u_x)$ for exactly one $x \in X$. Since $\sum_{x \in X} \eta(u_x)$ is the sum of $|X|$ non-zero terms and altogether is a sum of $|X|$ primitive idempotents, each $\eta(u_x)$ must be a primitive idempotent u_y for some $y \in X$. Thus N embeds in $Perm(X)$ where η maps to the permutation defined by $\eta(x) = y$ if $\eta(u_x) = u_y$.

To show that N is a regular subgroup of $Perm(X)$: first, we have that $|X| = |N|$ since $EN = H$ and XE is H-Galois over E. Suppose $Nu_x = \{u_y | y \in Y\}$ where Y is a proper subset of X. If $z \in X \setminus Y$, define $e_{xz} \in End_E(EN)$ by $e_{xz}(u_x) = u_z$ and $e_{xz}(u_y) = 0$ for $y \neq x$. If

$$j : XE \otimes EN \to End_E(EN)$$

is the usual map (2.7), then e_{xz} is not in $Im(j)$, for $(\sum_{w\in X,\eta\in N} \ell_{w,\eta}u_w\eta)(u_x) \in \sum_{y\in Y} Eu_y$. But this contradicts the assumption that XE/E is EN-Galois. So N acts transitively on X, and so is a regular subgroup of $Perm(X)$.

Conversely, if N is a regular subgroup of $Perm(X)$, then XE is EN-Galois over E. For if e_{xz} is defined as above for x, $z \in X$, then $\{e_{xz}|x, z \in X\}$ is an E-basis of $End_E(XE)$. If N is regular, then there exists $\eta \in N$ with $\eta(x) = z$, so $\eta(u_x) = u_z$ and

$$j(u_z \otimes \eta) = e_{xz}\ .$$

Thus $j : XE \otimes EN \to End_E(XE)$ is onto, hence, since $|X| = |N|$, a bijection. That completes the proof. □

We wish to apply this description of Hopf Galois structures on XE/E to obtain information on Hopf Galois structures on field extensions.

Suppose L is a field extension of K with normal closure E. Let $G = Gal(E/K), G' = Gal(E/L), X = G/G'$. Suppose L/K has a H-Hopf Galois structure

$$H \otimes_K L \to L.$$

Base change to E to get

$$E \otimes_K (H \otimes_K L) \cong (E \otimes_K H) \otimes_E (E \otimes_K L) \to E \otimes_K L \tag{6.4}$$

which is isomorphic to

$$EN \otimes_E XE \to XE.$$

By (6.1) the map $\gamma : E \otimes_K L \cong XE$, $\gamma(m \otimes \ell)(\overline{\sigma}) = m\sigma(\ell)$, is an isomorphism. Then by (6.3) there is a regular subgroup N of $Perm(X)$ so that $E \otimes H \cong EN$, and (6.4) becomes isomorphic to an action

$$EN \otimes_E XE \to XE. \tag{6.5}$$

The action of G on E yields actions of G on EN and XE to make (6.5) equivariant. To explain this action we need the left translation map

$$\lambda : G \to Perm(X),$$

defined by $\lambda(\tau)(\overline{\sigma}) = \overline{\tau\sigma}$.

(6.6) LEMMA. *λ is 1-1.*

PROOF. If $M = ker(\lambda)$, then

$$M \subset G' = \{\sigma \in G|\lambda(\sigma) \text{ fixes the identity coset } G' \text{ of } X\}.$$

But then E^M is a normal field extension of K containing L. Since E was the normal closure of L/K, $E = E^M$ and M is trivial. □

Now we show that the G-action on $E \otimes_K H$ induced by G acting on E translates into an action of G on EN by $\tau(\ell\eta) = \tau(\ell)(\lambda(\tau)\eta\lambda(\tau^{-1}))$ for $\ell \in E, \eta \in N, \tau \in G$: that is, the action of G on N is given by conjugation in $Perm(X)$ by $\lambda(G)$.

To see this, recall that $X = G/G'$, so we may write the basis $\{u_x | x \in X\}$ of orthogonal idempotents of $XE = Hom_E(EX, E)$ as

$$\{u_{\overline{\sigma}} | \overline{\sigma} \in X = G/G'\}:$$

for $\overline{\sigma}$, $\overline{\tau} \in G/G'$, $u_{\overline{\sigma}}(\overline{\tau}) = \delta_{\overline{\sigma},\overline{\tau}}$. Since G acts on EX and on E, G acts on $Hom_E(EX, E)$ by:

$$\tau(f)(y) = \tau(f(\tau^{-1}(y)))$$

for $\tau \in G, f \in Hom_E(EX, E), y \in EX$. Thus

$$\tau(u_{\overline{\sigma}})(\overline{\rho}) = \tau(u_{\overline{\sigma}}(\overline{\tau^{-1}\rho})) = u_{\overline{\sigma}}(\overline{\tau^{-1}\rho}) = u_{\overline{\tau\sigma}}(\overline{\rho}) = u_{\lambda(\tau)\overline{\sigma}}(\overline{\rho}).$$

Hence the action of $\tau \in G$ on the idempotents $\{u_{\overline{\sigma}} | \overline{\sigma} \in X\}$ corresponds to left translation of the cosets.

The action of N on $\{u_{\overline{\sigma}} | \overline{\sigma} \in X\}$ corresponds to an embedding of N in $Perm(X)$: $\eta(u_{\overline{\sigma}}) = u_{\eta(\overline{\sigma})}$.

Now the action

$$(E \otimes_K H) \otimes_E (E \otimes_K L) \to E \otimes_K L$$

is G-equivariant (where G acts on E). Therefore, after identifying $E \otimes H$ with EN and recalling that $\overline{\gamma'} : E \otimes_K L \to Map(X, E)$ is a G-module homomorphism, the action

$$EN \otimes XE \to XE$$

is G-equivariant. Since G respects the Hopf algebra structure of EN, G acts on N, the set of grouplike elements of EN, and we have for $\eta \in N, \overline{\sigma} \in X, \tau \in G$,

$$\tau(\eta)\tau(u_{\overline{\sigma}}) = \tau(\eta u_{\overline{\sigma}}) = \tau(u_{\eta(\overline{\sigma})}).$$

Since

$$\tau(\eta)\tau(u_{\overline{\sigma}}) = u_{\tau(\eta)(\lambda(\tau)(\overline{\sigma}))}$$

while

$$\tau(u_{\eta(\overline{\sigma})}) = u_{\lambda(\tau)\eta(\overline{\sigma})},$$

we have

$$\tau(\eta)(\lambda(\tau)(\overline{\sigma})) = \lambda(\tau)\eta(\overline{\sigma}),$$

or

$$\tau(\eta)(\overline{\sigma}) = \lambda(\tau)\eta\lambda(\tau^{-1})(\overline{\sigma}):$$

the action of τ in G on η in N is via conjugation by $\lambda(\tau)$ in $Perm(G)$.

We summarize:

(6.7) PROPOSITION. *Let L/K be an H-Galois extension with normal closure E, let $G = Gal(E/K)$, $G' = Gal(E/L)$ and $X = G/G'$. Then the base changed Galois action of H on L:*

$$(E \otimes_K H) \otimes_E (E \otimes_K L) \to E \otimes_K L$$

is equivalent to an action

$$EN \otimes_E XE \to XE$$

of $E \otimes H \cong EN$ on XE which corresponds to a regular embedding of N into $Perm(X)$ with the property that if $\lambda : G \to Perm(X)$ is left translation, then $\lambda(G)$ normalizes the image of N in $Perm(X)$. □

The theorem of Greither and Pareigis asserts that the above proposition has a converse, namely: given $N \subset Perm(X)$ normalized by $\lambda(G)$, there is a unique Galois structure on L/K which yields N:

(6.8) THEOREM. *Let L/K be a separable extension of fields with normal closure E, let $G = Gal(E/K)$, $G' = Gal(E/L)$ and $X = G/G'$. Then there is a bijection between regular subgroups N of $Perm(X)$ normalized by $\lambda(G)$ and Hopf Galois structures on L/K.*

PROOF. We already know that if N is a regular subgroup of $Perm(X)$ then XE is an EN-Galois extension of E. Now EN and XE are E-modules and the action

$$\phi : EN \otimes_E XE \to XE$$

is an E-module homomorphism. Suppose N is normalized by $\lambda(G)$. In order to show that ϕ lifts from an action over K, by Morita theory we need to show that ϕ is an $End_K(E)$-module homomorphism. Since E/K is a Galois extension with group G, $End_K(E) \cong D(E, G) = \sum_{\sigma \in G} Eu_\sigma$. Thus we need to show that $D(E, G)$ acts on EN and XE in such a way that ϕ is a $D(E, G)$-module homomorphism.

Now G acts on $XE = Map(X, E)$ by

$$\sigma(f)(\overline{\tau}) = \sigma(f(\overline{\sigma^{-1}\tau})) = \sigma(f(\lambda(\sigma^{-1})(\overline{\tau})))$$

for $f \in XE, \sigma, \tau \in G$. Also, E acts on XE by $(ef)(\overline{\tau}) = e(f(\overline{\tau}))$. Then

$$\sigma e(f)(\overline{\tau}) = \sigma(e)\sigma(f(\lambda(\sigma^{-1}))(\overline{\tau})) = \sigma(e)\sigma(f)(\overline{\tau})$$

so XE is a $D(E, G)$-module.

Also, G acts on EN by

$$\sigma(a\eta) = \sigma(a)\sigma(\eta) = \sigma(a)\lambda(\sigma)\eta\lambda(\sigma^{-1})$$

for $a \in E, \eta \in N \subset Perm(X)$, and $e(a\eta) = ea(\eta)$ for $e \in E$. Hence $(\sigma e)(a\eta) = \sigma(ea\eta) = \sigma(e)\sigma(a)\sigma(\eta) = \sigma(e)\sigma(a\eta)$, and EN is a $D(E, G)$-module.

Finally, the map ϕ is a $D(E, G)$-module homomorphism, where G acts on $EN \otimes XE$ diagonally: for $\sigma \in G, a \in E, \eta \in N, f \in XE, \overline{\tau} \in X$, we have

$$\begin{aligned}\phi(\sigma(a\eta \otimes f))(\overline{\tau}) &= \phi(\sigma(a\eta) \otimes \sigma(f))(\overline{\tau}) \\ &= \sigma(a)\sigma(\eta)\sigma(f)(\overline{\tau}) \\ &= \sigma(a)\sigma(f)(\sigma(\eta)^{-1}\overline{\tau}) \\ &= \sigma(a)\sigma(f(\lambda(\sigma^{-1})(\sigma(\eta)^{-1}\overline{\tau})))\end{aligned}$$

while

$$\begin{aligned}\sigma(\phi((a\eta\otimes f))(\overline{\tau}) &= \sigma(a\eta f)(\overline{\tau}) \\ &= \sigma(a)\sigma(\eta f)(\lambda(\sigma^{-1}))(\overline{\tau}) \\ &= \sigma(a)\sigma(f(\eta^{-1}(\lambda(\sigma^{-1}))(\overline{\tau}))\end{aligned}$$

But $\sigma(\eta) = \lambda(\sigma)\eta\lambda(\sigma^{-1})$ where $\lambda(\sigma)$ is left translation by σ on X. So $\lambda(\sigma^{-1})\sigma(\eta)^{-1} = \eta^{-1}\lambda(\sigma^{-1})$. Hence ϕ is a $D(E,G)$-module homomorphism.

Now, since E/K is a Galois extension with Galois group G, Morita theory yields an equivalence between the category of K-modules and the category of $D(E,G)$-modules, where (see (2.13)) the functors giving the equivalence are base change

$$M \mapsto E \otimes_K M$$

and G-invariants

$$N \mapsto N^G = \{n \in N | \sigma(n) = n \text{ for all } \sigma \in G\}.$$

Thus the $D(E,G)$-module map

$$EN \otimes_E XE \to XE$$

yields a unique $K = E^G$-module homomorphism

$$(EN)^G \otimes_K (XE)^G \to (XE)^G.$$

Since the G-action on EN respects the Hopf algebra structure on EN (i.e the Hopf algebra structure maps $\iota, \mu, \Delta, \varepsilon$ and λ (antipode) are $D(E,G)$-module homomorphisms), EN^G is a K-Hopf algebra. Since the G-action on XE respects the algebra structure on XE, XE^G is a K-algebra. And since $EN \otimes_E XE \to XE$ is an EN-module algebra action, so is

$$(EN)^G \otimes_K (XE)^G \to (XE)^G.$$

Finally, since $E \otimes_K (EN)^G \cong EN$, $E \otimes_K (XE)^G \cong XE$ (again by Morita theory) and the map $j : XE \otimes_E EN \to End_E(XE)$ is an isomorphism, it follows that

$$j : (XE)^G \otimes_K (EN)^G \to End_K((XE)^G)$$

is an isomorphism, and so $(XE)^G$ is an H-Galois extension of K, where $H = (EN)^G$.

Now $L \cong (XE)^G$ via the map $s \mapsto \sum_\tau \tau(s)u_{\overline{\tau}}$. To see this, observe that for $s_{\overline{\tau}}$ in E, we have

$$\sigma(\sum_{\overline{\tau}} s_{\overline{\tau}})u_{\overline{\tau}} = \sum_{\overline{\tau}} \sigma(s_{\overline{\tau}})u_{\overline{\sigma\tau}}.$$

If $\sum_{\overline{\tau}} s_{\overline{\tau}}u_{\overline{\tau}} \in (XE)^G$, that is,

$$\sigma \sum_{\overline{\tau}} s_{\overline{\tau}}u_{\overline{\tau}} = \sum_{\overline{\tau}} s_{\overline{\tau}}u_{\overline{\tau}},$$

then

$$\sum_{\overline{\tau}} \sigma(s_{\overline{\tau}}) u_{\overline{\sigma\tau}} = \sum_{\overline{\tau}} s_{\overline{\sigma\tau}} u_{\overline{\sigma\tau}}$$

so for all $\sigma \in G$, $\overline{\tau} \in G/G'$, $\sigma(s_{\overline{\tau}}) = s_{\overline{\sigma\tau}}$. In particular, for $\overline{\tau} = \overline{1}$, $s_{\overline{\sigma}} = \sigma(s_{\overline{1}})$ for all $\sigma \in G$, hence, if $\overline{\sigma} = \overline{1}$, that is, $\sigma \in G'$, then $\sigma(s_{\overline{1}}) = s_{\overline{1}}$. Setting $s_{\overline{1}} = s$, we have:

$$s \in E^{G'} = L \ ,$$

and $f = \sum_{\overline{\tau}} s_{\overline{\tau}} u_{\overline{\tau}} \in (XE)^G$ iff $s_{\overline{1}} = s$ and $f = \sum_{\overline{\tau}} \tau(s) u_{\overline{\tau}}$. The map $L \to (XE)^G$ by $s \mapsto \sum_{\overline{\tau}} \tau(s) u_{\overline{\tau}}$ is thus an isomorphism.

The map $EN \otimes_E XE \to XE$ defined by a regular subgroup N of $Perm(X)$ corresponds to a unique H-module algebra structure $H \otimes_K L \to L$ with $H = (EN)^G$. Since we showed prior to the statement of the theorem that if L/K is an H-Galois extension then the action of H on L yields a regular subgroup of $Perm(X)$ normalized by $\lambda(G)$, we get a *bijection* between regular subgroups N of $Perm(X)$ normalized by $\lambda(G)$ and Hopf Galois structures on L/K, as claimed. □

(6.9) Example. Let L/K be a Galois extension with group G. Then $X = G$. Inside $Perm(G)$ we have G embedded as a regular subgroup in two ways:

$$\lambda : G \to Perm(G) \ ,$$

$\lambda(\sigma)(\tau) = \sigma\tau$, left translation, and

$$\rho : G \to Perm(G),$$

$\rho(\sigma)(\tau) = \tau\sigma^{-1}$, right translation:

$$(\rho(\sigma_1\sigma_2))(\tau) = \tau(\sigma_1\sigma_2)^{-1} = \tau\sigma_2^{-1}\sigma_1^{-1} = \rho(\sigma_1)(\tau\sigma_2^{-1}) = \rho(\sigma_1)\rho(\sigma_2)(\tau)$$

We claim: $\lambda(G) = \rho(G) \subset Perm(G)$ iff G is abelian. For if G is abelian, then $\rho(\sigma) = \lambda(\sigma^{-1})$. Conversely, if $\rho(\sigma) = \lambda(\pi)$ for some σ, π, then necessarily $\pi = \sigma^{-1}$, since $\rho(\sigma)e = \sigma^{-1} = \lambda(\sigma^{-1})e$. However, if $\sigma\tau \neq \tau\sigma$ for some σ, τ in G, then $\rho(\sigma) \neq \lambda(\sigma^{-1})$: for $\rho(\sigma)\tau = \tau\sigma^{-1} \neq \sigma^{-1}\tau = \lambda(\sigma^{-1})\tau$. So $\rho(\sigma) \notin \lambda(G)$.

Since $\lambda(G)$ normalizes itself and commutes with $\rho(G)$ in $Perm(G)$, both $\lambda(G)$ and $\rho(G)$ are regular subgroups of $Perm(G)$ normalized by $\lambda(G)$.

(6.10) Proposition. *$\rho(G) \subset Perm(G)$ corresponds to the classical action of G on L.*

Proof. Let $N = \rho(G)$. Since $\lambda(G)$ commutes with $\rho(G)$, the action of $\lambda(G)$ on N is trivial. Hence $(LN)^G = L^G N = KN$. The action of KN on L is induced from that of KN on $(GL)^{\lambda(G)} = \{\sum_{\tau} \tau(s)u_\tau | s \in L\}$. Then if $s \in L$ corresponds to $\sum \tau(s)u_\tau \in GL$, then for $\sigma \in G$, $\rho(\sigma)(\sum_\tau \tau(s)u_\tau) = \sum_\tau \tau(s)\rho(\sigma)(u_\tau) = \sum_\tau \tau(s)u_{\tau\sigma^{-1}} = \sum_\tau \tau\sigma(s)u_\tau$ which corresponds to $\sigma(s) \in L$. That is, the action of $\rho(G)$ on $(GL)^{\lambda(G)}$ corresponds to the action of G on L. □

(6.11) COROLLARY. *If L/K is a Galois extension with non-abelian group G, then L/K has a non-classical Hopf Galois structure, namely, the structure corresponding to the regular subgroup $N = \lambda(G) \neq \rho(G)$.*

PROOF. If G is non-abelian then $\lambda(G) \neq \rho(G)$ in $Perm(G)$. So $(L\rho(G))^{\lambda(G)} = K\rho(G)$, but $(L\lambda(G))^{\lambda(G)} \ncong K\lambda(G)$. □

(6.12) EXAMPLE. Let $G = S_3 = \{\langle \sigma, \tau\rangle | \sigma^3 = \tau^2 = 1,\ \tau\sigma = \sigma^2\tau\}$, and let $N = \lambda(G)$. Then

$$(LN)^{\lambda(G)} = \{\xi \in L\lambda(G) | \xi = \tau(\xi) = \sigma(\xi)\}.$$

We identify N with G, noting that $\rho \in \lambda(G)$ acts on $\pi \in N$ by conjugation: $\pi^\rho = \rho\pi\rho^{-1}$. Then given

$$\xi = a_0 + a_1\sigma + a_2\sigma^2 + b_0\tau + b_1\sigma\tau + b_2\sigma^2\tau \in (LG)^{\lambda(G)},$$

we apply $\lambda(\tau)$ and get

$$\lambda(\tau)(\xi) = \tau(a_0) + \tau(a_1)\sigma^2 + \tau(a_2)\sigma + \tau(b_0)\tau + \tau(b_1)\sigma^2\tau + \tau(b_2)\sigma\tau;$$

thus $\lambda(\tau)(\xi) = \xi$ iff

$$\tau(a_0) = a_0, \tau(a_1) = a_2, \tau(b_0) = b_0, \text{ and } \tau(b_1) = b_2.$$

Also, we apply $\lambda(\sigma)$ to ξ and get

$$\lambda(\sigma)(\xi) = \sigma(a_0) + \sigma(a_1)\sigma + \sigma(a_2)\sigma^2 + \sigma(b_0)\sigma^2\tau + \sigma(b_1)\tau + \sigma(b_2)\sigma\tau$$

which implies that $\lambda(\sigma)(\xi) = \xi$ iff

$$\sigma(a_0) = a_0, \sigma(a_1) = a_1, \sigma(a_2) = a_2, \sigma(b_0) = b_2, \sigma(b_1) = b_0, \sigma(b_2) = b_1.$$

So

$$\begin{aligned}(LN)^{\lambda(G)} &= \{a_0 + a_1\sigma + \tau(a_1)\sigma^2 + b_0\tau + \sigma^2(b_0)\sigma\tau + \sigma(b_0)\sigma^2\tau \\ &\qquad \text{with } a_0 \in K, a_1 \in L^{\langle\sigma\rangle}, b_0 \in L^{\langle\tau\rangle}\} \\ &\cong K \times L^{\langle\sigma\rangle} \times L^{\langle\tau\rangle}\end{aligned}$$

as K-algebras.

(6.13) PROPOSITION [**GP87**]. *Let L/K be a separable field extension with normal closure E. Let $G = Gal(E/K)$ and $\overline{G} = Gal(E/L)$. If $[L : K] \leq 4$ then L/K is Hopf Galois.*

PROOF. We need to find a regular subgroup N of $Perm(G/G')$ normalized by $\lambda(G)$.

If L/K is normal, then choose $N = \rho(G)$. If $[L : K] = 3$ or 4 and $L \neq E$, we choose regular subgroups $N \subset \lambda(G) \subset Perm(G)$ as follows:

For $[L : K] = 3$, $G = S_3$, $\overline{G}$ is cyclic of order 2, $Perm(G/\overline{G}) \cong S_3$, and $\lambda : G \to Perm(G/G')$ is 1-1 (6.6), hence an isomorphism. We let $N = A_3 \subset \lambda(G)$, a regular subgroup of $Perm(G/\overline{G})$ which is normalized by $\lambda(G)$.

For $[L : K] = 4$, $G \cong D_4, A_4$ or S_4.

If $G = D_4$, $\overline{G} = C_2$ we let N be either subgroup of $\lambda(G)$ of index 2.

If $G = A_4$ or S_4, let $N = V_4 \subset \lambda(G)$ be the Klein 4-group. Then V_4 acts transitively on $G/\overline{G}$ and is normal in $Perm(G/\overline{G}) \cong S_4$. □

§7. **Byott's translation.** One difficulty with the Greither-Pareigis criterion is that, applied directly, we need to find out which regular subgroups of $Perm(G/G')$ are normalized by G, and for n much above 4, $Perm(G/G')$ has a large number of regular subgroups (for example, S_n has $\frac{(n-1)!}{\phi(n)}$ cyclic subgroups of order n). Thus it is useful to reverse the relationship between G and N. This reversal of relationship was implicit in [**GP87**], explicit in [**Ch89**], and was made precise by Byott [**By96a**].

Suppose L/K is Galois with group G. Then we seek regular subgroups N of $Perm(G)$ normalized by G. If N is a regular subgroup, then the map $N \xrightarrow{b} G$ by $b(\eta) = \eta \cdot e_G$, e_G =identity of G, is bijective and induces an isomorphism φ from $Perm(G)$ to $Perm(N)$ by $\varphi(\pi) = b^{-1}\pi b$. Under φ, N is mapped to $\lambda(N)$ in $Perm(N)$: for $\mu, \eta \in N$,

$$b^{-1}\eta b(\mu) = b^{-1}(\eta \cdot \mu \cdot e_G) = \eta \cdot \mu = \lambda(\eta)\mu \ ,$$

$\lambda(G)$ is mapped to some group $G_0 \cong G$ in $Perm(N)$, and, since $\lambda(G)$ normalizes N in $Perm(G)$, G_0 normalizes $\lambda(N)$ in $Perm(N)$.

By this translation, we shall see that to find regular subgroups $N' \cong N$ of $Perm(G)$ normalized by $\lambda(G)$ becomes a question of finding regular embeddings of G into the normalizer $Hol(N)$ of $\lambda(N)$ in $Perm(N)$, and $Hol(N)$, the holomorph of N, is far smaller than $Perm(G)$, and easy to describe (for example, $|Hol(C_n)| = n\phi(n)$, while $|Perm(C_n)| = n!$).

Once we have such an embedding of G into $Hol(N)$ it is easy to describe the corresponding K-Hopf algebra H for which L is H-Galois: see (7.7).

(7.1) Definition. The holomorph of N, $Hol(N)$, is the normalizer of $\lambda(N)$ in $Perm(N)$:

$$Hol(N) = \{\pi \in Perm(N) | \pi \text{ normalizes } \lambda(N)\}.$$

(7.2) Proposition. $Hol(N) = \rho(N) \cdot Aut(N)$

Proof. View $Aut(N) \subset Perm(N)$ in the obvious way. Then $Aut(N)$ normalizes $\lambda(N)$: for given $\gamma \in Aut(N), \eta, \mu \in N$,

$$(\gamma\lambda(\eta))(\mu) = \gamma(\eta\mu) = \gamma(\eta)\gamma(\mu) = (\lambda(\gamma(\eta))\gamma)(\mu) \ ,$$

hence $\gamma\lambda(\eta) = \lambda(\gamma(\eta))\gamma$, or

$$\gamma\lambda(\eta)\gamma^{-1} = \lambda(\gamma(\eta)) \in \lambda(N).$$

Also, $\rho(N)$ centralizes $\lambda(N)$: thus both $Aut(N)$ and $\rho(N)$ are subsets of $Hol(N)$.

Now $Aut(N) \cap \rho(N) = \{1\} \in Perm(N)$, since $Aut(N)$ fixes e_N, the identity element of N, and $\rho(N)$ is regular, i.e. the stabilizer in $\rho(N)$ of any element of N is trivial. Also, for $\gamma \in Aut(N)$ and $\eta, \mu \in N$,

$$\gamma\rho(\eta)(\mu) = \gamma(\mu\eta^{-1}) = \gamma(\mu)\gamma(\eta^{-1}) = \rho(\gamma(\eta))\gamma(\mu)$$

hence $\rho(\gamma(\eta))\gamma = \gamma\rho(\eta)$. Thus $\rho(N) \cdot Aut(N)$ is a subgroup of $Perm(N)$ which is contained in $Hol(N)$.

For the opposite inclusion, let $\pi \in Hol(N)$. Then for $\eta \in N$, $\pi\lambda(\eta)\pi^{-1} \in \lambda(N)$, hence $\pi\lambda(\eta)\pi^{-1} = \lambda(\gamma(\eta))$ for some $\gamma(\eta) \in N$. The map $\gamma : N \to N$ is easily seen to be an automorphism of N. Then

$$\begin{aligned} \pi(\eta) = \pi\lambda(\eta)(e) &= (\lambda(\gamma(\eta))\pi)(e) \\ &= \lambda(\gamma(\eta))\pi(e) \\ &= \gamma(\eta)\pi(e) \\ &= (\rho(\pi(e)^{-1})\gamma)(\eta) , \end{aligned}$$

hence $\pi = \rho(\pi(e)^{-1})\gamma \in \rho(N) \cdot Aut(N)$. Thus $Hol(N) \subseteq \rho(N) \cdot Aut(N)$. □

Here is Byott's Theorem, from [**By96a**]. Recall that to count Hopf Galois structures on L/K with normal closure E, and $G = Gal(E/K), G' = Gal(E/L)$, we seek regular subgroups of $Perm(G/G')$ normalized by $\lambda(G)$.

(7.3) THEOREM (BYOTT). *Let $G' \subset G$ be finite groups, let $X = G/G'$ and let N be an abstract group of order $|X|$. Then there is a bijection between $\mathcal{N} = \{\alpha : N \to Perm(X)$ a 1-1 homomorphism such that $\alpha(N)$ is regular$\}$ and $\mathcal{G} = \{\beta : G \to Perm(N)$ a 1-1 homomorphism such that $\beta(G')$ is the stabilizer of e_N, the identity of $N\}$.*

Under this bijection, if $\alpha, \alpha' \in \mathcal{N}$ correspond to $\beta, \beta' \in \mathcal{G}$, respectively, then :

$\alpha(N) = \alpha'(N)$ iff $\beta(G)$ and $\beta'(G)$ are conjugate by an element of $Aut(N)$; and

$\alpha(N)$ is normalized by $\lambda(G) \subset Perm(X)$ iff $\beta(G)$ is contained in $Hol(N)$, the normalizer of N in $Perm(N)$.

Call a homomorphism $\alpha : N \to Perm(X)$ so that $\alpha(N)$ is regular, a regular embedding.

PROOF. Let $\alpha \in \mathcal{N}$, i.e. $\alpha : N \to Perm(X)$ is a regular embedding, then $X = \alpha(N)\bar{e}$ where $\bar{e}$ is the coset in $X = G/G'$ of the identity e of G. Then α induces a bijection $a : N \to X$ by $a(\eta) = \alpha(\eta)\bar{e}$. The map a in turn yields an isomorphism

$$C(a) : Perm(N) \to Perm(X)$$

by $C(a)(\pi) = a\pi a^{-1}$ for $\pi \in Perm(N)$.

Let $\lambda_X : G \to Perm(X)$ be the left translation map, then

$$C(a)^{-1}\lambda_X : G \to Perm(N)$$

is an embedding. We show it is in $\mathcal{G}$: for e_N the identity element of N, we have

$$(C(a)^{-1}\lambda_X(\sigma))(e_N) = e_N$$

iff

$$(a^{-1}\lambda_X(\sigma)a)(e_N) = e_N$$

iff

$$\lambda_X(\sigma)(a(e_N)) = a(e_N)$$

iff

$$\lambda_X(\sigma)(\bar{e}) = \bar{e}$$

iff

$$\bar{\sigma} = \bar{e}$$

iff

$$\sigma \in G'.$$

So $C(a)^{-1}\lambda_X \in \mathcal{G}$.

The bijection we seek from $\mathcal{N}$ to $\mathcal{G}$ is $\Phi : \mathcal{N} \to \mathcal{G}$, defined by $\Phi(\alpha) = C(a)^{-1}\lambda_X$. For future use we note that

(7.4). $C(a)^{-1}\alpha = \lambda_N$:

for $C(a)^{-1}\alpha(\eta) = a^{-1}\alpha(\eta)a$, and

$$\begin{aligned}(a^{-1}\alpha(\eta)a)(\mu) &= (a^{-1}\alpha(\eta)(\alpha(\mu))\bar{e}) \\ &= a^{-1}((\alpha(\eta)\alpha(\mu))\bar{e}) \\ &= a^{-1}(\alpha(\eta\mu)\bar{e}) \\ &= \eta\mu \\ &= \lambda_N(\eta)\mu \quad .\end{aligned}$$

Thus $C(a)^{-1}\alpha = \lambda_N$.

Now we define the inverse Ψ of Φ.

If $\beta : G \to Perm(N)$ is in $\mathcal{G}$ then β yields a bijection $b : X \to N$ by $b(\bar{\sigma}) = \beta(\sigma)e_N$. Then b is 1-1 and well-defined on cosets because $G' = \{\sigma \in G | \beta(\sigma)e_N = e_N\}$, hence b is onto since $|X| = |N|$. Then

$$C(b) : Perm(X) \to Perm(N)$$

is an isomorphism, and

$$C(b^{-1})\lambda_N : N \to Perm(X)$$

is then a regular embedding of N in $Perm(X)$, hence in $\mathcal{N}$. Let $\Psi(\beta) = C(b^{-1})\lambda_N$.

Claim. Ψ and Φ are inverse maps. For given $\alpha \in \mathcal{N}$, let $\beta = \Phi(\alpha) = C(a)^{-1}\lambda$, then $b : X \to N$ is defined by

$$\begin{aligned}b(\bar{\sigma}) &= (C(a)^{-1}\lambda_X(\sigma))(e_N) \\ &= (a^{-1}\lambda_X(\sigma)a)(e_N) \\ &= a^{-1}(\bar{\sigma}) \quad ,\end{aligned}$$

hence $\Psi(\beta) = C(b)^{-1}\lambda_N = C(a)\lambda_N = \alpha$ by (7.4), and so $\Psi \circ \Phi$ is the identity on $\mathcal{N}$. The opposite composition is the identity on $\mathcal{G}$.

If $\alpha(N)$ is normalized by $\lambda_X(G)$ in $Perm(X)$ and $\beta = \Phi(\alpha)$ then $\beta(G)$ normalizes $\lambda_N(N) \subset Perm(N)$. For $\lambda_X(\sigma)\alpha(\eta)\lambda_X(\sigma^{-1}) \in \alpha(N) \subset Perm(X)$ for all $\sigma \in G$, $\eta \in N$. Mapping over to $Perm(N)$ via $C(a)^{-1}$, we have

$$C(a)^{-1}(\lambda_X(\sigma)\alpha(\eta)\lambda_X(\sigma^{-1})) \in C(a)^{-1}\alpha(N) \subset Perm(N) \ .$$

But $C(a)^{-1}\alpha(\eta) = a^{-1}\alpha(\eta)a = \lambda_N(\eta)$ as we observed above; set

$$\beta = C(a)^{-1}\lambda_X = \Phi(\alpha)$$

then

$$\begin{aligned} C(a)^{-1}(\lambda_X(\sigma)\alpha(\eta)\lambda_X(\sigma^{-1})) &= a^{-1}\lambda_X(\sigma)aa^{-1}\alpha(\eta)aa^{-1}\lambda_X(\sigma^{-1})a \\ &= C(a)^{-1}(\lambda_X(\sigma))\lambda_N(\eta)C(a)^{-1}(\lambda_X)(\sigma^{-1})) \\ &= \beta(\sigma)\lambda_N(\eta)\beta(\sigma^{-1}) . \end{aligned}$$

So $\beta(G)$ normalizes $\lambda_N(N) \subset Perm(N)$. Reversing the argument shows that if $\Phi(\alpha) = \beta$, then $\alpha(N)$ is normalized by $\lambda_X(G)$ iff $\beta(G)$ is contained in the normalizer of $\lambda_N(N)$.

Finally $\alpha(N) = \alpha'(N)$ iff $\gamma = \alpha^{-1}\alpha' : N \to N$ is an automorphism of N, hence $\alpha' = \alpha\gamma$. Now α yields $\beta = C(a)^{-1}\lambda_X : G \to Perm(N)$ and $\lambda_N = C(a)^{-1}\alpha$. If we replace α by $\alpha\gamma$, $\gamma \in Aut(N)$, then

$$C(\alpha\gamma)^{-1} = C(\gamma)^{-1}C(\alpha)^{-1} : Perm(X) \to Perm(N) ,$$

So if $\Phi(\alpha) = C(a)^{-1}\lambda_X = \beta$, then

$$\beta' = \Phi(\alpha\gamma) = C(\gamma)^{-1}C(a)^{-1}\lambda_X = C(\gamma)^{-1}\beta .$$

So β and $\beta' : G \to Perm(N)$ are embeddings which are conjugate by an automorphism of N. That completes the proof. □

As an initial application of Byott's theorem, we have

(7.5) PROPOSITION ([**GP87**], 4.8; [**Ch89**], THEOREM 2). *Let L/K be a field extension of prime order. Then L/K is Hopf Galois iff $G = Gal(E/K)$ is solvable.*

PROOF. Recall that E is the normal closure of L/K.

If N is a regular subgroup of $Perm(G/G')$ of order p, prime, then N is of course cyclic, so $Hol(N) \cong C_p \cdot C_{p-1}$ is solvable. Thus if $\beta : G \to Hol(N)$ is a regular embedding, then G is solvable.

For the converse, note that $L = K[a_0] \cong K[x]/(f(x))$ where $f(x)$ is an irreducible polynomial in $K[x]$ of degree p, prime, with roots $\{a_0, a_1, \dots , a_{p-1}\}$ in E, and G is a subgroup of the group of permutations of the roots. If G is solvable, then by a theorem of Galois, G is isomorphic to a transitive subgroup of the group Γ of permutations of the roots:

$$\Gamma = \{\pi_{r,s} | r, s \pmod p, r \not\equiv 0 \pmod p\},$$

where $\pi_{r,s}(a_i) = a_{ri+s}$. Then $\Gamma = \langle\pi_{1,1}\rangle \cdot \langle\pi_{b,0}\rangle \cong C_p \cdot C_{p-1} \cong Hol(C_p)$, where b is a primitive element modulo p, and $G' = \{\sigma \in G | \sigma(a_0) = a_0\}$ embeds in $\langle\pi_{b,0}\rangle$. Thus the embedding of G in Γ yields a Hopf Galois structure on L/K. □

Byott's theorem gives a way of counting Hopf Galois structures on a separable extension L/K:

(7.6) COROLLARY. *Let L/K be a separable extension with normal closure E; let $G = Gal(E/K)$, $G' = Gal(E/L)$. Let $\mathcal{S}$ be the set of isomorphism classes of groups N with $|N| = |G/G'|$. The number of Hopf Galois structures on L/K is $s(G,G') = \sum_{\{N\}\in\mathcal{S}} e(G,N)$ with $e(G,N) = |\mathcal{G}_N/\sim|$ = the cardinality of the set of equivalence classes of regular embeddings β of G into $Hol(N)$ such that $\beta(G')$ is the stabilizer of e_N, modulo conjugation by elements of $Aut(N) \subset Hol(N)$.*

PROOF. This is just putting Greither and Pareigis' theorem together with Byott's translation: Greither and Pareigis show that the number of Hopf Galois structures is in 1-1 correspondence with regular subgroups N of $Perm(X)$ normalized by $\lambda(G)$. The latter consists of all subgroups $\alpha(N)$ for $\alpha \in \mathcal{N}_N$ and $N \in \mathcal{S}$, and by Byott's theorem, the $\alpha(N)$ for $N \in \mathcal{N}$ are in 1-1 correspondence with equivalence classes of embeddings of G into $Hol(N)$ modulo conjugation by elements of $Aut(N)$. □

For each $\{N\} \in (S)$, the number $e(G,N)$ is the number of Hopf Galois structures on L/K by K-Hopf algebras H so that $L \otimes_K H \cong LN$.

Corollary (7.6) often facilitates obtaining a partial count of the number of Hopf Galois structures on L/K. Of course the cardinality of $\mathcal{S}$ can get large! For example, for $[L:K] = n$ we have the following mini-table:

n:	12	16	24	32	36	48	64	96	128	720
$\lvert\mathcal{S}\rvert$:	5	14	15	21	14	52	267	230	2328	840

Given $\beta : G \to Hol(N)$, the corresponding K-Hopf algebra which acts on L is easy to identify:

(7.7) PROPOSITION. *Let L/K be a Galois extension with Galois group G, and let $L \otimes L$ be a Galois extension of L with Galois group $\alpha(N)$ where α is a regular embedding of N into $Perm(G)$ normalized by $\lambda(G)$. Let $\beta : G \to Hol(N)$ be the corresponding regular embedding of G in $Hol(N)$. Let $\beta_2 : G \to Aut(N)$ be β followed by the canonical map from $Hol(N)$ to $Aut(N)$. Then the Hopf Galois structure on L/K corresponding to $\alpha(N)$ is given by the Hopf algebra $H = LN^G$ where G acts on L via the Galois action and on N via β_2.*

PROOF. $H = (L\alpha(N))^{\lambda(G)}$, where $\lambda(G)$ acts on $\alpha(N)$ in $Perm(G)$ by

$$\alpha(\eta)^{\lambda(\sigma)}(\tau) = (\lambda(\sigma)\alpha(\eta)\lambda(\sigma^{-1}))(\tau).$$

The map $C(a)^{-1} : Perm(G) \to Perm(N)$ satisfies $\lambda(\eta) = C(a)^{-1}\alpha(\eta)$ for $\eta \in N, \sigma, \tau \in G$ (7.4), hence maps $\alpha(N)$ onto $\lambda(N)$. The map $\beta : G \to Perm(N)$ corresponding to α is defined by $\beta(\sigma) = C(a)^{-1}\lambda(\sigma) \in Hol(\lambda(N)) = \rho(N)\cdot Aut(N)$. The action of $\lambda(G)$ on $\alpha(N)$ in $Perm(G)$ translates, under $C(a)^{-1}$, to an action of $\beta(G)$ on $\lambda(N)$: for

$$\begin{aligned} C(a)^{-1}(\lambda(\sigma)\alpha(\eta)\lambda(\sigma^{-1})) &= C(a)^{-1}(\lambda(\sigma))C(a)^{-1}(\alpha(\eta))C(a)^{-1}(\lambda(\sigma^{-1})) \\ &= \beta(\sigma)\lambda(\eta)\beta(\sigma^{-1}) \end{aligned}$$

So

$$H = (L\alpha(N))^{\lambda(G)} \cong (L\lambda(N))^{\beta(G)}.$$

Let $\beta(\sigma) = \rho(\mu)\delta \in \rho(N) \cdot Aut(N)$. Then $\beta_2(\sigma) = \delta$. Given $\theta \in N$,

$$\begin{aligned}\beta(\sigma)\lambda(\eta)\beta(\sigma^{-1})(\theta) &= [\rho(\mu)\delta \cdot \lambda(\eta) \cdot (\rho(\mu)\delta)^{-1}](\theta) \\ &= [\rho(\mu)\delta\lambda(\eta)\delta^{-1}\rho(\mu^{-1})](\theta) \\ &= [\rho(\mu)\delta\lambda(\eta)](\delta^{-1}(\theta\mu)) \\ &= \delta(\eta\delta^{-1}(\theta\mu))\mu^{-1} \\ &= \delta(\eta)(\theta)\end{aligned}$$

Hence for $\sigma \in G$, $\beta(\sigma)$ acts on $\lambda(N)$ via $\beta_2(\sigma)$, and so $H \cong LN^G$ where G acts on L via the Galois action and on N via $\beta_2 : G \to Aut(N)$. □

§**8. Byott's Uniqueness Theorem.** This theorem, from [**By96a**] generalizes observations for p-groups in [**Pa90**] and [**Ch89**].

(8.1) THEOREM (BYOTT). *A Galois extension L/K of fields with Galois group G has unique Hopf Galois structure (namely, that given by KG), iff $|G|$ is a Burnside number.*

A number g is Burnside if $(g, \phi(g)) = 1$.

(8.2) EXAMPLES.
$g = p$, prime;
$g = \ell p$ with $p > \ell$ primes, $p \not\equiv 1 \pmod{\ell}$;
$g = 3 \cdot 5 \cdot 17 \cdot 23 \cdot 53 \cdot 83 \cdot 257$.

(8.3) FACT (BURNSIDE). If G is a group and $|G|$ is a Burnside number, then G is cyclic.

For a proof see Robinson [**Ro82**].

PROOF OF BYOTT'S THEOREM. If $g = |G|$ is a Burnside number, then G is cyclic and any regular subgroup N of $Perm(G)$ is isomorphic to G. So the number of regular subgroups N of $Perm(G)$ normalized by $\lambda(G)$ is equal to the number of regular embeddings of G into $Hol(G)$, modulo conjugation by $Aut(G)$. Now G is cyclic, $G = C_g = \mathbb{Z}/g\mathbb{Z}$, so $Aut(G) = (\mathbb{Z}/g\mathbb{Z})^\times$ and $(|Aut(G)|, g) = 1$. We have

$$\rho(G) = ker(Hol(G) \xrightarrow{\pi} Aut(G));$$

if $\beta : G \to Hol(G)$ is any embedding, then $\pi\beta(G) = \{1\}$, and $\beta(G) = \rho(G)$. Thus modulo $Aut(G)$, any embedding β of G in $Hol(G)$ is equivalent to ρ, the embedding which corresponds to the classical KG-Galois structure on L/K.

The rest of the proof involves showing that if g is not a Burnside number, then L/K has a non-classical Hopf Galois structure.

To do this, we show that if $|G| = g$ is not a Burnside number, then there exist at least two regular subgroups of $Perm(G)$ normalized by G. By (6.11) we know that if G is not abelian, then $\lambda(G) \neq \rho(G)$ in $Perm(G)$ and both are regular and normalized by $\lambda(G)$. So we may restrict consideration to G abelian.

Suppose $G = G_1 \times G_2$, and let N_1, N_1' be distinct regular subgroups of $Perm(G_1)$ normalized by $\lambda(G_1)$. Then $N_1 \times \rho(G_2)$ and $N_1' \times \rho(G_2) \subset Perm(G_1) \times Perm(G_2) \subset Perm(G_1 \times G_2)$ and are normalized by $\lambda(G_1) \times \lambda(G_2) = \lambda(G_1 \times G_2)$. Thus, if we find that some abelian group G_1 has the property that $Perm(G_1)$ has at least two regular subgroups normalized by $\lambda(G_1)$, then the same will be true for any group G containing G_1 as a direct factor.

(8.4) CLAIM. *If G is abelian and $|G|$ is not Burnside then for some primes p, q, G has a direct factor of one of the following types*
I. $G_1 = C_{p^e}, e \geq 2$
II. $G_1 = C_p \times C_p$
III. $G_1 = C_p \times C_q$ *with q dividing $p - 1$.*

PROOF OF (8.4). For each prime p dividing $|G|$, the p-torsion subgroup of G either has order p or has a direct factor of type I or II. So suppose for each prime p dividing $|G|$, the p-torsion subgroup has order p. Then

$$G = C_{p_1} \times \ldots \times C_{p_r}$$

with $p_1 \ldots p_r$ distinct primes. If $|G| = g = p_1 \ldots p_r$ is not Burnside then $(g, (p_1 - 1) \cdot \ldots \cdot (p_r - 1)) \neq 1$, so p_j divides $p_i - 1$ for some i, j and G has a direct factor $C_{p_i} \times C_{p_j}$ of type III. Thus the claim is true.

To finish the proof of Byott's uniqueness theorem we apply (7.3) and show:

(8.5). If G is of type I-III, then either there are at least two equivalent classes of regular embeddings of G into $Hol(G)$, or there is a group $N \not\cong G$ of order $|G|$ which admits a regular embedding of G into $Hol(N)$.

We consider each case in turn.

I. $G = C_{p^e}$, $e \geq 2$.

First consider Case I with p odd.

(8.6) PROPOSITION. *Let $G = C_{p^e}$, $e \geq 2$, p odd. There are p^{e-1} equivalence classes of embeddings of C_{p^e} into $Hol(C_{p^e})$ modulo $Aut(C_{p^e})$.*

PROOF. We will determine $e(G, G)$, the number of equivalence classes of regular embeddings of G into $Hol(G)$. Now $Hol(G) = C_{p^e} \cdot Aut(C_{p^e}) = \langle \sigma, \delta \rangle \cong \mathbb{Z}_{p^e} \times \mathbb{Z}_{p^e}^{\times}$ where σ has order p^e, δ has order $p^{e-1}(p-1)$ and $\delta\sigma = \sigma^a \delta$ for some primitive element a modulo $p^e\mathbb{Z}$. If $\gamma = \delta^{p-1}$, then $\gamma\sigma = \sigma^q \gamma$ for some $q \equiv 1 \pmod{p}$.

(8.7) CLAIM. *$\sigma^r \gamma^s$ has order p^e iff $(r, p) = 1$.*

Let $t = q^s$. Then

$$\begin{aligned}(\sigma^r \gamma^s)^c &= (\sigma^r)^{1+t+\cdots+t^{c-1}} \gamma^{sc} \\ &= \sigma^{r(\frac{t^c - 1}{t-1})} \gamma^{sc}\end{aligned}$$

So $(\sigma^r \gamma^s)^c = 1$ iff p^{e-1} divides sc and p^e divides $r(\frac{t^c-1}{t-1})$. Let $t = 1 + p^m b$ for $(b, p) = 1$. Then

$$\begin{aligned}r(\frac{t^c - 1}{t-1}) &= r\frac{cp^m b + \binom{c}{2} p^{2m} b^2 + \ldots}{p^m b} \\ &= rc + r\binom{c}{2} p^m b + \ldots\end{aligned}$$

So p^e divides $r(\frac{t^c-1}{t-1})$ iff p^e divides rc. Thus if $\sigma^r\gamma^s$ has order p^e then the minimal c with p^e dividing rc is $c = p^e$, so $(r,p) = 1$. Conversely, if $(r,p) = 1$ and $(\sigma^r\gamma^s)^c = 1$ then p^e divides c, so $\sigma^r\gamma^s$ has order p^e. That verifies the claim.

Let $\beta_{r,s} : G \to Hol(G)$ by $\beta_{r,s}(\sigma) = \sigma^r\gamma^s$, where $(r,p) = 1$. Then $\beta_{r,s}$ is regular. For $\beta_{r,s}(\sigma^c)$ is in the stabilizer of the identity element e_G of the set G on which $Hol(G) \subset Perm(G)$ acts, iff $(\sigma^r\gamma^s)^c(e_G) = e_G$, iff

$$\sigma^{r(\frac{t^c-1}{t-1})} = 1$$

iff p^e divides rc, iff p^e divides c.

Now $\beta_{r,s} \sim \beta_{r',s'}$ iff there is some $\theta \in Aut(G)$ so that

$$\theta\sigma^r\gamma^s\theta^{-1} = \sigma^{r'}\gamma^{s'} .$$

But since $Aut(G)$ is abelian, thus holds iff $s = s'$ and $\theta(\sigma^r) = \sigma^{r'}$. Since $(rr',p) = 1$, the map $\theta : G \to G$ by $\theta(\sigma) = \sigma^{r'n}$ with $nr \equiv 1 \ (mod\ p^e)$ is an automorphism of G and $\theta(\sigma^r) = \sigma^{r'}$. So for any r, r' with $(p, rr') = 1$, $\beta_{r,s} \sim \beta_{r',s'}$ iff $s = s'$. Thus the equivalence classes of regular embeddings are parametrized by s, $0 \leq s < p^{e-1}$. □

We will show in (9.2) that for p odd there are exactly p^{e-1} Hopf Galois structures on a cyclic Galois extension L/K of degree p^e.

Now consider case I with $p = 2$.

(8.8). $G = C_{2^e}$, $e \geq 3$.

Then $Hol(C_{2^e}) \supseteq \langle\sigma, \alpha\rangle$ where $\alpha \in Aut(C_{2^e})$ satisfies $\alpha(\sigma) = \sigma^5$. Then $\sigma\alpha$ has order 2^e, just as in the odd prime case. So

$$\beta_0 : \sigma \mapsto \sigma$$
$$\beta_1 : \sigma \mapsto \sigma\alpha$$

are regular embeddings of C_{2^e} into $Hol(C_{2^e})$, and β_0 and β_1 are not equivalent under $Aut(C_{2^e})$.

(8.9). $G = C_4$.

Let $C_4 = \langle\tau\rangle$. We have $\rho : C_4 \to Hol(C_4)$ the standard embedding, and

$$\beta : C_4 \to Hol(V_4) = \mathbb{F}_2^2 \times GL_2(\mathbb{F}_2)$$

by $\beta(\tau) = \sigma\alpha$, where

$$\sigma = \begin{pmatrix}1\\0\end{pmatrix}, \quad \alpha = \begin{pmatrix}0 & 1\\1 & 0\end{pmatrix} .$$

(8.10). Now consider Case II: $G = C_p \times C_p \cong \mathbb{F}_p^2$.

Then

$$Hol(G) = \mathbb{F}_p^2 \cdot GL_2(\mathbb{F}_p) .$$

Let $\rho : G \to Hol(G)$ be right translation and let $\beta : G \to Hol(G)$ by

$$\beta\left(\begin{pmatrix}1\\0\end{pmatrix}\right) = \left(\begin{pmatrix}1\\0\end{pmatrix}, I\right)$$
$$\beta\left(\begin{pmatrix}0\\1\end{pmatrix}\right) = \left(\begin{pmatrix}0\\1\end{pmatrix}, \begin{pmatrix}1 & 1\\0 & 1\end{pmatrix}\right)$$

Then β is a regular embedding of G into $Hol(G)$ which is not conjugate to ρ under the action of $GL_2(\mathbb{F}_p)$. In fact, Tse [**Ts97**] shows that the number s_G of regular subgroups of $Hol(G)$ normalized by G is p^2: $e(G,G) = p^2$ and $e(G,N) = 0$ for $N = C_{p^2}$.

(8.11). Finally, consider Case III: $G = C_{pq} = \langle \tau \rangle$ where q divides $p-1$.

Consider the semidirect product $N = C_p \times C_q = \langle \sigma, \alpha \rangle$ where $\sigma^p = \alpha^q = 1$ and $\alpha\sigma = \sigma^s \alpha$, where s has order q in $(\mathbb{Z}/p\mathbb{Z})^\times$. If $C(\sigma) \in Aut(N)$ is conjugation by σ, then $\theta = \alpha C(\sigma) \in N \cdot Aut(N)$ has order pq and so the map

$$\beta : G \to Hol(N)$$
$$\beta(\tau) = \alpha C(\sigma)$$

is a regular embedding. Since $\rho : G \to Hol(G)$ is also a regular embedding, this completes the proof of the uniqueness theorem. □

§9. **Prime power cyclic Galois extensions.** Much remains unknown about how many Hopf Galois structures there are on a given field extension. The Greither-Pareigis theorem reduces the question to a purely group-theoretic question. If L/K is a separable extension with normal closure E, $Gal(E/K) = G$, and $Gal(E/L) = G'$, then the question depends only on G and G', and not on L/K (although, as we shall see, the structure of the Hopf algebras acting on L/K of course does depend on L/K).

If $G' = \{1\}$ i.e. L/K is Galois, then it appears that for G a p-group the number of structures $s(G, \{1\}) = s(G)$ is at least a linear function of $|G|$. For G a cyclic p-group of order p^n we showed (8.6) that the number of structures is at least p^{n-1}. In fact, we have

(9.1) THEOREM [**Ko98**]. *Let p be an odd prime and G be cyclic of order p^n. If $|N| = p^n$ but N is not cyclic, then there is no embedding of G into $Hol(N)$.*

This implies

(9.2) COROLLARY. *There are exactly p^{n-1} Hopf Galois structures on a Galois extension L/K cyclic of order p^n.*

PROOF OF (9.1). Proving the theorem is equivalent to showing that if N is a non-cyclic group of order p^n, then $Hol(N)$ contains no element of order p^n.

Now $Hol(N)$ is the semidirect product of N and $Aut(N)$. For any $\eta \in N, \alpha \in Aut(N)$, $\alpha\eta = \eta^\alpha \alpha$, so for all $e \geq 1$

$$(\eta\alpha)^e = \eta^{1+\alpha+\cdots+\alpha^{e-1}} \alpha^e \ ,$$

hence if $(\eta\alpha)^e = 1$ then $\alpha^e = 1$. Thus if we seek an element $\eta\alpha$ of $Hol(N)$ of order p^n, we can assume α has order a power of p.

Given an automorphism α of N of order a power of p, there exists a composition series for N as follows:

$$\{e\} = N_0 < N_1 < N_2 < \cdots < N_n = N \ ,$$

where:

1) N_i is a normal subgroup of N for all i
2) $\alpha(N_i) = N_i$ for all i, and
3) if N is non-cyclic, then $N_2 \cong C_p \times C_p$.

This is classical – see [**MBD16, p.134-5**].

Such a composition series allows us to describe η^α. We switch from exponential to functional notation for α acting on η: $\eta^\alpha = \alpha(\eta)$.

First, since $\alpha(N_i) = N_i$ for all i, α induces an automorphism on $N_i/N_{i-1} \cong C_p$. But since α has order a power of p, then α induces the identity on N_i/N_{i-1}, hence for all $\eta_i \in N_i$ there exists $\eta_{i-1} \in N_{i-1}$ so that $\alpha(\eta_i) = \eta_i\eta_{i-1}$

Claim. for any $\eta \in N$, $\alpha^{p^{s-1}}(\eta) = \eta\eta_{n-s}$ for some $\eta_{n-s} \in N_{n-s}$.

Let $\eta \in N$, then $\alpha(\eta) = \eta\eta_{n-1}$ for some $\eta_{n-1} \in N_{n-1}$ so the claim holds for $s = 1$. For $s = 2$: Suppose $\alpha^r(\eta) = \eta\eta_{n-1}^r\eta_{n-2}$ for some $\eta_{n-2} \in N_{n-2}$, then, since $\alpha(\eta_{n-1}) = \eta_{n-1}\eta'_{n-2}$ for some $\eta'_{n-2} \in N_{n-2}$,

$$\alpha^{r+1}(\eta) = \eta\eta_{n-1} \cdot (\eta_{n-1} \cdot \eta'_{n-2})^r \cdot \alpha(\eta_{n-2});$$

since N_{n-2} is a normal subgroup of N_{n-1}, $(\eta_{n-1}\eta'_{n-2})^r = \eta_{n-1}^r\eta''_{n-2}$ for some $\eta''_{n-2} \in N_{n-2}$; hence $\alpha^{r+1}(\eta) = \eta\eta_{n-1}^{r+1}\eta'''_{n-2}$ for some $\eta'''_{n-2} \in N_{n-2}$. Since $\eta_{n-1}^p \in N_{n-2}$, it follows that $\alpha^p(\eta) = \eta\eta_{n-2}$ for some $\eta_{n-2} \in N_{n-2}$, so the claim is true for $s = 2$. Repeating this argument, we find, for each $s \geq 1$, $\alpha^{p^s}(\eta) = \eta\eta_{n-s-1}$ for some $\eta_{n-s-1} \in N_{n-s-1}$.

In particular, $\alpha^{p^{n-1}}(\eta) = \eta\eta_0$ for some $\eta_0 \in N_0 = \{1\}$, that is, $\alpha^{p^{n-1}} = 1$: that is, no element of $Aut(N)$ has order p^n.

Now consider $\eta\alpha \in N \cdot Aut(N)$ for any $\eta \in N$ and $\alpha \in Aut(N)$ of order a power of p. Then one finds that

$$\begin{aligned}(\eta\alpha)^2 &= \eta\alpha\eta\alpha \\ &= \eta \cdot \eta\eta_{n-1} \cdot \alpha^2 = \eta^2\eta_{n-1}\alpha^2 \ , \\ (\eta\alpha)^3 &= \eta\alpha(\eta^2\eta_{n-1})\alpha^3 \\ &= \eta(\eta\eta_{n-1})^2\alpha(\eta_{n-1})\alpha^3 \\ &= \eta^3\eta'_{n-1}\alpha^3\end{aligned}$$

for some $\eta'_{n-1} \in N_{r-1}$. Continuing, one gets

$$(\eta\alpha)^p = \eta^p n''_{n-1}\alpha^p \in N_{n-1}\langle\alpha^p\rangle \ .$$

By induction, one finds that $(\eta\alpha)^{p^{n-2}} \in N_2\langle\alpha^{p^{n-2}}\rangle$. Let $\gamma = \alpha^{p^{n-2}}$ and $\eta_2 \in N_2$. Then $\gamma^p = 1$ and

$$\begin{aligned}(\eta_2\gamma)^p &= \eta_2\gamma(\eta_2) \cdot \ldots \cdot \gamma^{p-1}(\eta_2)\gamma^p \\ &= \eta_2\gamma(\eta_2) \cdot \ldots \cdot \gamma^{p-1}(\eta_2) \ .\end{aligned}$$

Now $\gamma(\eta_2) = \eta_2\eta_1$; also $\gamma|_{N_1} = id$ since γ has order p and $|N_1| = p$. So

$$\gamma^2(\eta_2) = \eta_2\eta_1 \cdot \eta_1 = \eta_2\eta_1^2 \text{ etc.}$$

$$\ldots,$$

$$\gamma^{p-1}(\eta_2) = \eta_2\eta_1^{p-1}.$$

So
$$(\eta_2\gamma)^p = \eta_2 \cdot \eta_2\eta_1 \cdot \eta_2\eta_1^2 \cdot \ldots \cdot \eta_2\eta_1^{p-1}.$$
Since N_2 is abelian of exponent p,
$$(\eta_2\gamma)^p = \eta_2^p \cdot \eta_1^{\frac{(p-1)p}{2}} = 1 .$$
Hence $(\eta\alpha)^{p^{n-1}} = 1$, completing the proof. □

§**10. Non-abelian extensions.** In this section we obtain a partial count of the number of Hopf Galois structures on Galois extensions L/K when the Galois group G is the symmetric group S_n for $n \geq 5$, and for non-abelian simple groups, such as the alternating group A_n for $n \geq 5$. For $K = \mathbb{Q}$, these include the Galois groups of "most"polynomials of degree $n > 4$. These results come from [**CC99**],

Recall (7.6) that the number of Hopf Galois structures
$$H \otimes L \to L$$
where $L \otimes H \cong LN$, is equal to the number of embeddings
$$\beta : G \to Hol(N)$$
such that the stabilizer in G of the identity element of the set N is trivial (i.e. β is *regular*), modulo conjugation by elements of $Aut(N)$. If we denote the number of equivalence classes of regular embeddings of G into $Hol(N)$ by $e(G, N)$, then the number $s(G)$ of Hopf Galois structures on a Galois extension with group G is the sum over the set of isomorphism classes of groups N of cardinality $|G|$, of $e(G, N)$.

In this section we examine $e(G, G)$ for simple groups and symmetric groups.

Simple groups. We begin with G simple:

(10.1) THEOREM. *Let A be a non-abelian simple group. Then $e(A, A) = 2$.*

Recall that $Inn(A) \subset Aut(A)$ is the subgroup of inner automorphisms, automorphisms of the form $C(\tau)$ for τ in A, where for σ in A, $C(\tau)(\sigma) = \tau\sigma\tau^{-1}$.

We first note the following unwinding lemma:

(10.2) PROPOSITION. *Let N be a group with trivial center and*
$$\beta : G \to Hol(N) = N \cdot Aut(N)$$
be a regular embedding. Let $j_1, j_2 : Hol(N) \to Aut(N)$ by $j_1(\eta\delta) = C(\eta)\delta$ and $j_2(\eta\delta) = \delta$ (then j_2 is the canonical quotient map). If the image of $j_2\beta$ is contained in $Inn(N)$, then $j_1\beta(\sigma) = C(\beta_1(\sigma))$, $j_2\beta(\sigma) = C(\beta_2(\sigma))$ for homomorphisms $\beta_1, \beta_2 : G \to N$, and
$$\beta(\sigma) = \beta_1(\sigma)\beta_2(\sigma^{-1})C(\beta_2(\sigma)).$$

PROOF. The map $j_1 : Hol(N) \to Aut(N)$ is a homomorphism because $\delta C(\mu) = C(\delta(\mu))\delta$ for all $\mu \in N, \delta \in Aut(N)$. The homomorphisms β_1, β_2 are well-defined because N has trivial center. If $\beta(\sigma) = \eta\delta$ and $\delta = C(\mu)$, then $\beta_2(\sigma) = \mu$ and $\beta_1(\sigma) = \eta\mu$, so
$$\beta(\sigma) = \eta\mu\mu^{-1}C(\mu) = \beta_1(\sigma)\beta_2(\sigma^{-1})C(\beta_2(\sigma)).$$
□

Returning to the case $G = N = A$ simple, we have:

(10.3) LEMMA. *If A is non-abelian and simple, then $\beta_i(A) \subset Inn(A)$.*

PROOF. Consider the composite θ of $j_2\beta$ with the canonical map from $Aut(A)$ to $Aut(A)/Inn(A)$. Since A is simple, θ is 1-1 or trivial. But if A is non-abelian, then θ cannot be 1-1 because $Aut(A)/Inn(A)$ is solvable, by the affirmative solution of Schreier's conjecture, which follows from the classification of finite simple groups [**Go82, p. 51**]. So $j_2\beta(A) \subset Inn(A)$.

Hence there are homomorphisms $\beta_1, \beta_2 : A \to A$ so that

$$\beta(\sigma) = \beta_1(\sigma)\beta_2(\sigma^{-1})C(\beta_2(\sigma)).$$

There are now three possibilities:
1. β_1 1-1, $\beta_2 = 0$;
2. $\beta_1 = 0$, β_2 1-1;
3. β_1 1-1, β_2 1-1.

Case 1. β_1 1-1, $\beta_2 = 0$. Then $\beta = \beta_1$ and is an isomorphism from A onto A, hence an automorphism of A. But then β is equivalent to β' where $\beta'(\mu) = \beta_1^{-1}(\beta(\mu))\beta_1 = \mu$ in $Hol(A)$. Thus in Case 1 any embedding is equivalent to the right regular embedding ρ. (Recall $Hol(A) = \rho(A) \cdot Aut(A)$.)

Case 2. $\beta_1 = 0$, β_2 is 1-1. Then β_2 is an isomorphism from A onto A. Then $\beta(\mu) = \beta_2(\mu^{-1})C(\beta_2(\mu))$ for all μ, and β is equivalent to β' where $\beta'(\mu) = \beta_2^{-1} \cdot (\beta(\mu)) \cdot \beta_2$ in $Hol(A)$: that is,

$$\begin{aligned} \beta'(\mu) &= \beta_2^{-1} \cdot (\beta_2(\mu^{-1})C(\beta_2(\mu))) \cdot \beta_2 \\ &= \beta_2^{-1}(\beta_2(\mu^{-1})) \cdot \beta_2^{-1}C(\beta_2(\mu))\beta_2 \\ &= \mu^{-1}C(\mu). \end{aligned}$$

and $(\mu^{-1}C(\mu))(\alpha) = \rho(\mu^{-1})C(\mu)\alpha = (\mu\alpha\mu^{-1})\mu = \mu\alpha$. Thus in Case 2 any embedding is equivalent to the left regular embedding $\lambda : A \to Perm(A)$.

Case 3. β_1 and β_2 are both 1-1. Then as in Case 1, we can assume $\beta_1 =$ identity, and β_2 is an automorphism of A, so that

$$\beta(\mu) = \mu\beta_2(\mu^{-1})C(\beta_2(\mu))$$

for $\mu \in A$.

Now β is regular iff the function $\mu \mapsto \mu\beta_2(\mu^{-1})$ is 1-1 (consider $\beta(A)e_A$). But this function is 1-1 iff the automorphism β_2 is fixed-point free: that is, $\beta_2(\mu) = \mu$ iff μ is the identity element of A. But since A is non-abelian and simple, A has no fixed-point free automorphisms, another consequence of the classification of finite simple groups [**Go82, p. 51**]. Thus Case 3 yields no regular embeddings of A into $Hol(A)$ and the proof of Theorem (10.1) is complete. □

There are additional results in [**CC99**] which suggest the possibility that $e(A, N) = 0$ for A simple, $N \not\cong A$.

Symmetric groups. Now we consider symmetric groups.

Let $S = S_n$ for $n \geq 5$. Then $A = A_n$ is simple. We compute $e(S, S)$.

(10.4) THEOREM. *For $S = S_n$, $n \geq 5$, $e(S, S) =$ two times the number of even permutations in S_n of order dividing 2.*

PROOF. Let $\beta : S \to Hol(S) = S \cdot Aut(S)$ be a regular embedding. Let $j_1\beta, j_2\beta : S \to Aut(S)$ be the corresponding homomorphisms: if $\beta(\sigma) = \tau\delta$, then $j_1\beta(\sigma) = C(\tau)\delta, j_2\beta(\sigma) = \delta$. If both $j_1\beta(A) = 1$ and $j_2\beta(A) = 1$, then $\beta(A) = 1$ and β is not an embedding. So at least one of $j_1\beta$ and $j_2\beta$ is 1-1.

Now if $n \neq 6, Aut(S) = Inn(S)$. If $n = 6$ and $j_i\beta$ is 1-1, then $j_i\beta(S) \subset Inn(S)$ by [**LL93**]. In either case, Proposition (10.2) applies, to yield homomorphisms $\beta_1, \beta_2 : S \to S$ so that

$$\beta(\sigma) = \beta_1(\sigma)\beta_2(\sigma^{-1})C(\beta_2(\sigma))$$

for all $\sigma \in S$, where β_1 or β_2 is 1-1.

If β_1 is 1-1, then β_1 is an automorphism of S, hence, as in Case 1 of Theorem (10.3), we can assume β_1 is the identity. Similarly if β_2 is 1-1. So we can assume β has one of the following forms:

$$\beta(\sigma) = \sigma\beta_2(\sigma^{-1})C(\beta_2(\sigma)),$$

or

$$\beta(\sigma) = \beta_1(\sigma)\sigma^{-1}C(\sigma).$$

If β_1 is the identity and β_2 is 1-1, then $\beta_2(A) = A$, so β restricted to A is of the form

$$\beta(\sigma) = \sigma\beta_2(\sigma^{-1})C(\beta_2(\sigma)).$$

But β is regular iff β_2 is a fixed-point free automorphism of A. None exist. So β_1 and β_2 cannot both be 1-1.

Thus for β on A a regular embedding, if β_1 is the identity, then β_2 is trivial on A, and similarly if β_2 is the identity. If β_i is trivial on A, then β_i maps every odd permutation to a single element τ of S of order dividing 2, and is trivial on all even permutations.

Thus for all $\sigma \in S$, $\beta(\sigma)$ can be assumed to have one of the two following forms:

1. $\beta(\sigma) = \sigma\tau^{-1}C(\tau)$ for σ odd, $\beta(\sigma) = \sigma$ for σ even; or
2. $\beta(\sigma) = \tau\sigma^{-1}C(\sigma)$ for σ odd, $\beta(\sigma) = \sigma^{-1}C(\sigma)$ for σ even.

where τ is a fixed element of S of order dividing 2. (Hence $\tau = \tau^{-1}$.)

The only further restriction on τ is that it must be even, for if τ were odd, then $\sigma\tau$ would be even for all odd σ in S, and so $\beta(S)e_S$ would be a subset of A and β would not be regular. On the other hand, if τ is even, then $\{\sigma\tau|\sigma \text{ odd}\}$ contains all odd permutations of S, and $\{\sigma|\sigma \text{ even}\}$, resp $\{\sigma^{-1}|\sigma \text{ even}\}$ contains all even permutations of S, and so in either case β is regular.

Thus to determine $e(S, S)$ it suffices to observe that if $\beta(\sigma) = \sigma\tau^{-1}C(\tau)$ for σ odd, $\beta(\sigma) = \sigma$ for σ even, or $\beta(\sigma) = \tau\sigma^{-1}C(\sigma)$ for σ odd, $\beta(\sigma) = \sigma^{-1}C(\sigma)$ for σ even, and β' is similarly of one of those two forms for some τ', then β and β' are not equivalent: that is, there exists no element $C(\pi)$ of $Inn(S) = Aut(S)$ so that $C(\pi)\beta(\sigma)C(\pi^{-1}) = \beta'(\sigma)$ for all σ. We have three cases.

Case I. $\beta(\sigma) = \sigma\tau^{-1}C(\tau)$ for σ odd, $\beta(\sigma) = \sigma$ for σ even; $\beta'(\sigma) = \sigma\tau'^{-1}C(\tau')$ for σ odd, $\beta'(\sigma) = \sigma$ for σ even. If $C(\pi)\beta(\sigma)C(\pi)^{-1} = \beta'(\sigma)$ in $Hol(S)$ for all σ, then for all σ even,

$$C(\pi)\sigma C(\pi^{-1}) = \sigma,$$

hence

$$C(\pi)\sigma = \sigma C(\pi),$$

so

$$\pi\sigma\pi^{-1}C(\pi) = \sigma C(\pi),$$

hence

$$\pi\sigma\pi^{-1} = \sigma.$$

But only the identity commutes with all even permutations. So $\pi = 1$ and $\beta = \beta'$

Case II. $\beta(\sigma) = \sigma\tau^{-1}C(\tau)$ for σ odd, $\beta(\sigma) = \sigma$ for σ even; $\beta'(\sigma) = \tau'\sigma^{-1}C(\sigma)$ for σ odd, $\beta'(\sigma) = \sigma^{-1}C(\sigma)$ for σ even. If $C(\pi)\beta(\sigma)C(\pi)^{-1} = \beta'(\sigma)$ in $Hol(S)$ for all σ, then for all σ even,

$$C(\pi)\sigma C(\pi^{-1}) = \sigma^{-1}C(\sigma)$$

or

$$\pi\sigma\pi^{-1}C(1) = \sigma^{-1}C(\sigma).$$

This never holds in $Hol(S)$ for $\sigma \neq 1$.

Case III. $\beta(\sigma) = \tau\sigma^{-1}C(\sigma)$ for σ odd, $\beta(\sigma) = \sigma^{-1}C(\sigma)$ for σ even; $\beta'(\sigma) = \tau'\sigma^{-1}C(\sigma)$ for σ odd, $\beta'(\sigma) = \sigma^{-1}C(\sigma)$ for σ even. If $C(\pi)\beta(\sigma)C(\pi)^{-1} = \beta'(\sigma)$ in $Hol(S)$ for all σ, then for all σ even,

$$C(\pi)\sigma^{-1}C(\sigma)C(\pi^{-1}) = \sigma^{-1}C(\sigma)$$

or

$$\pi\sigma^{-1}\pi^{-1}C(\pi\sigma\pi^{-1}) = \sigma^{-1}C(\sigma).$$

But then π commutes with all even σ, so π is trivial and $\beta = \beta'$.

Thus $e(S, S)$ is twice the number of even permutations in S of order dividing 2, as we wished to show. □

(10.5) COROLLARY. $e(S, S) = 2\sum_{k=0}^{\lfloor n/4 \rfloor} \frac{n!}{(n-4k)!2^{2k}(2k)!}$

PROOF. Any permutation of A_n of order dividing 2 is the product of an even number of disjoint transpositions. To find all products of $2k$ disjoint transpositions for $0 \leq k \leq \lfloor n/4 \rfloor$, pick two numbers from the original n, then two from the remaining $n-2$ numbers, then two from the remaining, etc.: the number of choices is $\binom{n}{2} \cdot \binom{n-2}{2} \cdot \ldots \cdot \binom{n-(4k-2)}{2}$. That gives $\frac{n!}{(n-4k)!2^{2k}}$ choices. But since the order of the $2k$ transpositions doesn't matter, we divide by $(2k)!$. The result is the number of ways of choosing an element which is a product of $2k$ disjoint transpositions. □

In [**CC99**] is shown that for $n \geq 5$, $e(S_n, A_n \times C_2) =$ twice the number of odd permutations in S of order 2, by a similar argument. Thus the number of Hopf Galois structures on a Galois extension L/K with Galois group S_n is at least equal to twice the number of permutations in S_n of order dividing 2. The number of such permutations is at least $\sqrt{n!}$.

There is a lower bound for $s(G)$ when $G = S_6$ in [**CC99**].

§11. Pure extensions. For an application of the theory to non-normal field extensions, consider a pure extension L/K of fields of degree p^n, p an odd prime, that is, an extension $L = K[w]$ with $w^{p^n} \in K$. Suppose K contains a primitive p^rth root of unity and L does not contain a primitive p^{r+1}st root of unity, so that if ζ is a primitive p^nth root of unity then $K[\zeta]/K$ is Galois of degree p^{n-r} with Galois group $\Delta = \langle\delta\rangle$, $\delta(\zeta) = \zeta^{a^{p^{r-1}}}$ where a is some integer $\equiv 1 \pmod p$ whose image generates the p-Sylow subgroup of $(\mathbb{Z}/p^n\mathbb{Z})^\times$ = the group of units of $\mathbb{Z}_{p^n}\mathbb{Z}$ congruent to 1 modulo p. Then the normal closure of L/K is $E = K[w, \zeta]$, $K[\zeta] \cap L = K$, and $\Gamma = Gal(E/K) = \langle\sigma, \delta\rangle$ where $\sigma(w) = \zeta w$, $\sigma(\zeta) = \zeta$, and δ fixes w; then $\Delta = Gal(E/L)$ and $\delta\sigma = \sigma^{a^{p^{r-1}}}\delta$.

(11.1) THEOREM [**Ko98**]. *If $r < n$ there are p^r Hopf Galois structures on L/K.*

The Galois case $r = n$ was treated in (9.2).

PROOF. By Byott's theorem (7.5), the number of Hopf Galois structures on L/K is the sum $s(\Gamma, \Delta) = \sum_N e(\Gamma, N)$ where N runs through the isomorphism classes of abstract groups of order p^n and $e(\Gamma, N)$ is the number of embeddings of Γ into $Hol(N) \subset Perm(N)$ such that the image of Δ in $Perm(N)$ is the stabilizer of the identity of N, modulo conjugation by elements of $Aut(N) \subset Hol(N)$.

Now Γ contains $\langle\sigma\rangle$, a cyclic group of order p^n. By (9.1), if N is not cyclic of order p^n, then $Hol(N)$ contains no element of order p^n, and so there is no embedding of Γ in $Hol(N)$. Thus $s(\Gamma, \Delta) = e(\Gamma, C)$ where C is the cyclic group of order p^n: any K-Hopf algebra H such that L/K is H-Galois must be a form of the group ring of a cyclic group of order p^n.

Let $Hol(C) = \langle\eta, \gamma\rangle$ where η has order p^n, γ has order $p^{n-1}(p-1)$ and $\gamma\eta = \eta^\pi\gamma$ where π is a primitive element modulo p^n. Then $\gamma^{p-1} = \theta$ satisfies $\theta\eta = \eta^b\theta$ where the image of b modulo p^n generates the group of units congruent to 1 modulo p^n. By replacing γ by a suitable power we can assume $b = a$, above.

The stabilizer of the identity of C in $Hol(C) \subset Perm(C)$ is $Aut(C)$. So any embedding $\beta : \Gamma = \langle\sigma, \delta\rangle \to Hol(C)$ of interest satisfies $\beta(\delta) \in Aut(C)$. Thus β has the form

$$\beta(\delta) = \theta^m \ , \quad \beta(\sigma) = \eta^i\theta^j$$

for some m, i, j.

Since σ has order p^n, we must have $\eta^i\theta^j$ of order p^n. By (8.7) this is true iff i and p are relatively prime. If we conjugate β by γ, the generator of $Aut(C)$, we have

$$\gamma(\beta(\delta))\gamma^{-1} = \gamma\theta^m\gamma^{-1} = \theta^m$$

since $\theta = \gamma^{p-1}$, and

$$\gamma(\beta(\sigma))\gamma^{-1} = \gamma(\eta^i\theta^j)\gamma^{-1} = \eta^{i\pi}\theta^j;$$

more generally,

$$\gamma^t(\beta(\sigma))\gamma^{-t} = \eta^{i\pi^t}\theta^j.$$

Since π is a primitive element modulo p^n and $(i,p)=1$, we can find a unique t so that $i\pi^t \equiv 1 \pmod{p^n}$. Thus we can assume β is of the form

$$\beta(\sigma) = \eta\theta^\ell, \; \beta(\delta) = \theta^m$$

for some ℓ, m, and each conjugacy class of β modulo $Aut(C)$ yields a unique pair (ℓ, m).

We need to count the possible pairs (ℓ, m), given the relation between σ and δ.

Now $\delta\sigma = \sigma^{a^{p^{r-1}}}\delta$, hence

$$\beta(\delta)\beta(\sigma) = \beta(\sigma)^{a^{p^{r-1}}}\beta(\delta)$$

or

$$\theta^m\eta\theta^\ell = (\eta\theta)^{\ell a^{p^{r-1}}}\theta^m. \tag{11.2}$$

We have $\theta^m\eta = \eta^{a^m}\theta^m$, while for any $e>0$,

$$(\eta\theta^\ell)^e = \eta^{1+a^\ell+\ldots+a^{\ell(e-1)}}\theta^{\ell e}.$$

Hence (11.2) becomes

$$\eta^{a^m}\theta^{\ell+m} = \eta^{1+a^\ell+\ldots+a^{\ell(a^{p^{r-1}}-1)}}\theta^{\ell a^{p^{r-1}}+m}.$$

For this to be valid, we must have

$$\ell + m \equiv \ell a^{p^{r-1}} + m \pmod{p^{n-1}} \tag{i}$$

and

$$a^m \equiv 1 + a^\ell + \ldots + a^{\ell(a^{p^{r-1}}-1)} \pmod{p^n}. \tag{ii}$$

For (i), since a generates the group of units congruent to 1 modulo p of $\mathbb{Z}/p^n\mathbb{Z}$, we have for all $s>0$,

$$a^{p^{s-1}} = 1 + p^s u$$

with $(u,p)=1$. Hence (i) becomes

$$p^{n-1} \text{ divides } \ell(a^{p^{r-1}} - 1) = \ell(p^r u)$$

or p^{n-1-r} divides ℓ, or $\ell \equiv p^{n-1-r}d \pmod{p^{n-1}}$. That is, the number of possible exponents ℓ is

$$p^{n-1}/p^{n-1-r} = p^r.$$

For each ℓ, congruence (ii) is valid for a unique m modulo p^{n-1}. For since $a \equiv 1 \pmod{p}$, $a^{\ell s} \equiv 1 \pmod{p}$ for all s, so the right side of (ii) is

$$1 + a^\ell + \ldots + a^{\ell(p^r u)} \equiv 1 + 1 + \ldots + 1 \; (p^r u + 1 \text{ copies}) \equiv 1 \pmod{p}.$$

So the right hand side of (ii) is a unit $\equiv 1 \pmod{p}$ of $\mathbb{Z}/p^n\mathbb{Z}$ and so for each ℓ there is a unique m so that

$$a^m \equiv \text{ right side of (ii)} \mod p^n.$$

Thus the number of pairs (ℓ, m) is exactly p^r, which completes the proof. □

CHAPTER 3

Tame Extensions and Noether's Theorem

This chapter introduces the basic setting of local Galois module theory and presents the concept of tame extension.

§**12. Associated orders.** Suppose L is a Galois extension of K, with Galois group G. Then L is a KG-module and the Normal Basis Theorem implies that L is a free KG-module of rank one. (For a proof of the Normal Basis Theorem see (2.17) or any text on Galois theory of fields.)

Now suppose that L, K are local fields. Let S, R be the valuation rings of L, K, respectively. Then G acts on S, with fixed ring R, so one can view S as an RG-module. Since R is a discrete valuation ring, S is a free R-module of rank equal to the cardinality of G, so we can ask if S has a normal basis over R, or equivalently, S is free as an RG-module. This question was answered by Emmy Noether, using the notion of tame ramification:

(12.1) Definition. An extension L/K of local fields is tamely ramified ("tame") if the ramification index $e_{L/K}$ of L over K is not divisible by the characteristic of the residue field of K.

An extension which is not tamely ramified is, of course, wildly ramified, or wild. An equivalent criterion is given by:

(12.2) Proposition. *L/K is tamely ramified iff the trace map $tr : S \to R$ defined by $tr(a) = \sum_{\sigma \in G} \sigma(a)$ is onto, that is, the integral $\theta = \sum_{\sigma \in G} \sigma$ of RG satisfies $\theta S = R$.*

(For a proof of (12.2), see [**Fr67**]).

Noether's theorem is:

(12.3) Theorem [**No31**]. *With K, L, R, S, G as above, S has a normal basis over R iff L/K is tamely ramified.*

Noether's theorem is the foundation of tame Galois module theory. The freeness of S over RG in the local tame case implies that for L/K a tamely ramified Galois extension of algebraic number fields, the ring of integers $\mathcal{O}_L$ of L is a locally free module over both $\mathcal{O}_K G$ and $\mathbb{Z}G$, and therefore defines a class in the class group of locally free rank one $\mathcal{O}_K G$-modules, resp. locally free $\mathbb{Z}G$-modules. The class of $\mathcal{O}_L$ is a subtle invariant, which relates to the Artin L-functions of the extension L/K. See [**Fr83, Section 1.1**] for an exposition of this theory.

Wild extensions are easy to find. For example, any ramified Galois extension L/K over a local field K containing $\mathbb{Q}_p$ with Galois group G of degree a power of p is wildly ramified.

To try to understand wild extensions, two approaches have been tried. One is to study the structure of the non-projective RG-module S, either directly (see **[EM94]**, **[El95]**) or, for global extensions, via a K-theoretic approach which allows one to bypass the non-projectivity at certain primes (see **[De84]**, work of Queyrut **[Fr83, p. 252]**, or, most influentially, work of Chinburg **[Cn85]**.

The other, due to Leopoldt **[Le59]**, is to replace the group ring RG by a larger order over R in KG, namely the associated order,

$$\mathfrak{A} = \{\alpha \in KG | \alpha s \in S \text{ for all } s \in S\}$$

with the idea that S may have better properties as a module over $\mathfrak{A}$ than over RG.

From the viewpoint of Hopf Galois extensions, Leopoldt's approach seems natural. In general, inside a K-Hopf algebra A there is no obvious analogue of the order RG in the group ring KG. But the notion of associated order extends easily:

(12.4) DEFINITION. Suppose A is a finite K-Hopf algebra and L/K is an A-Hopf Galois extension of global or local fields. Let R be the ring of integers (= valuation ring, if K is local) of K. Let S be the integral closure of R in L. The *associated order of S in A* is

$$\mathfrak{A} = \{\alpha \in A | \alpha s \in S \text{ for all } s \in S\}.$$

We note that $\mathfrak{A}$ is an R-algebra, for $1 \in \mathfrak{A}$, and if $\alpha, \beta \in \mathfrak{A}$, then for all $s \in S, (\beta\alpha)(s) = \beta(\alpha(s)) \in S$ since $\alpha(s) \in S$. However, $\mathfrak{A}$ need not be a Hopf algebra. (For example, if $K = \mathbb{Q}$ and L is a Galois extension of $\mathbb{Q}$ with abelian Galois group G, then $\mathfrak{A}$ is not Hopf if any odd prime ramifies wildly in L: see **[Ch87, Section 5]**.)

To justify the choice of $\mathfrak{A}$ as the "correct" algebra to act on S, we note:

(12.5) PROPOSITION. *Let L/K be an A-Hopf Galois extension, $H \subset \mathfrak{A}$ an order over R in A, and suppose S is H-free of rank one. Then $H = \mathfrak{A}$.*

PROOF. If $S = Ht$ is free of rank one as an H-module with basis t for some t in S, then $L = At$ is a free A-module with basis t. Let $\alpha \in \mathfrak{A}$, then $\alpha t \in S$, so $\alpha t = ht$ for some h in H. Since L is A-free with basis t, $\alpha = h$. Hence $\mathfrak{A} = H$. □

Here is an initial example where S is free over its associated order $\mathfrak{A}$ where $\mathfrak{A} \neq RG$:

(12.6) EXAMPLE. Let $K = \mathbb{Q}, R = \mathbb{Z}, w = \sqrt{2}$ and $L = K[w]$. Then L/K is Galois with group $G = \{1, \sigma\}$ where $\sigma(w) = -w$. The ring of integers of L is $S = R[w]$. The group ring RG clearly acts on S, but the trace map $tr : S \to R$ is not onto: $tr(a + bw) = 2a$, hence $tr(S) = 2R$: the extension locally is wildly ramified at 2.

Let $e_1 = \frac{1+\sigma}{2}, e_{-1} = \frac{1-\sigma}{2}$. Then e_1 and e_{-1} map S to S, so are in the associated order $\mathfrak{A}$. Since $R[e_1, e_{-1}]$ is the maximal order of RG in KG, $\mathfrak{A} = R[e_1, e_{-1}]$. Moreover, S is a free $\mathfrak{A}$-module with $\mathfrak{A}$-basis $\alpha = 1 + w$ since $e_1\alpha = 1, e_{-1}\alpha = w$.

The most striking success achieved by replacing RG by the associated order $\mathfrak{A}$ as the algebra acting on the valuation ring (ring of integers) of an extension L of a local (global) field K was Leopoldt's original result **[Le59]**, that if L is any abelian extension of $\mathbb{Q}$, then the ring of integers S of L is a free module of rank one over its associated order.

Leopoldt's result does not generalize easily, however: subsequent investigation (Bergé [**Be78**], and Bertrandias and Ferton [**BF72**], [**BBF72**], [**Fe74**]) showed that for L/K an extension of local fields containing $\mathbb{Q}_p$, S need not be free over its associated order if $K \neq \mathbb{Q}_p$ or if G is non-abelian. Finding criteria for S to be free over its associated order has been difficult.

The main point of this chapter is to present the following general criterion:

(12.7) THEOREM [**Ch87**], [**CM94**]. *Suppose L/K is a finite A-Hopf Galois extension of local fields. If $\mathfrak{A}$ is a Hopf order in A, then S is $\mathfrak{A}$-free of rank one.*

To obtain this result, we need to extend the notion of tameness to Hopf algebras.

(12.8). The converse of (12.7) is false: there are many wildly ramified Galois extensions L/K whose valuation rings are are free over their associated order $\mathfrak{A}$ but $\mathfrak{A}$ is not a Hopf order. If L is an abelian extension of $\mathbb{Q}$ with Galois group G, then $\mathcal{O}_L$ is free over $\mathfrak{A}$ by Leopoldt's theorem, but $\mathfrak{A}$ is not a Hopf order in $\mathbb{Q}G$ if any odd prime of $\mathbb{Z}$ ramifies wildly in $\mathcal{O}_L$: see [**Ch87, Corollary 5.6**].

There is a significant body of literature related to proving or disproving the freeness of the ring of integers of a Galois extension of global or local fields over its associated order when the associated order is not Hopf. The introduction to [**Lt98**] reviews some of these results.

[**Ta95**] describes an extension of Leopoldt's theorem to arbitrary wild abelian extensions by replacing the valuation ring $\mathcal{O}_L$ by a possibly non-maximal order in L.

§13. Tame extensions. Suppose L is an A-Hopf Galois extension of K, local fields, for some K-Hopf algebra A, necessarily of rank equal to the dimension of L over K. Suppose H is a Hopf order over R in A: that is, H is an R-submodule of A of finite rank over R, H is an R-Hopf algebra and the map $K \otimes_R H \to A$ induced by the inclusion map $H \subset A$ is an isomorphism of K-Hopf algebras. Suppose the valuation ring S of R is an H-module via restriction of the action of A on L. Then S is an H-module algebra (since L is an A-module algebra). We can then ask for conditions under which S is free as an H-module.

Motivated by (12.2), we call S a tame H-extension of R if $\theta \cdot S = R$ for some left integral θ of H.

If $H = RG, A = KG$, then S is a tame H-extension of R iff L/K is tamely ramified.

More generally, let R be a commutative ring, H a cocommutative R-Hopf algebra which is a finitely generated projective R-module, and let I be the module of left integrals of H.

(13.1) DEFINITION [**CH86**]. Let S be an R-algebra which is a finitely generated projective R-module, and suppose S is an H-module algebra with $S^H = R$. We say that S is a *tame* H-extension of R if

1. $rank_R(S) = rank_R(H)$;
2. S is a faithful H-module (i.e. if $hS = 0$ then $h = 0$); and
3. $IS = R$.

To help understand condition (3), we have

(13.2) LEMMA. *Let H be a finite cocommutative R-Hopf algebra with module of left integrals I, and S an H-module algebra. Then $IS \subset S^H$.*

PROOF. Recall that

$$S^H = \{s \in S | h \cdot s = \varepsilon(h)s \text{ for all } h \in H\}.$$

Let $\xi = \sum \theta_i s_i \in IS$, where $\theta_i \in I$. Then for any $h \in H$,

$$h\xi = h(\sum \theta_i s_i) = \sum (h\theta_i)s_i = \sum \varepsilon(h)\theta_i s_i = \varepsilon(h)\xi.$$

So $\xi \in S^H$. □

When $H = RG$ this says that $tr(s) = \sum_{\sigma \in G} \sigma(s)$ is fixed by G for all $s \in S$.

Given (13.2) and $S^H = R$, tameness implies that IS is as large as possible.

For valuation rings of local fields, condition (3) is the only relevant condition for tameness. Suppose R, S are valuation rings of an A-Hopf Galois extension L/K of local fields. Let H be a Hopf order over R in A. Then $S^H = R$ and conditions (1) and (2) always hold. However condition (3) often fails. The simplest example is to take L/K to be Galois with Galois group G and let RG act on S. Then the condition, $IS = R$, is the same as the condition that the trace map $S \to R$ is onto, which holds iff L/K is tamely ramified.

The first observation related to (12.7) is that if $\mathfrak{A}$ is Hopf, then S is $\mathfrak{A}$-tame:

(13.3) THEOREM. *Suppose L/K is an A-Hopf Galois extension of local fields. Let R be the valuation ring of K, let S be the integral closure of R in L. If the associated order $\mathfrak{A}$ of S is a Hopf order in A, with module of integrals $I = R\theta$, then $\theta S = R$, hence S is a tame $\mathfrak{A}$-extension of R.*

PROOF. We know that since R is local, the module of integrals I of $\mathfrak{A}$ is a free R- module of rank one (3.4), so let θ be a generator of I.

Since L is an A-Hopf Galois extension of K, the fixed ring $L^A = K$, hence $S^{\mathfrak{A}} = R$. Hence $\theta S \subset S^{\mathfrak{A}} = R$ by (13.2).

Let π be a parameter for R. Since θS is an ideal of $R, \theta S = R\pi^i$ for some $i \geq 0$. Hence $(\theta/\pi^i)S = R$. Since θ/π^i maps S to $R \subset S$, $\theta/\pi^i \in \mathfrak{A}$, and is an integral of $\mathfrak{A}$. Since θ generates the module of integrals of $\mathfrak{A}, i = 0$ and $\theta S = R$.

The other conditions for S to be $\mathfrak{A}$ tame are true for $\mathfrak{A}$ any order over R in A. That completes the proof. □

(12.7) is then implied by the following generalization of Noether's theorem.

(13.4) THEOREM. *Let R be a local domain with quotient field K of characteristic zero. Let H be a finite cocommutative R-Hopf algebra and S be an R-algebra which is a finitely generated projective R-module. Suppose S is a tame H-extension of R. Then S is a free H-module of rank one.*

The proof of (13.4) is in two parts, one formal, one representation-theoretic. Here is the formal part.

(13.5) THEOREM. *Let R be a local ring, H a finite free cocommutative R-Hopf algebra with module of left integrals I, and S an R-algebra which is a finite free R-module and an H-module algebra. If $IS = R$ then S is H-projective.*

PROOF. Let $I = R\theta$, then $\theta S = R$, so there exists some z in S so that $\theta z = 1$.

Since S is R-projective, $H \otimes_R S$, with H-action via the action of H on H, is H-projective. The scalar multiplication map $\mu : H \otimes S \to S, \mu(h \otimes s) = hs$, is then an H-module homomorphism, so to show S is H-projective, it suffices to show that μ splits: there exists some H-module homomorphism $\nu : S \to H \otimes S$, so that $\mu \circ \nu$ is the identity on S.

If $h \in H$, write the antipode of h exponentially, viz: h^λ

Define ν by $\nu(s) = \sum_{(\theta)} \theta_{(1)} \otimes z(\theta^\lambda_{(2)} s)$. Then

$$\mu\nu(s) = \sum_{(\theta)} \theta_{(1)}(z(\theta^\lambda_{(2)} s)).$$

Since S is an H-module algebra, this

$$= \sum_{(\theta)} \theta_{(1)}(z)\theta_{(2)}(\theta^\lambda_{(3)} s)).$$

Since $\sum_{(\theta)} \theta_{(1)}\theta^\lambda_{(2)} = \varepsilon(\theta)$, this becomes

$$\begin{aligned} &= \sum_{(\theta)} \theta_{(1)}(z)\varepsilon(\theta_{(2)})s \\ &= (\sum_{(\theta)} \theta_{(1)}\varepsilon(\theta_{(2)}))(z)s \\ &= \theta(z)s = s. \end{aligned}$$

So ν is injective and splits the map μ.

It suffices to show that ν is an H-module homomorphism.

First we recall (3.7): for any $h \in H$,

$$\sum_{(\theta)} h\theta_{(1)} \otimes \theta^\lambda_{(2)} = \sum_{(\theta)} \theta_{(1)} \otimes \theta^\lambda_{(2)} h.$$

Using this identity, we show that ν is an H-module homomorphism, as follows:

$$\begin{aligned} h\nu(s) &= \sum_{(\theta)} h\theta_{(1)} \otimes z(\theta^\lambda_{(2)} s) \\ &= (1 \otimes z)((\sum_{(\theta)} h\theta_{(1)} \otimes \theta^\lambda_{(2)})(1 \otimes s)) \\ &= (1 \otimes z)((\sum_{(\theta)} \theta_{(1)} \otimes \theta^\lambda_{(2)} h)(1 \otimes s)) \\ &= \sum_{(\theta)} \theta_{(1)} \otimes z\theta^\lambda_{(2)}(h(s)) \\ &= \nu(hs). \end{aligned}$$

Thus ν is an H-module homomorphism, and so S is a direct summand as H-module of the projective H-module $H \otimes S$, hence is H-projective. □

(13.4) now follows from:

(13.6) PROPOSITION. *Let K be a local field with valuation ring R, let A be a finite cocommutative K-Hopf algebra. Let S be an order over R in an A-Galois extension L of K. If H is a Hopf order over R in A which is contained in the associated order $\mathfrak{A}$ of S and S is H-projective, then $S \cong H$ as left H- modules.*

(13.6) is a consequence of the following general result:

(13.7) THEOREM (SCHNEIDER [**Sch77**]). *Let R be a local domain with quotient field K of characteristic zero. Let H be a finite cocommutative R-Hopf algebra. Let P, Q be finitely generated projective left H-modules. If $K \otimes_R P \cong K \otimes_R Q$ as $K \otimes H$-modules, then $P \cong Q$ as H-modules.*

PROOF OF (13.6). S is a projective left H-module, $K \otimes_R S \cong L$, $K \otimes_R H \cong A$, and $L \cong A$ as left A-modules since L is A-Galois over K (2.16). To conclude that $S \cong H$ as left H-modules we apply (13.7). □

Putting (13.6) together with (13.5) and (13.3) yields Theorem (12.7), that if L/K is an A-Galois extension of local fields, and the associated order $\mathfrak{A}$ of the valuation ring S of L is Hopf, then S is $\mathfrak{A}$-free.

We note that (13.4), tameness implies freeness, has a converse. See (14.5) below.

Remarks on Schneider's Theorem. When $H = RG$, Schneider's theorem (13.7) is a well-known result of Swan [**Sw60**] which provided the first really satisfactory proof of Noether's theorem.

Swan's theorem is proved in four places in [**CR81**]: (18.16), Exercise 18.14, (21.21) and (32.5). The second proof uses the invertibility of the Cartan matrix, defined as follows: let $k = R/m$ and $\overline{H} = k \otimes RH$. If $\{E_i\}$ represents the isomorphism classes of simple $\overline{H}$- modules, and P_i is the projective cover of E_i, then $\{P_i\}$ represents the isomorphism classes of indecomposable projective H-modules. Let $c_{i,j}$ be the number of composition factors of P_i which are isomorphic to E_j, then $C = \{c_{i,j}\}$ is the Cartan matrix of $\overline{H}$.

The main theorem of [**Sch77**] is that C is invertible.

Schneider's proof involves a reduction to the case where $K \otimes H \cong KG$, G a finite group, and then an extension of the K-theoretic induction machinery of Swan [**Sw60**] as formulated by Lam [**Lm68**] to Hopf orders in KG' for subgroups G' of G. Presenting the details would carry us rather far afield. Readers interested in investigating K-theory for Hopf algebras may wish to consult Schneider's paper [**Sch77**] as well as Pareigis [**Pa73**] and Weinraub [**We94**], [**We95**].

If A is a finite commutative K-Hopf algebra, where K has characteristic zero, then A is a separable K-algebra, by [**Wa79, §11.4**]. The abelian version of Schneider's theorem then follows from the following exercise [**CR81, sec. 35, Ex 9**]:

Let R be a discrete valuation ring and Λ be an order in a finite, commutative, separable K-algebra A. Let M, N be finitely generated projective Λ- modules. Then $M \cong N$ as Λ-modules iff $KM \cong KN$ as A-modules.

On tameness. The concept of tameness for the action of a Hopf algebra on an extension of rings, as in (13.1), has been generalized in [**CEPT96**] to actions of affine group schemes on affine schemes. The key step is to translate the notion of tame extension to comodule algebra structures, which is accomplished by the following result.

(13.8) PROPOSITION. *Let H be a finite cocommutative R-Hopf algebra, R a local ring, and S a finite faithful H-module algebra which is a finite R-algebra of rank equal to the rank of H. Then S is a tame H-extension of R iff there is a left H-module homomorphism $g : H^* \to S$ such that $g(1) = 1$.*

Such a map g is an H^*-comodule homomorphism from H^* to S.

PROOF. Recall that H^* is an H^*-comodule via the comultiplication $\Delta : H^* \to H^* \otimes H^*$, so H^* is an H-module. From (3.3) there is an integral t of H^* which is a basis of H^* as a free H-module: $H^* = Ht$. Hence $1 = \theta \cdot t$ for some $\theta \in H$. Then θ generates the module of left integrals of H. For using (2.3), we have for any h in H that $h\theta \cdot t = h \cdot (\theta \cdot t) = h \cdot 1 = \langle h, 1\rangle = \varepsilon(h) = \varepsilon(h)\theta \cdot t$.

If S is H-tame, then there exists z in S with $\theta \cdot z = 1$; if we define an H- module homomorphism $g : H^* \to S$ by $g(t) = z$, then $g(1) = g(\theta \cdot t) = \theta \cdot z = 1$.

Conversely, if $g : H^* \to S$ is a left H-module homomorphism with $g(1) = 1$, then $z = g(t)$ satisfies $\theta \cdot z = \theta \cdot g(t) = g(\theta \cdot t) = g(1) = 1$, so S is H- tame. □

Chinburg, Erez, Pappus and Taylor define an H-comodule algebra structure $\alpha : S \to S \otimes_R H$ to be tame if there is an H-comodule map $g : H \to S$ such that $g(1) = 1$. Thus (13.8) implies that CEPT-tameness is equivalent to tameness as defined in (13.1) (which those authors call CH-tameness). CEPT-tameness, translated to schemes, allows an extension of the notion of tame action to the case of actions (X, G) of an affine but not necessarily finite group scheme G on an affine S-scheme X. This setting permits a far-reaching generalization of the main result of classical tame Galois module theory [**Fr83**], namely Taylor's proof of Frohlich's conjecture relating analytic invariants (Artin root numbers) to global Galois module structure. We defer to [**CEPT96**] and its sequels for further details and consequences.

§**14. Galois extensions.** Section 2 contains a generalization (2.16) of the Normal Basis Theorem for Hopf Galois extensions: if R is local and S/R is H-Galois, then S is H-free of rank one. The proof in the Galois case did not use the full strength of Schneider's theorem, but was a direct application of the Krull-Schmidt theorem. Thus it is of interest to understand the relationship between H-Galois and H-tame.

We first show that H-Galois implies H-tame.

(14.1) PROPOSITION. *Let S be an H-Hopf Galois extension of R. Then S is an H-tame extension of R.*

This should be expected, since in the case $H = RG$ and S the valuation ring of an extension of local fields, Galois is equivalent to unramified and tame to tamely ramified.

To prove this, we need to reexamine the Morita correspondence (c.f. (2.13)) between R-modules and $S\#H$-modules for S an H-Galois extension of R. Recall that multiplication in $S\#H$ is by

$$(s\#h)(t\#k) = \sum_{(h)} sh_{(1)}(t)\#h_{(2)}k.$$

Let $E = End_R(S)$. For any left $S\#H$-module M,

$$M \cong S \otimes_R Hom_E(S, E) \otimes_E M \tag{14.2}$$

since $S \otimes_R Hom_E(S, E) \cong E$. We note that $E \cong S\#H$ is a projective left H-module with $g \in H$ acting via left multiplication by $1\#g$: since

$$\begin{aligned} s\#h &= s\#\sum \varepsilon(h_{(1)})h_{(2)} \\ &= \sum_{(h)} \varepsilon(h_{(1)})s\#h_{(2)} \\ &= \sum_{(h)} h_{(2)} \cdot \lambda(h_{(1)})(s)\#h_{(3)} \\ &= \sum_{(h)} (1\#h_{(2)})(\lambda(h_{(1)})(s)\#1), \end{aligned}$$

any element of $S\#H$ can be written as a sum of terms of the form $(1\#h)(s\#1)$, on which $g \in H$ acts by

$$g((1\#h)(s\#1)) = (1\#gh)(s\#1).$$

Thus, as left H-module, $S\#H = H \otimes S$. Since S is R-projective, $S\#H = H \otimes S$ is left H-projective.

Also, S is a projective left E-module since S, being a faithful finitely generated projective R-module, is a generator in the category of left R- modules.

Now we have

(14.3) PROPOSITION. *Let S be an H-Galois extension of R and $E = End_R(S) \cong S\#H$. Then for any left E-module M,*

$$Hom_E(S, E) \otimes_E M \cong Hom_E(S, M) \cong M^H \cong IM$$

where $I = H^H$. Hence $M \cong S \otimes IM$ as left E-modules.

PROOF. The first isomorphism holds for $S = E$, hence for S a free E- module, hence for S a direct summand of a free E-module, i. e. for S a projective E-module.

The second isomorphism is by $f \mapsto f(1)$, as noted in (2.13).

For the last, $M^H \cong E^H \otimes_E M$ since $Hom_E(S, E) \cong E^H$, so it suffices to show that $E^H \cong H^H E$. But as left H-modules, $E \cong H \otimes S$, where S is a projective R-module. If $P = R^n = \sum Rv_i$ is a free R-module, then

$$H^H(H \otimes P) = \sum H^H(Hv_i) = \sum H^H v_i = (\sum Hv_i)^H = (H \otimes P)^H$$

since H^H is a right ideal of H. If $H^H(H \otimes P) = (H \otimes P)^H$ for P a free R-module, it also holds for P projective, e. g. $P = S$.

The isomorphism $M \cong S \otimes IM$ then follows from (14.2). □

PROOF OF (14.1). The isomorphism $M^H \cong IM$ with $M = S$ shows that $R = IS$, and (3) of (13.1) holds. The other conditions (1) (equality of ranks) and (2) (faithfulness) of (13.1) follow from the isomorphisms of (2.9). □

The proof of (14.3) yields the following useful result:

(14.4) COROLLARY. *Let S be an H-Galois extension of R. Let T be an H-module algebra. and $\phi : S \to T$ be an R-algebra, H-module homomorphism. Let $I = H^H$, then $IT = T^H$. If $IT = T^H = R$, then ϕ is an isomorphism.*

PROOF. View T as an $S\#H$-module via $(s\#h)(t) = \phi(s)h(t)$. Then by (14.3), $T^H = IT$ and $S \otimes IT \cong T$ via $s \otimes ft \mapsto \phi(s)ft$. If $T^H = IT = R$ then $S \cong S \otimes IT$ by $s \mapsto s \otimes 1$ and the composite isomorphism $S \to S \otimes IT \to T$ is ϕ. □

From (14.1) we have that H^*, the trivial H-Galois extension (2.10a), is H-tame. This observation yields a converse to (13.4), tame implies free:

(14.5) COROLLARY. *Let S be an H-module algebra such that $S^H = R$ and $S \cong H^*$ as H-modules. Then S is H-tame.*

PROOF. H^* is a faithful H-module, so S is, and $rank_R(H^*) = rank_R(S)$. We need to show that $IS = R$. If $\phi : S \to H^*$ is an H-module isomorphism then $\phi(IS) = IH^*$, so S is H-tame if H^* is. But H^* is H-Galois, hence H-tame by (14.1). □

The rest of this section is devoted to showing that when H is local, then H-tame implies H-Galois, so that when H is local, all three conditions on an H-module algebra S: H-tame, H-Galois, H-free, are equivalent.

(14.6) DEFINITION. A ring A is local if A has a unique maximal left ideal.

If A is a local ring then the Jacobson radical $rad(A)$ is the maximal left ideal. If A is a local R-Hopf algebra, R a field, then $rad(A) = ker(\varepsilon)$ since $ker(\varepsilon)$ is a maximal left ideal. One class of examples is $A = RG$ where R is a field of characteristic p and G is a p-group ([**CR81, (5.25)**]).

(14.7) THEOREM. *Let R be a local ring with maximal ideal m, let H be a local, cocommutative R-Hopf algebra with module of integrals $R\theta$, S a finite R- algebra and a faithful H-module algebra. Suppose $rank_R H = rank_R S$. Then the following are equivalent*

(1) *θS is a tame H-extension of R: that is, there is some $t \in S$ so that $\theta t = 1$;*
(2) *$S = Ht$ for some t in S;*
(3) *S is an H-Galois extension of R.*

To obtain (2) implies (3), we first do that case over a field.

(14.8) THEOREM. *If H is a finite local cocommutative K-Hopf algebra, K a field, and S is an H-module algebra which is free of rank one as an H- module, then S is an H-Galois extension of K.*

For H commutative and cocommutative this result appears in [**CH86**]. The proof here is due to Waterhouse [**Wa92**].

We need some preliminary information.

First, recall that if S is an H^*-comodule via $\rho : S \to S \otimes H^*$,

$$\rho(s) = \sum_{(s)} s_{(0)} \otimes s_{(1)}$$

with $s_{(1)} \in H^*$, then the H-module action on S is by

$$h \cdot s = \sum_{(s)} s_{(0)} \langle s_{(1)}, h \rangle.$$

Conversely, given an H-module action on S, we get a coaction $\rho : S \to S \otimes H^*$ by $\rho(s) = \sum_i h_i s \otimes f_i$, where $\{h_i, f_i\}_{i \in I}$ is a projective coordinate system for H (2.2).

We also recall (3.5): If H be a finite R-Hopf algebra, R a principal ideal domain and θ generates the module of left integrals of H^*, then $\beta(h, k) = \langle \theta, hk \rangle$ is a non-degenerate, associative bilinear form on H.

Using β, we have:

(14.9) PROPOSITION. *Let H be a finite local K-Hopf algebra, K a field. If $M \supset N$ are left ideals of $H, M \neq N$, then there exists some $g \in H$ so that $Mg \neq 0, Ng = 0$.*

PROOF. For M a left ideal of H, set

$$M^\perp = \{f \in H | \beta(M, f) = 0\}.$$

Then $M^\perp = \{f \in H | Mf = 0\}$, a right ideal of H. For by associativity of β,

$$\beta(M, f) = \beta(HM, f) = \beta(H, Mf).$$

Since β is non-degenerate, $\beta(M, f) = 0$ iff $Mf = 0$.

Similarly, we can define the left ideal $N^\perp$ for N a right ideal of H. Thus $M^{\perp\perp}$ is a left ideal of H and $M^{\perp\perp} = M$. For clearly $M^{\perp\perp} \supset M$; to get equality, we apply non-degeneracy of β to show, by a standard dual basis argument, that $dim M^\perp + dim M = dim H$, hence $dim M = dim M^{\perp\perp}$, hence $M^{\perp\perp} = M$.

Now if $N \subset M, N \neq M$, then $M^\perp \subset N^\perp, M^\perp \neq N^\perp$, so for $g \in N^\perp$ and $g \notin M^\perp$, then $Ng = 0, Mg \neq 0$, as we wished to show. $\square$

Now we proceed to the proof of (14.8), that for H is finite and local, if S is H-free of rank one, then S is K-Galois.

PROOF OF (14.8). First, if $S = Ht$ for some generator t in S, then t is multiplied to 1 by a left integral of H. For $1 = \theta t$ for some θ in H. But then for any $h \in H$,

$$h\theta t = h \cdot 1 = \varepsilon(h)1 = \varepsilon(h)\theta t;$$

since S is H-free with basis t, we have $h\theta = \varepsilon(h)\theta$, so θ is a left integral of H.

To show S is H-Galois, it suffices, by a dimension argument, to show that

$$\gamma : S \otimes S \to S \otimes H^*$$

is onto.

Pick a K-basis $\{f_0, ..., f_{n-1}\}$ of H as follows:

$f_0 = 1$;

$f_1, \ldots, f_{m(1)}$ in J yield a K-basis of J/J^2;

$f_{m(1)+1}, \cdots, f_{m(2)}$ in J^2 yield a K-basis of J^2/J^3;

etc. Let $a_0, ..., a_n$ be the dual basis in H^* of $\{f_0, ..., f_{n-1}\}$; then $a_0 = \varepsilon$ is the identity element of H^*.

The basis of H and corresponding dual basis of H^* define the coaction of H^* on S, namely, $\rho : S \to S \otimes H^*$ by $\rho(s) = \sum f_i s \otimes a_i$. Then for any $s_1, s_2 \in S$,

$$\gamma(s_1 \otimes s_2) = \sum s_1(f_i s_2) \otimes a_i$$

so to show γ is onto, it suffices to show that $1 \otimes a_i$ is in the image of γ for all i.

First, $\rho(1) = 1 \otimes 1 = 1 \otimes a_0$ is in the image of γ.

Assume $1 \otimes a_i$ is in the image of γ for $i < r$.

Now Jf_r is a K-linear combination of the basis elements f_j only for $j > r$. So

$$V_r = \sum_{i \geq r} K f_i$$

is a left ideal of H. By (14.9), there is some $g \in H$ so that $V_r g \neq 0$, while $V_{r+1} g = 0$. Hence $f_j g = 0$ for $j > r$, but $f_r g \neq 0$. Since $Jf_r \subset \sum_{j>r} K f_j$, we have $J f_r g = 0$. Since $J = ker(\varepsilon)$, for any $h \in H$, $h - \varepsilon(h) \in J$, so $h f_r g = \varepsilon(h) f_r g$. Thus $f_r g$ is a non-zero left integral of H. Adjusting g by a scalar, we can assume $f_r g = \theta$. Then for t the H-basis element of S,

$$\begin{aligned} \gamma(1 \otimes gt) &= \rho(gt) \\ &= \sum_i f_i(gt) \otimes a_i. \end{aligned}$$

Since $f_j g = 0$ for $j > r$ this

$$\begin{aligned} &= f_r gt \otimes a_r + \sum_{i<r} f_i(gt) \otimes a_i \\ &= \theta t \otimes a_r + \sum_{i<r} f_i(gt) \otimes a_i \\ &= 1 \otimes a_r + \sum_{i<r} f_i(gt) \otimes a_i. \end{aligned}$$

Since $\sum_{i<r} f_i(gt) \otimes a_i$ is in the image of γ, by induction, $1 \otimes a_r$ is in the image of γ. Hence γ is onto. □

(14.8) need not be true if H is not local. For example, let $R = \mathbb{Q}_p, G = C_p$, cyclic of order p with generator σ and $H = RG^* = \sum_{i=0}^{p-1} Re_i$ where $e_i(\sigma^j) = \delta_{i,j}$. Let $S = R[z]$ with $z^p = 0$. Define an H-module algebra structure on S by $e_i(z^j) = \delta_{i,j}$; equivalently, S is a G-graded R-algebra, hence an H^*-comodule algebra via $\alpha : S \to S \otimes H^*$ where $\alpha(z^i) = z^i \otimes \sigma^i$. Then S is tame: for $\theta = \varepsilon = e_0$ is a integral of H and $e_0(1) = 1$, so $e_0 S = R$. However S is not H-Galois since z^p is not a unit of R. In fact, if

$$\gamma : S \otimes S \to S \otimes H^*$$

is the map such that $\gamma(s \otimes f) = (s \otimes 1)\alpha(t)$, then γ is not 1-1:

$$\gamma(z^i \otimes z^j) = (z^i \otimes 1)(z^j \otimes \sigma^j) = z^{i+j} \otimes \sigma^j = 0$$

if $i + j \geq p$.

Now we proceed to the proof of (14.7), that if H is local then H-tame, H-Galois and H- free are equivalent.

PROOF OF (14.7). 3) implies 1), H-Galois implies H-tame, was shown in (14.1).

2) implies 3), H-free implies H-Galois. If S is isomorphic to H as H-module, then S/mS is isomorphic to H/mH as H/mH-modules, hence, since H/mH is local, S/mS is a Galois H/mH- module by (14.8); but then, using Nakayama's lemma, the map $S \otimes S \to S \otimes H^*$ is an isomorphism because it is true modulo m. (The converse, H-Galois implies H-free, was shown in (2.16).)

1) implies 2), H-tame implies H-free. This is (13.4), which follows from Schneider's theorem (13.7). But when H is local we can find an explicit normal basis element for S.

Suppose θ is an integral of H so that $\theta S = R$. Let $t \in S$ so that $\theta t = 1$. Let $j : H \to S$ by $j(h) = ht$. We show that j is an isomorphism. By Nakayama's Lemma, it suffices to show that j is an isomorphism modulo the maximal ideal m. Now over $k = R/m$, if j is 1-1, then j is onto, hence an isomorphism, since $rank(S) = rank(H)$.

Working over k, let $M = ker(j)$. Now $M = (M^{\perp})^{\perp}$ as in (14.9), where

$$M^{\perp} = \{f \in H | Mf = 0\}$$

is a two-sided ideal, since M is, and is proper if $M \neq 0$. Thus if $M \neq 0$, then $M^{\perp} \subset ker(\varepsilon)$, the maximal ideal of H, and so for any $h \in M^{\perp}, h\theta = \varepsilon(h)\theta = 0$. Hence $\theta \in (M^{\perp})^{\perp} = M = ker(j)$. But then

$$0 = j(\theta) = \theta t = 1,$$

impossible. So $M = 0$ and j is 1-1.

That completes the proof of (14.7). □

See Byott [**By95c, Theorem 5**] or (29.2) for conditions on when H is local.

CHAPTER 4

Hopf Algebras of Rank p

In this chapter we describe the classification of Hopf algebras of rank p over the valuation ring R of K, a finite extension of $\mathbb{Q}_p$, due to Tate and Oort [**TO70**]. The description demonstrates that classifying abelian R-Hopf algebras involves the arithmetic of R is an essential way.

Fix the prime p throughout. All R-Hopf algebras are finite (that is, finitely generated and projective as R-modules).

§15. Rank p Hopf orders over algebraically closed fields.

(15.1) PROPOSITION. *Let R be the valuation ring of a local field K of characteristic zero. Then any cocommutative R-Hopf algebra H of rank p is also commutative.*

PROOF. Let $\overline{K}$ be the algebraic closure of the field of fractions of R. Then since $\overline{K}$ has characteristic zero, any finite $\overline{K}$-Hopf algebra is separable ([**Wa79, (11.4)**]). Since H^* is commutative, therefore $\overline{K}H^* \cong \overline{K} \times \overline{K} \times \ldots \times \overline{K}$ (p copies), and so $\overline{K}H = \overline{K}C_p$ by the argument of (6.3), hence is commutative. Since the map from H to $\overline{K}H$ is 1-1, H is commutative. □

As a prelude to the classification of rank p Hopf algebras over valuation rings R, we consider the same classification over an algebraically closed field L. Recall, "abelian" means commutative and cocommutative.

(15.2) THEOREM. *Let L be an algebraically closed field. There are, up to isomorphism, at most three abelian L-Hopf algebras of rank p: LC_p, LC_p^* and $L[x]/(x^p)$ with x primitive.*

Recall: x is primitive if $\Delta x = x \otimes 1 + 1 \otimes x$.

PROOF. If L is algebraically closed and the L-Hopf algebra H of rank p is separable, then $H \cong LC_p^*$ as noted in the proof of (15.1), hence $H^* \cong LC_p$. In general, by Waterhouse ([**Wa79, (6.8)**]), any finite abelian group scheme is the direct product of a separable group scheme and a connected group scheme. Translating to Hopf algebras, a L-Hopf algebra H has the form $H = H_\ell \otimes_L H_{et}$, where H_ℓ is local and H_{et} is separable. The dual H^* decomposes similarly. If H has prime rank, then $H = H_\ell$ or H_{et} and similarly for the dual.

By [**Wa79, (14.4)**], if the characteristic of L is $q \neq p$ then H is separable, as is H^*. Thus $H = LC_p^*$.

Suppose char $K = p$. Then there are four possible cases.

If H is separable, then $H \cong LC_p^*$, in which case $H^* \cong LC_p$, which is local. Similarly if H^* is separable. Thus the separable with separable dual case cannot occur. The remaining case is where both H and H^* are local.

By [**Wa79, (14.4)**], if H is local, $H \cong L[x_1, \dots, x_n]/(J)$ where J is generated by pth powers of $x_1, \dots, x_n$. Thus if H is local of rank p, with local dual, then $H \cong L[x]/(x^p)$ and $H^* \cong L[t]/(t^p)$. In this case, we show that x is primitive, as follows.

Let $\{x_0, x_1, \dots, x_{p-1}\}$ be a dual basis for $\{1, t, t^2, \dots, t^{p-1}\}$. Since $t^p = 0$, $\varepsilon_{H^*}(t^r)^p = \varepsilon_{H^*}(t^{rp}) = 0$ for $r > 0$, hence $\varepsilon_{H^*}(t^r) = 0 = \langle x_0, t^r \rangle$. Since $\varepsilon_{H^*}(1) = 1 = \langle x_0, 1 \rangle$, $x_0 = \varepsilon_{H^*}$, the identity element of H^*. Set $x = x_1$. Then x is primitive: for suppose

$$\Delta(x) = \sum a_{i,j} x_i \otimes x_j.$$

Then

$$\delta_{r+s,1} = \langle x, t^{r+s} \rangle = \sum a_{i,j} \langle x_i, t^r \rangle \langle x_j, t^s \rangle = a_{r,s}.$$

Hence $\Delta(x) = x \otimes 1 + 1 \otimes x$. Also

$$0 = \langle x, 1 \rangle = \varepsilon(x) = \mu(1 \otimes \lambda)\Delta(x) = x + \lambda(x)$$

so $\lambda(x) = -x$.

Finally, $H = L[x]$. We show that if $\Delta(x) = x \otimes 1 + 1 \otimes x$, then $x^s = s! x_s$: for we have

$$\Delta(x^r) = \Delta(x)^r = \sum \binom{r}{k} x^k \otimes x^{r-k},$$

hence for any i, j with $1 \leq i, j \leq r - 1$,

$$\langle x^r, t^{i+j} \rangle = \sum_{k=0}^{r} \binom{r}{k} \langle x^k, t^i \rangle \langle x^{r-k}, t^j \rangle = \sum_{k=1}^{r-1} \binom{r}{k} \langle x^k, t^i \rangle \langle x^{r-k}, t^j \rangle$$

since $\langle 1, t^j \rangle = 0$ for $j > 0$. If $x^k = k! x_k$ for $k < r$, then

$$\langle x^r, t^{i+j} \rangle = r! \sum_{k=1}^{r-1} \langle x_k, t^i \rangle \langle x_{r-k}, t^j \rangle = r! \delta_{k,i} \delta_{r-k,j}.$$

So $\langle x^r, t^{i+j} \rangle = 0$ if $i + j \neq r$, while for $i + j = r$ with $i, j, \geq 1$,

$$\langle x^r, t^r \rangle = r! \sum \langle x_k, t^i \rangle \langle x_{r-k}, t^{r-i} \rangle = r! \ .$$

Thus $x^r = r! x_r$ for $r \geq 1$ and H has L-basis $1, x, x^2, \dots x^{p-1}$. □

REMARK. If H is a Hopf algebra, not necessarily commutative or cocommutative, of rank p, prime, over an algebraically closed field of characteristic prime to p, then Y. Zhu [**Zu94**] showed that H is a group algebra. The difficult question of the number of isomorphism types of Hopf algebras of rank p^n, $n > 1$, over an algebraically closed field has been of interest since Kaplansky conjectured in 1975 that there exist only finitely many non-isomorphic Hopf algebras of a given finite dimension q over an algebraically closed field K of characteristic not dividing q. However, in 1998-99 three papers independently disproved Kaplansky's conjecture by showing that there are infinitely many non-isomorphic K-Hopf algebras of dimension p^4: [**AS98**], [**Ge98**] and [**BDG99**].

§16. Eigenspace decompositions. The key to understanding rank p R-Hopf algebras H for R a $\mathbb{Z}_p$-algebra is an eigenspace decomposition of the augmentation ideal $H^+ = ker(\varepsilon)$.

For H an abelian R-Hopf algebra, recall (1.10) that $End_{R-Hopf}(H)$ is a ring under convolution $*$ and composition. Define $[n] : H \to H$ for $n \in \mathbb{Z}$ by

$$\begin{aligned}
[0] &= \varepsilon \\
[1] &= I, \text{ the identity homomorphism,} \\
[n+1] &= \mu([n] \otimes 1)\Delta = [n] * [1], \text{ for } n > 0, \\
[-1] &= \lambda, \text{ the antipode,} \\
[n-1] &= [n] * [-1] \text{ for } n < 0.
\end{aligned}$$

Then $[n]$ is a Hopf algebra endomorphism of H for all $n \in \mathbb{Z}$ since $End_{R-Hopf}(H)$ is closed under $*$.

(16.1) PROPOSITION.
(1) $[\ ]$ *is a ring homomorphism from* $\mathbb{Z}$ *to* $End_{R-Hopf}(H)$.
(2) *If* $H = RG$, G *abelian, then* $[n](\sigma) = \sigma^n$ *for* $\sigma \in G$.
(3) *If* R *is a domain and* H *has rank* p *then* $[p] = [0] = \varepsilon_H$

PROOF. To show $[m+n] = [m] * [n]$ is easy by induction on $[n]$: first,

$$[m] * [0](x) = \sum [m](x_{(1)})\varepsilon(x_{(2)}) = [m](x) \text{ for all } x \in H;$$

and

$$[m] * [n \pm 1] = [m] * ([n] * [\pm 1]) = [m+n] * [\pm 1] = [(m+n) \pm 1].$$

For $[mn] = [m] \circ [n]$, first note that by (1.10),

$$([m] \circ [-1]) * ([m] \circ [1]) = [m] \circ ([-1] * [1]) = [m] \circ \varepsilon = [0],$$

hence $[m] \circ [-1] = [-m]$. Then

$$\begin{aligned}
[m] \circ [n \pm 1] &= [m] \circ ([n] * [\pm 1]) \\
&= ([m] \circ [n]) * ([m] \circ [\pm 1]) \\
&= [mn] * [\pm m] = [m(n \pm 1)].
\end{aligned}$$

For (2): if $[k](\sigma) = \sigma^k$ then $[k+1](\sigma) = \mu(\sigma \otimes \sigma^k) = \sigma^{k+1}$.

For (3), pass to the algebraic closure $\overline{K}$ of the quotient field K of R and use (15.2). If H is a rank p R-Hopf algebra, then H embeds in $\overline{K}H$.

If $\overline{K}H = \overline{K}G$, $G = \langle\sigma\rangle$ cyclic of order p, then $[p](\sigma^i) = \sigma^{ip} = 1 = \varepsilon(\sigma^i)$, so $[p] = [0]$.

If $\overline{K}H = \overline{K}G^* = \oplus_{i=0}^{p-1}\overline{K}e_i$, then $\Delta(e_k) = \sum_{i+j=k} e_i \otimes e_j$, so it is easy to see that

$$[m](e_k) = \sum e_{i_1} e_{i_2} \cdot \ldots \cdot e_{i_m}$$

where the sum is over all $(i_1, i_2, ..., i_m)$ with $i_1 + i_2 + \ldots + i_m \equiv k \pmod{p}$. Since each summand $= 0$ unless $i_1 = i_2 = \ldots = i_m$, we have

$$[m](e_k) = \sum_{mi \equiv k \pmod{p}} e_i.$$

This equals $e_{km^{-1}}$ for $1 \leq m \leq p-1$, while for $m = p$ we get $[p](e_k) = 0$ unless $k = 0$, in which case $[p](e_0) = \sum_{i=0}^{p-1} e_i = 1$. So $[p] = \varepsilon$.

If $\overline{K}H = \overline{K}[x]$ with x primitive, then $[m](x) = mx$, $[m](x^r) = (mx)^r$ for $r > 0$, so $[p](x^r) = 0$ for $r > 0$. Thus $[p] = \varepsilon = [0]$ in this case also. □

If H has rank p, so that $[p] = [0]$, then the homomorphism

$$[\ \ ] : \mathbb{Z} \to End_{R-Hopf}(H)$$

induces a homomorphism, also denoted $[\ \ ]$, from $\mathbb{F}_p = \mathbb{Z}/p\mathbb{Z}$ to $End_{R-Hopf}(H)$. Thus the group of units $\mathbb{F}_p^\times$ of $\mathbb{Z}/p\mathbb{Z}$ acts as Hopf algebra automorphisms of H, and so, since $\mathbb{Z}_p \subset R$, H becomes a module over the group ring $\mathbb{Z}_p[\mathbb{F}_p^\times]$.

There is a unique character

$$\chi : \mathbb{F}_p \to \mathbb{Z}_p$$

which maps $\mathbb{F}_p^\times$ into the group of $p-1$st roots of unity in $\mathbb{Z}_p$ and is a multiplicative section of the residue class map (χ is called the Teichmuller character). Using χ, we obtain in $\mathbb{Z}_p[\mathbb{F}_p^\times]$ a set of pairwise orthogonal minimal idempotents $\eta_1, \ldots, \eta_{p-1}$, defined by

$$\eta_j = \frac{1}{p-1} \sum_{m \in \mathbb{F}_p^\times} \chi^j(m)^{-1}[m]$$

for $j = 1, \ldots, p-1$, so that

$$\mathbb{Z}_p[\mathbb{F}_p^\times] = \sum_{j=1}^{p-1} \mathbb{Z}_p \eta_j.$$

Any $\mathbb{Z}_p[\mathbb{F}_p^\times]$-module, such as an R-Hopf algebra H of rank p, (where $\mathbb{Z}_p \subset R$) decomposes similarly.

(16.2) LEMMA. *H^+ is a $\mathbb{Z}_p[\mathbb{F}_p^\times]$-submodule of H.*

PROOF. Since $\varepsilon([m](x)) = [m](\varepsilon(x)) = \varepsilon(x)[m](1) = \varepsilon(x)$, if $x \in H^+ = ker(\varepsilon)$ then $[m](x) \in H^+$ for all m. □

Let $H_i = \eta_i H^+$. Then $H^+ = \oplus_{i=1}^{p-1} H_i$.

Note that since $[m]\eta_i = \chi^i(m)\eta_i$, we have easily:

$$H_i = \{h \in H^+ | [m](h) = \chi^i(m)h\}.$$

(16.3) LEMMA. *$H_i H_j \subset H_{i+j}$*

PROOF. If $h_i \in H_i$, $h_j \in H_j$, then

$$\begin{aligned} [m](h_i h_j) &= [m](h_i)[m](h_j) \\ &= \chi^i(m)\chi^j(m) h_i h_j \\ &= \chi^{i+j}(m) h_i h_j \in H_{i+j}. \end{aligned}$$

□

Now $H \subset \overline{K}H = \overline{K}G$, and the inclusion respects the $\mathbb{Z}_p[\mathbb{F}_p^\times]$-structures. Thus to understand the structure of an arbitrary rank p R-Hopf algebra, it is helpful to see what $\mathbb{Z}_pG$ and $\mathbb{Z}_pG^*$ look like as $\mathbb{Z}_p[\mathbb{F}_p^\times]$-modules.

Until (16.10), $R = \mathbb{Z}_p$ and $H = RG$ where $G = \langle\sigma\rangle$ is of order p.

Recall

$$\eta_i = \frac{1}{p-1}\sum_{m\in\mathbb{F}_p^\times} \chi^i(m)^{-1}[m]$$

for $i = 1, \dots, p-1$, are the pairwise orthogonal idempotents of $\mathbb{Z}_p[\mathbb{F}_p^\times]$. Set

$$x_i = (p-1)\eta_i(1-\sigma) = \sum_{m\in\mathbb{F}_p^\times} \chi^i(m)^{-1}(1-\sigma^m). \tag{16.4}$$

Thus since $\chi^{p-1}(m)^{-1} = 1$ for all m,

$$x_{p-1} = p-1-\sum_{m\in\mathbb{F}_p^\times} \sigma^m \text{ for } i = p-1,$$

while

$$x_i = -\sum_{m\in\mathbb{F}_p^\times} \chi^{-i}(m)\sigma^m \text{ for } 1 \le i < p-1.$$

(16.5) PROPOSITION. *H^+ has an R-basis $x_1, ..., x_{p-1}$, hence $H_i = Rx_i$.*

PROOF. H^+ has an R-basis $\{\sigma^m - 1 | m \in \mathbb{F}_p^\times\}$, as is easily checked. Hence $x_i \in H^+$ for all i. But by orthogonality, for $m \in \mathbb{F}_p^\times$,

$$1-\sigma^m = \frac{1}{p-1}\sum_{i=1}^{p-1} \chi^i(m)x_i.$$

Hence $\{x_1, \dots, x_{p-1}\}$ generates H^+, so is an R-basis of H^+.

Evidently $x_i \in H_i$ and $H^+ = H_1 \oplus \dots \oplus H_{p-1}$. So $H_i = Rx_i$ for $i = 1, \dots, p-1$.□

(16.6) PROPOSITION. *Let $x = x_1$. Then $H = R[x]$.*

PROOF. Consider kH, where k is the residue field of R, of characteristic p. Set $t = \sigma - 1$. Then $kH = k[t]$ with $t^p = 0$. We have

$$x = -\sum_{m\in\mathbb{F}_p^\times} \chi(m)^{-1}(\sigma^m - 1).$$

Over k, $\chi(m)^{-1} = m^{-1}$, since χ is a multiplicative section of the canonical map from $\mathbb{Z}_p$ to $\mathbb{F}_p = k$. So over k,

$$x = -\sum_{m\in\mathbb{F}_p^\times} m^{-1}((1+t)^m - 1).$$

Modulo t^2,

$$x \equiv -\sum_{m\in\mathbb{F}_p^\times} m^{-1}((1+mt-1) = t.$$

Hence $1, x, x^2, \dots, x^{p-1}$ is a k-basis of kG and $x^p \equiv 0$.

Since $x^i \in (kH)_i$ and is nonzero, $(kH)_i = kx^i$ has rank one. Hence, over R, $H_i = Rx^i$ for $i = 1, \dots, p-1$, and $RG = R[x]$ where $x^p = w_p x$ for some $w_p \in pR$ since $x^p \in H_1 = Rx$ and x^p maps to 0 in kG.

Now $H_i = Rx^i = Rx_i$ for $i = 1, \dots, p-1$, so define units w_i of $R = \mathbb{Z}_p$ by $x^i = w_i x_i$. Since $x_1 = x$, $w_1 = 1$. Recall also that $x^p = w_p x$.

Now we consider the comultiplication of RG.

We have, for $m \in \mathbb{F}_p^\times$,

$$1 - \sigma^m = \frac{1}{p-1} \sum_{i=1}^{p-1} \chi^i(m) x_i,$$

so in particular,

$$1 - \sigma = \frac{1}{p-1} \sum_{i=1}^{p-1} x_i.$$

Now by (16.4),

$$\Delta(x_i) = \sum_{i=1}^{p-1} \chi^{-i}(m)(1 \otimes 1 - \sigma^m \otimes \sigma^m),$$

while

$$x_i \otimes 1 = \sum_{i=1}^{p-1} \chi^{-i}(m)(1 \otimes 1 - \sigma^m \otimes 1)$$

and

$$1 \otimes x_i = \sum_{i=1}^{p-1} \chi^{-i}(m)(1 \otimes 1 - 1 \otimes \sigma^m).$$

So

$$\begin{aligned}
\Delta(x_i) - x_i \otimes 1 - 1 \otimes x_i &= -\sum_{i=1}^{p-1} \chi^{-i}(m)(1 - \sigma^m) \otimes (1 - \sigma^m) \\
&= -(\frac{1}{p-1})^2 \sum_{i=1}^{p-1} \chi^{-i}(m)(\sum_{j,k=1}^{p-1} \chi^j(m)\chi^k(m) x_j \otimes x_k) \\
&= -(\frac{1}{p-1})^2 (\sum_{j,k=1}^{p-1} \sum_{i=1}^{p-1} \chi^{-i+j+k}(m) x_j \otimes x_k) \\
&= -\frac{1}{p-1} \sum_{j+k \equiv i \pmod{p-1}} x_j \otimes x_k.
\end{aligned}$$

Hence for $i = 1, \dots p-1$,

$$\Delta(x_i) = x_i \otimes 1 + 1 \otimes x_i - \frac{1}{p-1} \sum_{j=1}^{p-1} x_j \otimes x_{i-j} \tag{16.7}$$

(subscripts modulo $p-1$).

(16.8) PROPOSITION. *We have*

(1) $w_i \equiv i! \pmod{p}$ *for* $1 \le i \le p-1$

(2) $w_p = p w_{p-1}$.

PROOF. We first show that $w_i \equiv i!$. Since

$$1-\sigma = \frac{1}{p-1}\sum x_i$$

and $x = x_1$ satisfies $x^i = w_i x_i$, we have

$$\sigma \equiv 1 + x + \frac{x^2}{w_2} + \cdots + \frac{x^{p-1}}{w_{p-1}} \pmod{p}$$

and $x^p \equiv 0 \pmod{p}$.

Now $\Delta(\sigma) = \sigma \otimes \sigma$. Substituting, we obtain

$$1 + \Delta x + \frac{(\Delta x)^2}{w_2} + \cdots + \frac{(\Delta x)^{p-1}}{w_{p-1}}$$

$$\equiv (1 + x + \frac{x^2}{w_2} + \cdots + \frac{x^{p-1}}{w_{p-1}}) \otimes (1 + x + \frac{x^2}{w_2} + \cdots + \frac{x^{p-1}}{w_{p-1}}) \pmod{p}.$$

Substitute in the left side

$$\Delta(x) \equiv x \otimes 1 + 1 \otimes x + \sum_{j=1}^{p-1} \frac{x^j}{w_j} \otimes \frac{x^{p-j}}{w_{p-j}} \pmod{p}.$$

We then consider the coefficient of $x^i \otimes x$ for $0 \leq i \leq p-2$. In the right side the coefficient is $\frac{1}{w_i}$. In the left side $x^i \otimes x$ occurs only in $\frac{(\Delta x)^{i+1}}{w_{i+1}}$ and then only in the term containing $(x \otimes 1 + 1 \otimes x)^{i+1}$: the coefficient is $\binom{i+1}{1}\frac{1}{w_{i+1}}$. Hence

$$\frac{1}{w_i} \equiv \frac{i+1}{w_{i+1}} \pmod{p}.$$

Since $w_1 = 1$, this yields formula (1).

For formula (2), we map R to $\overline{K}$ by $\sigma \mapsto \zeta$, a primitive pth root of unity. Then

$$x_i \mapsto \xi_i = -\sum \chi^{-i}(m)\zeta^m \text{ for } i \neq p-1,$$

$$x_{p-1} \mapsto \xi_{p-1} = p - \sum_{m=0}^{p-1} \zeta^m = p.$$

Now $x^i = w_i x_i$ for $i = 1, \ldots, p$, so $\xi_1^{p-1} = w_{p-1}p$, which means, since w_{p-1} is a unit, that $\xi_1 \neq 0$ and since $x^p = w_p x$,

$$w_p = \frac{\xi_1^p}{\xi_1} = \xi_1^{p-1} = w_{p-1}p.$$

□

To sum up our findings on RC_p, $R = \mathbb{Z}_p$:

(16.9) THEOREM. *$RC_p = R[x]$ with $x^p = w_p x$, where $w_p = pw_{p-1}$ and*

$$\Delta(x) = x \otimes 1 + 1 \otimes x - \frac{1}{p-1}\sum_{j=1}^{p-1} \frac{x^j}{w_j} \otimes \frac{x^{p-j}}{w_{p-j}}$$

where $w_i \equiv i! \pmod p$. The $\mathbb{Z}_p[\mathbb{F}_p^\times]$-eigenspaces of $(RC_p)^+$ are $Rx^i = Rx_i$, $i = 1, \dots, p-1$, with $x_i = (p-1)\eta_i(1-\sigma)$, $x = x_1$ and $x^i = w_i x_i$.

□

Now we look at RG^*.

(16.10) PROPOSITION. *$RG^* = R[\xi]$ with $\xi^p = \xi$ and the $\mathbb{Z}_p[\mathbb{F}_p^\times]$-eigenspaces of $(RG^*)^+$ are $(RG^*)_i = R\xi^i$, $i = 1, \dots, p-1$.*

PROOF. Let $\{e_0, \dots, e_{p-1}\}$ be the dual basis in RG^* to $\{1, \sigma, \dots, \sigma^{p-1}\}$ in RG. Then $RG^* = \bigoplus_{k=0}^{p-1} Re_k$ and $\{e_0, \dots, e_{p-1}\}$ are pairwise orthogonal idempotents.

Now $\overline{K}G^*$ contains $\hat{G} = Hom(G, \overline{K}^\times)$, which is cyclic of order p, generated by ϕ, and

$$e_k = \frac{1}{p}\sum_{j=0}^{p-1} \langle \phi^{-j}, \sigma^k \rangle \phi^j \ .$$

Since $[n](\phi^k) = \phi^{nk}$, then $[n](e_k) = e_{kn^{-1}}$ (subscripts modulo p), as is easily checked.

Let

$$\xi = \sum_{k=1}^{p-1} \chi(k) e_k.$$

Then

$$\xi^i = \sum_{k=1}^{p-1} \chi^i(k) e_k$$

and $\xi^p = \xi$. If we set

$$H_j = \{h \in H^+ | [n](h) = \chi^j(n)h\},$$

then $H_i \supset R\xi_i$. For

$$\begin{aligned}[n](\xi^i) &= \sum_{k=1}^{p-1} \chi^i(k) e_{kn^{-1}} \\ &= \sum_{k=1}^{p-1} \chi^i(kn) e_k \\ &= \chi^i(n)\xi^i.\end{aligned}$$

But from

$$\xi^i = \sum_{k=1}^{p-1} \chi^i(k) e_k$$

it follows by orthogonality that

$$e_j = \frac{1}{p-1}\sum_{i=1}^{p-1}\chi^{-i}(j)\xi_i$$

for $i = 1, \dots p-1$. Since $\{e_1, \dots, e_{p-1}\}$ is an R-basis of H^+, $\{\xi, \dots \xi^{p-1}\}$ is an R-basis of H^+, hence $R\xi^i = H_i$. □

(16.11) LEMMA. *Set $x_0 = 1$ and $\xi_0 = \varepsilon(1-p)$. Then the dual basis in RG^* of the basis $\{x_i\}_{i=0\dots p-1}$ of RG is $\{\frac{\xi_i}{1-p}\}_{i=0\dots p-1}$.*

PROOF. For $i, j = 0, \dots, p-1$ we have $\langle \sigma_i, e_j\rangle = \delta_{i,j}$. Therefore, for $i, j = 1, \dots, p-1$,

$$\begin{aligned}\langle x_i, \xi_j\rangle &= \langle \sum_{m=1}^{p-1}\chi^{-i}(m)(1-\sigma^m), \sum_{k=1}^{p-1}\chi^j(k)e_k\rangle \\ &= \sum_{m,k}\chi^{-i}(m)\chi^j(k)(\langle 1, e_k\rangle - \langle \sigma^m, e_k\rangle) \\ &= -\sum_{m,k}\chi^{-i}(m)\chi^j(k)\delta_{m,k} \\ &= -\sum_m \chi^{-i}(m)\chi^j(m) \\ &= -\sum_m \chi^{j-i}(m) \\ &= \begin{cases} 1-p \text{ if } i = j \\ 0 \text{ if } i \neq j \end{cases}\end{aligned}$$

Also $\langle 1, \xi_j\rangle = \langle x_i, \varepsilon\rangle = 0$ and $\langle x_0, \frac{\xi_0}{1-p}\rangle = 1$. Hence $\{x_i\}_{i=0\dots p-1}$ and $\{\frac{\xi_i}{1-p}\}_{i=0\dots p-1}$ are dual bases. □

Now we return to the general situation: K a finite extension of $\mathbb{Q}_p$ with valuation ring R, and H an arbitrary finite R-Hopf algebra of rank p.

(16.12) PROPOSITION. *Each H_i has rank one as an R-module.*

PROOF. $H^+ = \bigoplus_{i=1}^{p-1} H_i$, so each H_i is a free R-module. To find the rank of H_i we pass to the algebraic closure $\overline{K}$ of K, then $\overline{K}H = \overline{K}C_p$ and the action by $\mathbb{Z}_p[\mathbb{F}_p]$ on H extends to $\overline{K}C_p$. Since the proposition is true for $\overline{K}C_p$, it is true for H. □

(16.13) PROPOSITION. *If H is a rank p R-Hopf algebra and $H_1 = Rx$, then $H_i = Rx^i$ for $i = 1, \dots, p-1$, hence $H = R[x]$ with $x^p = bx$ for some $b \in R$.*

PROOF. We have $Rx^i \subset H_i$ for all i. To show that $Rx^i = H_i$, we may pass to the residue field k and then to $\overline{k}$, the algebraic closure of k, in which case H has one of three forms, as observed in (15.2).

If $\overline{k}H = \overline{k}G$, $G = C_p$, then (16.13) is true since it is true for $\mathbb{Z}_pG$ by (16.5) and (16.6).

If $\overline{k}H = (\overline{k}G)^*$, $G = C_p$, then (16.13) is true since it is true for $\mathbb{Z}_pG^*$ by (16.10).

If $\overline{k}H = \overline{k}[t], t^p = 0$ with t primitive, then $[n](t) = nt$ for $n = 1, \dots, p-1$, as one verifies quickly by induction using primitivity of t. Since $\chi(n) = n$ in $\overline{k}$, we have $[n](t) = \chi(n)t$, and so $(\overline{k}H)_1 = \overline{k}t$, hence $(\overline{k}H)_i = \overline{k}t^i$ for $i = 1, \dots, p-1$ by (15.2).

Since $H = R \cdot 1 \oplus H^+ = R \oplus H_1 \oplus \dots \oplus H_{p-1}$, we have $H = R[x]$. Since $H_1^p \subset H_1$, $x^p = bx$ for some $b \in R$. □

Denote $H = R[x]$ with $x^p = bx$ by H_b. Evidently b is only determined up to the $p-1$st power of a unit of u of R: if $H = R[x]$ with $x^p = bx$ then also $H = R[y]$ with $y = ux$ and $y^p = u^{p-1}by$.

(16.14) PROPOSITION. *The dual of H_b is H_a where $ab = w_p$.*

PROOF. Let $H_b = R[z]$ with $z^p = bz$, and let $H_a = R[\tau]$ with $\tau^p = a\tau$. Then $Rz = (H_b)_1$ and $R\tau = (H_a)_1$.

Pass to the valuation ring S of a field containing K and the $p-1$st roots of $1, a, b$ and w_p.

Let c and d in S satisfy $c^{p-1} = w_p/b$ and $d^{p-1} = a$. Then we may embed $SG = S[x]$ into SH_b by $x = cz$: for $x^p = w_px$ iff $z^p = \frac{x^p}{c^p} = \frac{w_p}{c^{p-1}}z = bz$. Then $SH_b \subset LG$. We may also embed SH_a in $SG^* = S[\xi]$ by $\tau = d\xi$, since $\xi^p = \xi$ iff $\tau^p = d^p\xi^p = d^{p-1}\tau = a\tau$. Then $SH_a \subset LG^*$. Viewed inside LG and LG^*, respectively, we have

CLAIM. *If $d = c$ then $\{1, \tau, \dots, \tau^{p-1}\}$ is, up to units, a dual basis over R in H^* for $\{1, z, \dots, z^{p-1}\}$ in H.*

For passing to S,

$$\begin{aligned}\langle z^i, \tau^j \rangle &= \langle \frac{x^i}{c^i}, d^j \xi^j \rangle \\ &= \frac{d^j w_i}{c^i} \langle x_i, \xi_j \rangle\end{aligned}$$

which $= 0$ if $i \neq j$, and $= \frac{d^j w_j}{c^j}$ if $i = j$. Thus if $c = d$, then

$$\{\frac{\tau^j}{w_j} | j = 0 \dots, p-1\}$$

in H_a is a dual basis for $\{z^i\}$ in H_b, and so $H_a = H_b^*$. But if $c = d$ then $ab = w_p$.□

(16.15) COROLLARY. *The comultiplication of $H_b = R[z], z^p = bz$ is*

$$\Delta(z) = z \otimes 1 + 1 \otimes z - \frac{a}{p-1} \sum_{j=1}^{p-1} \frac{z^j}{w_j} \otimes \frac{z^{p-j}}{w_{p-j}}$$

where $ab = w_p$.

PROOF. Over the valuation ring S used in the proof of (16.14) we have by (16.9),

$$\Delta(z) = \Delta(\frac{x}{c}) = \frac{1}{c}(x \otimes 1 + 1 \otimes x - \frac{1}{p-1}\sum_{j=1}^{p-1} \frac{x^j}{w_j} \otimes \frac{x^{p-j}}{w_{p-j}})$$
$$= \frac{1}{c}(cz \otimes 1 + 1 \otimes cz - \frac{1}{p-1}\sum_{j=1}^{p-1} \frac{(cz)^j}{w_j} \otimes \frac{(cz)^{p-j}}{w_{p-j}})$$
$$= z \otimes 1 + 1 \otimes z - \frac{c^{p-1}}{p-1}\sum_{j=1}^{p-1} \frac{z^j}{w_j} \otimes \frac{z^{p-j}}{w_{p-j}},$$

and $c^{p-1} = a \in R$. Hence this formula is valid over R. □

The Hopf algebras of rank p are thus of the form $H_b = R[z], z^p = bz$ with comultiplication given by (16.15). The Hopf orders over R in KC_p are the H_b with $w_p = c^{p-1}b$ for some $c \in R$; the Hopf orders over R in KC_p^* are the H_a with $a = c^{p-1}$ for some $c \in R$. Finally,

(16.16) COROLLARY. *$H_b \subseteq H_{b'}$ iff $b = c^{p-1}b'$ for some $c \in R$. Thus $H_b = H_{b'}$ iff $b = u^{p-1}b'$ for some unit u of R.*

PROOF. Let $H_b = R[x]$, $x^p = bx$, then $(H_b)_1 = Rx$. Let $H_{b'} = R[t]$, $t^p = b't$, then $(H_{b'})_1 = Rt$. If $c^{p-1}b' = b$, then

$$x \mapsto ct$$

is a well-defined algebra homomorphism ($x^p \mapsto c^pt^p = c^pb't$ while $bx \mapsto b'c^{p-1}ct$), and one checks that the comultiplication is preserved. Conversely, if $\phi : H_b \to H_{b'}$ is a homomorphism of R-Hopf algebras, then

$$\begin{array}{ccc} \mathbb{Z} & \xrightarrow{[\ ]} & H_b \\ \| & & \downarrow \phi \\ \mathbb{Z} & \xrightarrow{[\ ]} & H_{b'} \end{array}$$

commutes since $[\ ]$ is defined in terms of the Hopf algebra structure maps μ and Δ. So $(H_b)_1$ maps to $(H_{b'})_1$. If $(H_b)_1 = Rx$ and $(H_{b'})_1 = Rt$, then $\phi(x) = ct$ for some $c \in R$. But then $\phi(x^p) = c^pt^p = c^pb't$ while $\phi(bx) = cbt$, so $cbt = c^pb't$ and $b = c^{p-1}b'$. □

This completes the classification of rank p Hopf algebras over rings $R \supset \mathbb{Z}_p$.

(16.17) NOTES.

1) Tate and Oort obtain the classification under more general conditions. Namely, let Λ_p be the intersection in $\mathbb{Q}_p$ of $\mathbb{Z}_p$ with the $\mathbb{Z}$-algebra generated by $\frac{1}{p(p-1)}$ and the image in $\mathbb{Z}_p$ of the Teichmuller character $\chi : \mathbb{F}_p \to \mathbb{Z}_p$, that is, the $p-1$st roots of unity in $\mathbb{Z}_p$. Thus Λ_p is a ring obtained by adjoining to $\mathbb{Z}$ certain units of $\mathbb{Z}_p$. The Tate-Oort classification works over Λ_p-algebras. Evidently $\Lambda_p \subset \mathbb{Z}_p$, and developing the classification for $\mathbb{Z}_p$-algebras suffices for our purposes in these notes. For the more general setting, see [**TO70**] or [**Sh86**].

2) Since $\Lambda_2 = \mathbb{Z}$, the description of rank 2 Hopf algebras given in (1.5) follows from the Tate-Oort classification.

3) The Hopf orders in KG, G cyclic of order p, arise as Larson orders–see Chapter 5. The Larson description provides an attractive algebra structure for such Hopf orders, and the comultiplication is easier. The advantage of the Tate-Oort description is that it applies not just to orders in group rings, but to arbitrary rank p Hopf algebras over valuation rings, and, in particular, gives a description of all rank p Hopf algebras over fields of characteristic zero.

Hopf orders in KG, G cyclic of order p may also be constructed (and generalized) via dimension one polynomial formal groups–see Chapter 11.

4) Raynaud [**Ra74**] generalized the construction of Tate and Oort by working over valuation rings R whose residue field contains $\mathbb{F}_q$ where $q = p^n$, and classifying rank q R-Hopf algebras H for which H^+ decomposes into rank one eigenspaces with respect to the action of the multiplicative group of $\mathbb{F}_q$. Raynaud algebras which are orders in KG, G elementary abelian of order q, are studied as iterated extensions of Tate-Oort Hopf orders in [**CS98**] and [**GC98**].

5) The Tate-Oort classification shows that a Hopf algebra of rank p over the valuation ring of a local field involves a single parameter b (up to equivalence). In Chapter 9 we will show that the construction of Hopf algebras of rank p^2 involves three parameters. There is evidence to suggest that one should expect $n(n+1)/2$ parameters in the description of Hopf algebras of rank p^n. Most of that evidence for $n > 2$ is found in [**CS98**] and [**GP98**].

CHAPTER 5

Larson orders

This chapter is an exposition of results in Larson [**La76**], which contains the first general construction of Hopf orders inside group rings. Larson's construction remains the only method available for constructing orders in group rings of non-abelian groups.

§17. Group valuations. Let R be the valuation ring of a local field K containing $\mathbb{Q}_p$, and let π be a parameter (i.e. a generator of the maximal ideal of R). Let G be a finite group. Larson constructs Hopf orders using group valuations. To motivate the definition of a group valuation, we will start with a Hopf order and construct the corresponding valuation on G, then reverse the process.

For each $\sigma \neq 1 \in G$, let $I_\sigma = \{s \in K | s(\sigma - 1) \in H\}$. Then I_σ is a fractional ideal of R, containing 1 since $RG \subset H$. (I_σ is finitely generated because it is isomorphic, via $s \mapsto s(\sigma - 1)$, to an R-submodule of the finitely generated R-module H.) So I_σ^{-1} is an integral ideal, hence is generated by some power of π. Define $v(\sigma) \in \mathbb{Z}_{\geq 0}$ by $I_\sigma^{-1} = \pi^{v(\sigma)}R$. Set $v(1) = \infty$. In this way we define a map $v : G \to \mathbb{Z}_{\geq 0} \cup \{\infty\}$.

(17.1) THEOREM. *The map v satisfies the following properties. For all $\sigma, \tau \in G$:*

(i) $v(\sigma\tau) \geq min\{v(\sigma), v(\tau)\}$

(ii) $v([\sigma, \tau]) \geq v(\sigma) + v(\tau)$ *where* $[\sigma, \tau] = \sigma\tau\sigma^{-1}\tau^{-1}$ *is the commutator of* σ *and* τ;

(iii) $v(\sigma) = 0$ *if the order of* σ *is not a power of* p

(iv) $v(\sigma) \leq ord(p)/\phi(\text{order of } \sigma)$ *if* σ *has order a power of* p

(v) $v(\sigma^p) \geq pv(\sigma)$.

A function $v : G \to \mathbb{Z}_{\geq 0} \cup \infty$ satisfying properties (i) - (v) is called a *p-adic order-bounded group valuation on G*.

PROOF. (i). From the identity

$$\sigma\tau - 1 = (\sigma - 1)\tau + (\tau - 1)$$

we have that $I_\sigma \cap I_\tau \subset I_{\sigma\tau}$, hence

$$I_{\sigma\tau}^{-1} \subset (I_\sigma \cap I_\tau)^{-1} = I_\sigma^{-1} + I_\tau^{-1},$$

hence $v(\sigma\tau) \geq min\{v(\sigma), v(\tau)\}$.

(ii).

$$\begin{aligned}[\sigma, \tau] - 1 &= \sigma\tau\sigma^{-1}\tau^{-1} - 1 \\ &= (\sigma\tau - \tau\sigma)\sigma^{-1}\tau^{-1} \\ &= ((\sigma - 1)(\tau - 1) - (\tau - 1)(\sigma - 1))\sigma^{-1}\tau^{-1}.\end{aligned}$$

Hence $I_{[\sigma,\tau]} \supset I_\sigma \cdot I_\tau$, hence $I_\sigma^{-1} \cdot I_\tau^{-1} \supset I_{[\sigma,\tau]}^{-1}$, or $v(\sigma) + v(\tau) \leq v([\sigma, \tau])$.

(iii) and (iv). Let $\tau \in G$, let $G_0 = \langle\tau\rangle \subset G$ and let $g_0 = |G_0|$. If $L \supset K$ contains a primitive g_0th root of unity ζ, then $LG_0 \cong \sum_{i=0}^{g_0-1} Le_i$ where the e_i are pairwise orthogonal idempotents with the property that $\tau e_i = \zeta^i e_i$ for all i. Then LG is spanned as an L-vector space by $\{e_i\sigma | \sigma \in G, i = 0, \ldots, p-1\}$, and for each $\tau \in G_0$, $\tau(e_i\sigma) = \zeta^i(e_i\sigma)$. Pick a subset $\{a_1, \ldots, a_g\}$ of $\{e_i\sigma | \sigma \in G\}$ which is an L-basis of LG. Then for each j, there is a g_0th root of unity ζ_j such that

$$\frac{\tau - 1}{\pi_K^{v(\tau)}} a_j = \frac{\zeta_j - 1}{\pi_K^{v(\tau)}} a_j.$$

Now

$$LG = LG_0 \otimes_{LG_0} LG \equiv \bigoplus_{i=0}^{g_0-1} (Le_i \otimes_{LG_0} LG)$$

and each $(Le_i \otimes_{LG_0} LG)$ has dimension $[G : G_0]$. Hence among the basis elements $\{a_1, \ldots, a_g\}$, there are g/g_0 a_j's so that $\tau a_j = \zeta^i a_j$ for each $i, 0 \leq i \leq g_0 - 1$. Hence each g_0th root of unity occurs as a ζ_j.

Now $\frac{\tau-1}{\pi^{v(\tau)}} \in H$ and multiplies H to itself. So $\frac{\tau-1}{\pi^{v(\tau)}}$ is the root of a monic polynomial $m(x) \in S[x]$, S the valuation ring of L. But then

$$0 = m(\frac{\tau - 1}{\pi^{v(\tau)}}) a_j = m(\frac{\zeta_j - 1}{\pi^{v(\tau)}}) a_j.$$

Since the a_j are linearly independent over L, $m(\frac{\zeta_j-1}{\pi^{v(\tau)}}) = 0$, hence $\frac{\zeta_j-1}{\pi^{v(\tau)}}$ is in S, not just in L.

If $g_0 = p^s r$ with $(r, p) = 1$, $r > 1$, consider $\frac{\zeta_j-1}{\pi^{v(\tau)}}$ with ζ_j a primitive qth root of unity for q a prime dividing r. As is well-known, if q is a unit of R, then so is $\zeta - 1$ for ζ a primitive qth root of unity. Hence $v(\tau) = 0$, proving (iii).

If $g_0 = p^s$, then choosing ζ_j to be a primitive p^sth root of unity, the condition $\frac{\zeta_j-1}{\pi^{v(\tau)}}$ in S implies that

$$v(\tau) ord_L(\pi_K) \leq ord_L(\zeta_j - 1) = ord_L(p)/\phi(p^s)$$

or

$$v(\tau) \leq \frac{ord_L(p)}{ord_L(\pi_K)\phi(p^s)} = \frac{ord_K(p)}{(\phi(p^s))},$$

proving (iv).

(v). To show $v(\sigma^p) \geq pv(\sigma)$, by (iii) we may assume σ has order p^s. Then

$$(\sigma - 1)^p = \sigma^p + \sum_{r=0}^{p-1} \binom{p}{r} \sigma^r (-1)^r = \sigma^p - 1 + pb$$

for some $b \in RG$.

If $s = 1$ then $v(\sigma) \leq v(\sigma^p) = v(1) = \infty$, so $v(\sigma^p) \geq pv(\sigma)$.

If $s > 1$ then

$$v(\sigma) \leq \frac{ord(p)}{(p-1)p^{s-1}}$$

from (iv), so

$$pv(\sigma) \leq \frac{1}{(p-1)p^{s-2}} ord(p) < ord(p).$$

Suppose $t = pv(\sigma)$, then

$$\frac{(\sigma - 1)^p}{\pi^t} = \frac{\sigma^p - 1 + pb}{\pi^t} \in H.$$

Since $t < ord(p)$, $\frac{pb}{\pi^t} \in H$, hence $\frac{\sigma^p - 1}{\pi^t} \in H$, that is, $t \leq v(\sigma^p)$. Thus $pv(\sigma) \leq v(\sigma^p)$. □

(17.2) COROLLARY. *For any $\sigma \in G$,*

(vi) $$v(\sigma^r) = v(\sigma) \text{ if } (r, \text{order of } \sigma) = 1.$$

For by (i), $v(\sigma\tau) \geq min\{v(\sigma), v(\tau)\}$, which implies easily by induction that $v(\sigma^i) \geq v(\sigma)$ for all $i > 0$. But since $(\sigma^r)^s = \sigma$ for some s, $v(\sigma) = v((\sigma^r)^s) \geq v(\sigma^r)$.

Also,

(17.3) COROLLARY. *For all $\sigma, \tau \in G$,*

(vii) $$v(\tau\sigma\tau^{-1}) = v(\sigma).$$

For using (i) we have

$$v(\tau\sigma\tau^{-1}) \geq min\{v(\tau\sigma\tau^{-1}\sigma^{-1}), v(\sigma)\}.$$

By (ii),

$$v(\tau\sigma\tau^{-1}\sigma^{-1}) \geq v(\sigma) + v(\tau) \geq v(\sigma).$$

So

$$v(\tau\sigma\tau^{-1}) \geq v(\sigma).$$

But then also

$$v(\sigma) = v(\tau^{-1}(\tau\sigma\tau^{-1})\tau) \geq v(\tau\sigma\tau^{-1}).$$

(17.4) COROLLARY. *$G_+ = \{\sigma | v(\sigma) > 0\}$ is a p-group which is a normal subgroup of G.*

PROOF. G_+ is normal by (vii) and a p-group by (iii). □

Thus for a finite group G to have a non-trivial p-adic order bounded group valuation, G must have a non-trivial normal subgroup which is a p-group. Thus groups with no non-trivial p-adic order bounded group valuations for any p include the symmetric groups S_n for $n \geq 4$ and non-abelian simple groups. See Section 20 for some consequences of (17.4).

§18. Orders from valuations. Now we reverse the process: we start with a p-adic order-bounded group valuation on a finite group G and construct a Hopf order in KG, where, as before, $K \supset \mathbb{Q}_p$.

(18.1) THEOREM. *Let $v : G \to \mathbb{Z}_{\geq 0} \cup \{\infty\}$ be a p-adic order-bounded group valuation on G, a finite group. Then the R-algebra H with generators*

$$\{\frac{\sigma - 1}{\pi^{v(\sigma)}} | \sigma \in G\}$$

is a Hopf order over R in KG.

PROOF. By construction H is an R-subalgebra of KG. We need to show that H is an order, that is, is finitely generated as an R-module, and that the Hopf algebra structure maps Δ (comultiplication), λ (antipode) and ε (counit), when restricted to H, maps into $H \otimes H$, H and R, respectively. These last items are easy: if $u = \frac{\sigma-1}{\pi^{v(\sigma)}}$ then $\Delta(u) = u \otimes 1 + 1 \otimes u + \pi^{v(\sigma)} u \otimes u$, $\varepsilon(u) = 0$ and $\lambda(u) = -\sigma^{-1} u \in H$ since $RG \subset H$.

To show that H is finitely generated as an R-module, we reduce to the case $G = G_+$:

(18.2) PROPOSITION. *If $G_+ = \{\sigma \in G | v(\sigma) > 0\}$ and $H_{>0}$ is the R-algebra generated by $\{\frac{\sigma-1}{\pi^{v(\sigma)}} | \sigma \in G_+$, then H is a free left $H_{>0}$-module with basis consisting of a transversal of G_+ in G.*

PROOF. Since G_+ is a normal subgroup of G, if $\sigma \in G_+$ then

$$\tau(\frac{\sigma - 1}{\pi^{v(\sigma)}})\tau^{-1} = \frac{\tau\sigma\tau^{-1} - 1}{\pi^{v(\sigma)}} = \frac{\tau\sigma\tau^{-1} - 1}{\pi^{v(\tau\sigma\tau^{-1})}} \in H_{>0}$$

for all $\tau \in G$. So $\tau H_{>0} \subset H_{>0}\tau$ for all $\tau \in G$.

Now by definition, H is the R-algebra generated by

$$\{\frac{\sigma - 1}{\pi^{v(\sigma)}} | \sigma \in G_+\} \cup \{\tau - 1 | v(\tau) = 0\}.$$

So

$$\begin{aligned} H &= H_{>0} \cdot R[\{\tau - 1 | v(\tau) = 0\}] \\ &= H_{>0} \cdot R[\{\tau | v(\tau) = 0\}]. \end{aligned} \tag{18.3}$$

Since $RG_+ \subset H_{>0}$, we have $H = H_{>0} \cdot RG$. Since $H_{>0}\tau = \tau H_{>0}$ for all $\tau \in G$, every element of H has the form $\sum_{\tau \in G} h_\tau \tau$ for h_τ in $H_{>0}$. Thus H is generated as a left $H_{>0}$-module by the elements of G. But since $G_+ \subset H_{>0}$, if $\tau_1, \ldots, \tau_n$ is a transversal for G_+ in G (i.e. a set of representatives in G of the cosets of G_+), then $\tau_1, \ldots, \tau_n$ generate H as a left $H_{>0}$-module. Since $\tau_1, \ldots, \tau_n$ are a basis of KG as a free KG_+-module, $\tau_1, \ldots, \tau_n$ are therefore a basis of H as a free left $H_{>0}$-module.□

Given (18.2), to show that H is finitely generated as R-module, we must show that $H_{>0}$ is finitely generated as an R-module. In the remainder of the proof assume $G_+ = G$, which implies that G is a p-group and for every $\sigma \neq 1$ in $G, v(\sigma) > 0$.

Pick $\sigma \neq 1$ in G. We first show that $H_\sigma = R[\frac{\sigma-1}{\pi^{v(\sigma)}}]$ is finitely generated as an R-module.

Let $t = \frac{\sigma-1}{\pi^{v(\sigma)}}$, and let $v = v(\sigma)$. Let p^s be the order of σ. The order bounded property,

$$v(\sigma) \leq \frac{ord(p)}{p^{s-1}(p-1)}$$

implies that t satisfies a monic polynomial of degree p^s with coefficients in R, as follows: since $\sigma = 1 + t\pi^v$,

$$1 = \sigma^{p^s} = (1 + t\pi^v)^{p^s}$$

$$= 1 + \sum_{r=1}^{p^s - 1} \binom{p^s}{r} t^r \pi^{rv} + t^{p^s} \pi^{vp^s}. \tag{18.4}$$

Divide (18.4) by π^{vp^s} to get a monic polynomial of degree p^s satisfied by t, whose coefficients will be in R if π^{vp^s} divides $\binom{p^s}{r}\pi^{rv}$ for all $r \geq 1$.

For $s = 1$ we need that π^{vp} divides $p\pi^v$, that is, $vp \leq ord(p) + v$, that is, $v \leq \frac{ord(p)}{p-1}$, which is exactly the order bounded property for $s = 1$.

For $s > 1$ we need that π^{vp^s} divides $\binom{p^s}{1}\pi^v, \ldots, \binom{p^s}{p}\pi^{vp}, \ldots, \binom{p^s}{p^2}\pi^{vp^2}, \ldots$. Now $ord(\binom{p^s}{1}) = s\ ord(p)$, $ord(\binom{p^s}{p}) = (s-1)ord(p)$, $ord(\binom{p^s}{p^2}) = (s-2)ord(p)$, etc. So we need

$$vp^s \leq (s - r)ord(p) + vp^r$$

for $0 \leq r \leq s - 1$, or

$$v \leq \frac{(s-r)ord(p)}{p^s - p^r}.$$

But the order bounded property gives

$$v \leq \frac{ord(p)}{p^{s-1}(p-1)} = \frac{ord(p)}{p^s - p^{s-1}} \leq \frac{ord(p)}{p^s - p^r} \leq \frac{(s-r)ord(p)}{p^s - p^r}$$

for $r \leq s - 1$. This shows that each H_σ is finitely generated and free of rank p^s if σ has order p^s.

Now H is the algebra generated by $s_\sigma = \frac{\sigma - 1}{\pi^{v(\sigma)}}$ for all $\sigma \neq 1$ in G. Let $\mathcal{S} = \{s_\sigma | \sigma \neq 1, \sigma \in G\}$. Order the elements of $\mathcal{S}$ so that $s_1 < s_2 < \ldots < s_{n-1}$ if $v(\sigma_1) \geq v(\sigma_2) \geq \ldots \geq v(\sigma_{n-1})$ (the ordering on $\mathcal{S}$ is not unique). Let $H_i = R[s_i]$ and define the R-module homomorphism

$$m : H_1 \otimes \ldots \otimes H_{n-1} \to H$$

by multiplication.

Since each H_i is a finitely generated R-module, the R-algebra H with generators $s_1, \ldots, s_{n-1}$ will be finitely generated if m is onto.

If G is abelian, then the generators of H commute, so m is obviously onto.

In general, we show that m is onto as follows: as a module, H is spanned by $\alpha = s_{i_1}s_{i_2}\ldots s_{i_\ell}$ for $1 \leq i_k \leq n-1$ and $\ell \geq 0$. To show that α is in the image of m it suffices to show that α is a sum of ordered terms: terms of the form $s_{j_1}\ldots s_{j_\nu}$ with

$$j_1 \leq j_2 \leq \cdots \leq j_\nu.$$

This is obvious if all $i_k = 1$.

Assume it is true if all $i_k \leq h$. Let $\alpha = s_{i_1}s_{i_2}\ldots s_{i_\ell}$ be a term in which all $i_k \leq h+1$. Then α is a mixture of factors s_{h+1} and factors s_i with $i \leq h$. Call a factor s_{i_k} *misplaced* if $s_{i_k} = s_{h+1}$ and $s_{i_{k'}} = s_j$ with $j < h+1$ for some $k' > k$.

Suppose α has r misplaced factors and the rightmost misplaced factor has q factors $s_{i_{m+1}}, \dots, s_{i_{m+q}} \le s_h$ to its right:

$$\alpha = \dots s_{h+1} \dots s_{h+1} \dots s_{h+1} s_{i_{m+1}} \dots s_{i_{m+q}} s_{h+1}^e$$

(where there are r copies of s_{h+1} to the left of $s_{i_{m+1}}$).

(18.5) LEMMA. *For $j < h+1$ there is some $c < j$ and elements $r_1, \dots r_5$ of R so that*

$$s_{h+1}s_j = r_1 s_j s_{h+1} + r_2 s_c s_j s_{h+1} + r_3 s_c s_{h+1} + r_4 s_c s_j + r_5 s_c.$$

Using the lemma we can replace α by a sum of terms each of which has fewer than r misplaced factors, or which has r misplaced factors and the rightmost misplaced factor has $q-1$ factors $\le s_h$ to its right. Applying the lemma q times replaces α by a sum of terms all of which have $< r$ misplaced factors. By induction on r we can replace α by a sum of terms with no misplaced factors, that is, by a sum of terms of the form

$$s_{j_1} \dots s_{j_r} s_{h+1}^f \tag{18.6}$$

where each $s_{j_k} \le s_h$ for $1 \le k \le r$. By induction we can replace $s_{j_1}, \dots, s_{j_r}$ by a sum of ordered terms involving only $s_1, \dots, s_h$; multiplying these on the right by s_{h+1}^f means we can replace (18.6) by a sum of ordered terms. Thus α is a sum of ordered terms, and hence is in the image of m.

Repeating this argument for $2 \le h \le n-1$ shows that m is surjective.

To prove Lemma (18.5), let $s_{h+1} = \frac{\sigma-1}{\pi^{v(\sigma)}}, s_j = \frac{\tau-1}{\pi^{v(\tau)}}$ and $s_c = \frac{\rho-1}{\pi^{v(\rho)}}$ with $\rho = [\sigma, \tau]$. By property (ii) of the group valuation v,

$$v(\rho) = v(\sigma\tau\sigma^{-1}\tau^{-1}) \ge v(\sigma) + v(\tau).$$

By the choice of ordering on $\mathcal{S}$, $0 < v(\sigma) \le v(\tau)$, hence $v(\rho) > v(\tau)$, and hence $s_c < s_j < s_{h+1}$. Now

$$(\sigma-1)(\tau-1) = (\tau-1)(\sigma-1) + (\rho-1)(\tau-1)(\sigma-1)$$
$$+(\rho-1)(\sigma-1) + (\rho-1)(\tau-1) + (\rho-1)$$

as one verifies by multiplying out. Hence

$$s_{h+1}s_j = s_j s_{h+1} + \pi^{v(\rho)} s_c s_j s_{h+1}$$
$$+\pi^{v(\rho)-v(\tau)} s_c s_{h+1} + \pi^{v(\rho)-v(\sigma)} s_c s_j + \pi^{v(\rho)-v(\sigma)-v(\tau)} s_c,$$

and all powers of π are non-negative.

That proves Lemma (18.5), and completes the proof of ontoness of m and finite generation of H, thereby proving (18.1). □

§**19. Lattice Properties.** We have constructed a map $\mathcal{V}$ from the set of Hopf orders in KG to the set of p-adic order bounded group valuations on G and a map $\mathcal{A}$ from the set of p-adic order bounded group valuations on G to the set of Hopf orders over R in KG . For valuations v and w on G, define $w \le v$ iff $w(\sigma) \le v(\sigma)$ for all σ in G. Then we have

(19.1) PROPOSITION. *$\mathcal{A}$ and $\mathcal{V}$ satisfy the following lattice properties:*
(1) *If $v_1 \leq v_2$ are valuations on G, then $\mathcal{A}(v_1) \subset \mathcal{A}(v_2)$.*
(2) *If $H_1 \subset H_2$ are Hopf orders in KG, then $\mathcal{V}(H_1) \leq \mathcal{V}(H_2)$.*
(3) *For any Hopf order H in KG, $\mathcal{AV}(H) \subset H$.*
(4) *For any valuation w on G, $w \leq \mathcal{VA}(w)$.*

PROOF. Recall that if H is an R-Hopf order in KG, G a finite group, then $\mathcal{V}(H)$ is the valuation $v : G \to \mathbb{N} \cup \infty$ defined as follows: for $\sigma \in G$, $v(\sigma)$ is maximal such that $\frac{\sigma-1}{\pi^{v(\sigma)}} \in H$. If v is a p-adic order-bounded group valuation, then $\mathcal{A}(v)$ is the Hopf order generated as an R-module by $\{\frac{\sigma-1}{\pi^{v(\sigma)}} | \sigma \in G\}$.

If $v_1 \leq v_2$, then $\mathcal{A}(v_1) \subset \mathcal{A}(v_2)$, for

$$\frac{\sigma - 1}{\pi^{v_1(\sigma)}} = \pi^{v_2(\sigma)-v_1(\sigma)} \frac{\sigma - 1}{\pi^{v_2(\sigma)}} \subset \mathcal{A}(v_2)$$

for all σ. This proves (1).

To prove (2): if $H_1 \subset H_2$ with $v_i = \mathcal{V}(H_i)$, then

$$v_1(\sigma) = max\{t|\frac{\sigma - 1}{\pi^t} \in H_1\}$$

$$\leq max\{t|\frac{\sigma - 1}{\pi^t} \in H_2\} = v_2(\sigma)$$

so $\mathcal{V}(H_1) = v_1 \leq v_2 = \mathcal{V}(H_2)$.

To prove (3): given H, if $v(\sigma)$ is maximal so that $\frac{\sigma-1}{\pi^{v(\sigma)}} \in H$ for all $\sigma \in G$, then $\frac{\sigma-1}{\pi^{v(\sigma)}} \in H$ for all σ, and so

$$\mathcal{AV}(H) \subset H$$

for all H.

To prove (4): given a p-adic order bounded group valuation w, if H is spanned over R by $\frac{\sigma-1}{\pi^{w(\sigma)}}$ for all $\sigma \in G$, then

$$w(\sigma) \leq max\{t|\frac{\sigma - 1}{\pi^t} \in H\} = v(\sigma)$$

where $v = \mathcal{V}(H)$. So for all p-adic order bounded group valuations w on G,

$$w \leq \mathcal{VA}(w)$$

□

(19.2) DEFINITION. H is a *Larson order* if $H = \mathcal{A}(v)$ for some p-adic order bounded group valuation v on G.

Using the lattice properties of Proposition (19.1) we can begin to see the role of Larson orders among all Hopf orders in KG:

(19.3) COROLLARY.
(1) *For any Hopf order H, $\mathcal{AV}(H)$ is the largest Larson order contained in H.*
(2) *$\mathcal{AVA}(v) = \mathcal{A}(v)$: that is, the Larson order defined by the valuation on a Larson order is the original Larson order.*
(3) *$\mathcal{VAV}(H) = \mathcal{V}(H)$: that is, the valuation on G defined by a Hopf order H is the same as the valuation arising from the largest Larson order in H.*

PROOF. (1): The Larson order $\mathcal{AV}(H)$ is contained in H; if $\mathcal{A}(w) \subset H$ for some valuation w, then $w \leq \mathcal{VA}(w) \leq \mathcal{V}(H)$, so $\mathcal{A}(w) \subset \mathcal{AV}(H)$. Thus $\mathcal{AV}(H)$ is maximal among Larson orders in H.

(2) follows immediately from (1), since $A(v)$ is clearly the largest Larson order contained in itself.

(3): $\mathcal{AV}(H) \subset H$, hence $\mathcal{VAV}(H) \leq \mathcal{V}(H)$; on the other hand, if $w = \mathcal{V}(H)$, then by Proposition (19.1), (4), $w \leq \mathcal{VA}(w)$. □

If $|G| = p^n, n > 1$, then there exist (many) Hopf orders which are not Larson: see Chapter 9 for examples. On the other hand, every valuation on G is the valuation of a Larson order. This will follow from the next results, which give a set of algebra generators for a Larson order. That will complete the circle of ideas around $\mathcal{A}$ and $\mathcal{V}$.

Recall that G_+ is the subgroup of G of elements of G with positive valuation. If $H = \mathcal{A}(v)$ is a Larson order, let $H_1 = H \cap KG_+, \overline{H} = Im\{H \to K(G/G_+)\}$. Then $H_1 = H_{>0}$, so by (18.2) H has a basis as free left H_1-module consisting of $\{1 = \tau_1, \dots, \tau_m\}$, a transversal of G_+ in G. Thus to find a set of generators for H, we need to find a set of generators for H_1. So in the next results we assume $G = G_+$ is a p-group equipped with a valuation which is positive on all elements of G.

We first prove:

(19.4) LEMMA. *Let $G = \langle g \rangle$ be cyclic of order p. Let v be a p-adic order bounded group valuation on* G, *let $r = v(g)$ and let $u = \frac{g-1}{\pi^r}$. Then $\frac{g^k-1}{\pi^{v(g^k)}} \in R[u]$ and $\mathcal{A}(v) = R[u]$ is free of rank p over R.*

PROOF. We know by the proof of (18.1) that $R[u]$ is R-free of rank p. For $1 \leq k < p$, $v(g^k) = v(g) = r$ and

$$\begin{aligned}\frac{g^k - 1}{\pi^r} &= \frac{(g - 1 + 1)^k - 1}{\pi^r}\\ &= \frac{1}{\pi^r} \sum_{\ell=1}^{k} \binom{k}{l} (g-1)^\ell \\ &= \sum_{\ell=1}^{k} \binom{k}{l} \pi^{(\ell-1)r} (\frac{g-1}{\pi^r})^\ell \in R[u]. \quad \square\end{aligned}$$

Let

$$G_r = \{\sigma \in G | v(\sigma) \geq r\}.$$

Then G_r is a normal subgroup of G. If m is the maximum of the valuations of the non-identity elements of G, then

$$\{1\} \subset G_m \subset G_{m-1} \subset \dots \subset G_1 = G$$

is a normal chain for G. Refine this chain to a composition series for G:

$$\{1\} \subset N_s \subset N_{s-1} \subset \dots \subset N_1 \subset N_0 = G.$$

Choose elements $g_0, \ldots, g_s$ so that $g_i \in N_i$ and $g_i N_{i+1}$ generates the factor group $N_i/N_{i+1} \cong C_p$. Let $u_j = \frac{g_j - 1}{\pi^{v(g_j)}}$ for $j = 0, \ldots, s$.

(19.5) THEOREM. *$\mathcal{A}(v)$ is a free R-module with basis*

$$\{u_0^{e_0} \cdot \ldots \cdot u_s^{e_s} | 0 \leq e_i \leq p-1 \textit{ for } i = 0, \ldots, s\}.$$

PROOF. Let $A_i = \mathcal{A}(v) \cap K[N_i]$. Thus $A_{s+1} = R$, and A_s has an R-basis $\{1, u_s, \ldots, u_s^{p-1}\}$ by (19.4). Suppose for some $i \leq s+1$, $A_{i+1} = \mathcal{A}(v) \cap K[N_{i+1}]$ has an R-basis

$$\{u_{i+1}^{e_{i+1}} \cdot \ldots \cdot u_s^{e_s} | 0 \leq e_j \leq p-1 \text{ for } j = i+1, \ldots, s\}.$$

Now $A_i = \mathcal{A}(v) \cap K[N_i]$ is the R-subalgebra of $K[N_i]$ generated by

$$\{\frac{\sigma - 1}{\pi^{v(\sigma)}} | \sigma \in N_i\}.$$

Suppose $G_{r+1} \subseteq N_{i+1} \subset N_i \subseteq G_r$. If gN_{i+1} generates N_i/N_{i+1}, then $v(g) \geq r, v(g) \not\geq r+1$, so $v(g) = r$. Let $\sigma = g^k \tau$ with $1 \leq k < p$ and $\tau \in N_{i+1}$, then $v(\sigma) = r$ and $v(\tau) \geq r$.

Let $s_\sigma = \frac{\sigma - 1}{\pi^{v(\sigma)}}$, then

$$\begin{aligned} s_\sigma &= \frac{\sigma - 1}{\pi^{v(\sigma)}} \\ &= \frac{g^k \tau - \tau + \tau - 1}{\pi^{v(\sigma)}} \\ &= \tau(\frac{g^k - 1}{\pi^r}) + \frac{\tau - 1}{\pi^r} \\ &= \tau(\frac{g^k - 1}{\pi^{v(g^k)}}) + \pi^{v(\tau) - r} \frac{\tau - 1}{\pi^{v(\tau)}}. \end{aligned}$$

So

$$s_\sigma = \tau s_{g^k} + \pi^{v(\tau) - r} s_\tau.$$

Note that $\tau = \pi^{v(\tau)} s_\tau + 1$. Also, by the proof of (19.4), s_{g^k} is an R-linear combination of $1, s_g, \ldots, s_g^{p-1}$ for $k = 1, \ldots, p-1$. So if

$$s_{\sigma_1}^{f_1} \ldots s_{\sigma_\ell}^{f_\ell}$$

is an arbitrary monomial in A_i, where $\sigma_1, \ldots, \sigma_\ell \in N_i$, we may assume that $s_{\sigma_1}^{f_1} \ldots s_{\sigma_\ell}^{f_\ell}$ is an R-linear combination of monomials where either $\sigma_i \in N_{i+1}$ or $\sigma_i = g$.

Now for any $\tau \in N_{i+1}$,

$$\begin{aligned} s_\tau s_g &= \frac{\tau g - g - \tau + 1}{\pi^{r + v(\tau)}} \\ &= s_g(\frac{g^{-1} \tau g - 1}{\pi^{v(\tau)}}) + \tau(\frac{\tau^{-1} g^{-1} \tau g - 1}{\pi^{r + v(\tau)}}) \\ &= s_g s_{g^{-1} \tau g} + \tau \pi^c s_{\tau^{-1} g^{-1} \tau g} \end{aligned}$$

where $c \geq 0$ by (17.1 (ii)). Thus every monomial of the form

$$s_{\sigma_1}^{f_1} \ldots s_{\sigma_\ell}^{f_\ell}$$

is an R-linear combination of monomials $s_g^k e$ where e is in A_{i+1}. Set $s_g = u_i$. Now since $g_i = 1 + u_i\pi^r$, we have

$$g^p = 1 + \sum_{k=1}^{p-1} \binom{p}{k} u_i^k \pi^{rk} + u_i^p \pi^{rp}. \tag{19.6}$$

Since

$$r = v(g) \leq \frac{ord(p)}{\phi(\text{order of } g)} \leq \frac{ord(p)}{p-1},$$

and $v(g^p) \geq pv(g)$, dividing (19.6) by π^{rp} shows that $u_i^p = s_{g^p} +$ an R-linear combination of $u_i, \ldots, u_i^{p-1}$. Since $s_{g^p} \in A_{i+1}$, thus A_i is spanned as a right A_i-module by $1, u_i, \ldots, u_i^{p-1}$, which are linearly independent over A_{i+1} because they are over $K[N_{i+1}]$. Applying induction completes the proof. □

Recapitulating (18.2) and (19.5) we have

(19.7) Corollary. *If $H = \mathcal{A}(v)$ is a Larson order then H is a free R-module with basis*

$$\{u_0^{e_0} \cdot \ldots \cdot u_s^{e_s} \tau_i | 0 \leq e_j \leq p-1, i = 0, \ldots, m\},$$

where $u_j = \frac{g_i - 1}{\pi^{v(g_i)}}$, $g_0, \ldots, g_s \in G_+$ are defined above (19.5), and $\{\tau_1, \ldots \tau_m\}$ is a transversal for G_+ in G. □

Using this last result, we have

(19.8) Proposition. *If v is a p-adic order bounded group valuation on G, and G is a p-group, then $\mathcal{VA}(v) = v$.*

Proof. Let $w = \mathcal{VA}(v)$. We know $v \leq \mathcal{VA}(v) = w$. Let $\{N_i\}$ be the subgroups in the composition series for G_+ as exhibited above (19.5). If $v(\sigma) = 0$ then we may choose 1 and σ as part of a transversal of G_+ in G. If $\frac{\sigma-1}{\pi^s} \in \mathcal{A}(v)$ with $s \geq 0$, then $\frac{\sigma-1}{\pi^s}$ is an $H_1 = H \cap KG_+$-linear combination of $\tau_1, \ldots, \tau_m$, iff $\frac{\sigma-1}{\pi^s}$ is an H_1-linear combination of 1 and σ. But if $\frac{\sigma-1}{\pi^s} = h_1 + h_2\sigma$ in $\mathcal{A}(v)$ for $h_1, h_2 \in H_1$, then it also holds in KG, hence $h_1, h_2 \in K \cap H_1 = R$, hence $s = 0$. Thus $w(\sigma) = 0$.

Now suppose $0 < v(\sigma) < w(\sigma)$. Let $\sigma \in N_i$, σ not in N_{i+1}. Then $v(\sigma) = r$. Let $w(\sigma) = s > r$. Since the coset of any element of N_i which is not in N_{i+1} generates N_i/N_{i+1}, we may choose $g_i = \sigma$. Then A_i is free as a right A_{i+1}-module with basis $1, u, \ldots, u^{p-1}$, where $u = \frac{\sigma-1}{\pi^r}$. Then $\frac{\sigma-1}{\pi^s} \in A_i$ iff $\frac{\sigma-1}{\pi^s}$ is an A_{i+1}-multiple of $\frac{\sigma-1}{\pi^r}$, that is, $\frac{\sigma-1}{\pi^s} = a\frac{\sigma-1}{\pi^r}$, or $\pi^r(\sigma-1) = \pi^s a(\sigma-1)$. But in KG this is true iff $\pi^r = \pi^s a$, iff $a \in K \cap A_{i+1} = R$, iff $s \leq r = v(\sigma)$. So $w(\sigma) = v(\sigma)$. □

§**20. Uniqueness results.** We have already observed that if H is a Hopf order in KG, then the set of elements of G which have a positive valuation is a p-group which is a normal subgroup of G. We can strengthen this result somewhat.

(20.1) PROPOSITION. *Suppose $K \supset \mathbb{Q}_p$ and G is a group with no non-trivial normal subgroup which is a p-group. Then the only Hopf order in KG is RG.*

PROOF. Let H be a Hopf order in KG. Then $RG \subset H$. Pass to the residue field k to get

$$j : kG \to k \otimes_R H,$$

a homomorphism of Hopf algebras, and hence a group homomorphism

$$j : \Gamma(kG) \to \Gamma(k \otimes_R H)$$

where $\Gamma(k \otimes H)$ is the group of grouplike elements of $k \otimes H$, and $\Gamma(kG) = G$. We show $G_+ = ker(j)$. For let $x_1, \ldots, x_n$ be an R-basis of H; then their images in $k \otimes_R H$ are a k-basis of $k \otimes_R H$. Then $j(\sigma) = 1$ iff $j(\sigma - 1) = 0$ iff $\sigma - 1 = \sum r_i x_i$ in H and all $r_i \mapsto 0$ in $k \otimes_R H$, iff the parameter π divides r_i for all i, iff $\frac{\sigma-1}{\pi} \in H$, iff $v(\sigma) > 0$.

Since G_+ is a p-group and G has no non-trivial normal p-subgroups, $G_+ = (1)$ and j is 1-1.

If j is 1-1, then $\Gamma(k \otimes_R H)$ contains a subgroup $j(G) \equiv G$ whose order is equal to the dimension over k of $k \otimes_R H$. Since grouplikes are linearly independent over k (1.7), $j(G)$ is a k-basis of $k \otimes_R H$, and hence $k \otimes_R H = k[j(G)]$. But then by Nakayama's Lemma, $RG = H$. □

(20.2) COROLLARY. *Let H be an order over R in KG. If $H \neq RG$ then $v(H) > 0$.*

PROOF. By the last proof, if $H \neq RG$ then j is not 1-1, and if $j(\sigma) = 1$ for some $\sigma \neq 1$ in G, then $v(\sigma) > 0$. □

(20.3) COROLLARY. *Let $K \supset \mathbb{Q}_p$. If G has order prime to p, then the only order in KG is RG.* □

Now we consider orders over R in A, a cocommutative K-Hopf algebra, where $K \supset \mathbb{Q}_p$.

The following result extends an observation of Byott [**By98, Corollary 4.8**]:

(20.4) PROPOSITION. *Let $K \supset \mathbb{Q}_p$ and G be a finite group with no non-trivial normal p-subgroup. Let L/K be a Galois extension with Galois group Γ, and let A be a K-Hopf algebra with $L \otimes_K A \cong LG$. Then there is an R-Hopf algebra order H in A iff there is a subfield M of L containing K so that M/K is unramified and $M \otimes_K A \cong MG$.*

PROOF. We have $A = (LG)^\Gamma$ where Γ acts on G via a map $\beta : \Gamma \to Aut(G)$ and on L via acting as the Galois group. First assume β is 1-1. We will show that A has an R-Hopf order iff L/K is unramified.

If L/K is unramified and S is the valuation ring of R, then S/R is a Galois extension with Galois group Γ (2.15) and SG is an $S\#R\Gamma$-module. By Galois descent (2.12), $H = (SG)^\Gamma$ is an R-Hopf order in A.

Conversely, if H is an R-Hopf order in A, then H^* is an R-Hopf order in A^*, and $S \otimes_R H^*$ is an S-Hopf order in $L \otimes_K H^* = (LG)^*$. By uniqueness of orders in LG (20.1), $S \otimes_R H^* = (SG)^*$.

Let M/K be the maximal unramified subextension of L/K, and T the valuation ring of M. Then $T \otimes_R H$ is a Hopf order in $M \otimes_K A$, and L/M is totally ramified. Then $T \otimes_R H \cong TG$. For if $\overline{k}$ is the residue field of S, then $\overline{k} \otimes_R H \cong \overline{k} \otimes_S (S \otimes_R H) \cong \overline{k}G$. But $\overline{k}$ is the residue field of T, and so $\overline{k} \otimes_T (T \otimes_R H) \cong \overline{k}G$. Taking duals, we have $\overline{k} \otimes_T (T \otimes H^*) \cong (\overline{k}G)^*$. Since idempotents lift from $\overline{k}$ to T, $T \otimes_R H^* \cong (TG)^*$, and so $T \otimes_R H \cong TG$. But then $M \otimes_K A \cong MG$. Since β is 1-1, that implies that $M = L$ and L/K is unramified.

If β is not 1-1, then let M be the fixed field of the kernel of β. Then $A \otimes_K M = (LG)^{ker(\beta)} = MG$. By replacing L by M in the argument just given, we have that A has a Hopf order iff M/K is unramified. □

(20.5) COROLLARY. *Let L/K be a totally ramified Galois extension of local fields with Galois group Γ, and let $\beta : \Gamma \to Aut(G)$ be a non-trivial homomorphism. Let $A = (LG)^\Gamma$ and suppose G has no non-trivial normal p-subgroups. Then A has no Hopf orders.*

PROOF. $A \not\cong KG$ since β is non-trivial.

By (20.4), if H is a Hopf order in A then $E \otimes_K A \cong EG$ for some unramified extension E/K. Let M be the compositum of E and L. Let $u_1, \dots, u_n$ be a K-basis of A. Then in MG, each $\sigma \in G$ is a unique linear combination $\sigma = \sum a_{\sigma,i} u_i$ of the u_i, for $a_{\sigma,i} \in M$. Let F be the subfield of M generated over K by the $a_{\sigma,i}$. Since $E \otimes A = EG, F \subset E$. Since $L \otimes A = LG, F \subset L$. Thus $F \subset E \cap L$, which $= K$ since $E \cap L$ is both totally ramified and unramified over K. Hence $A \cong KG$, impossible. □

(20.4) allows an easy construction of Hopf algebras A with no Hopf orders. For example, let $K \supset \mathbb{Q}_3$ and L/K be a totally ramified Galois extension with Galois group $\Gamma = C_3$, cyclic of order 3. Let $A = (LG)^\Gamma$ where G is cyclic of order 7 and $\beta : C_3 \to Aut(C_7) \cong C_6$ is an embedding. Then A has no Hopf orders.

Here is a globalization of (20.5).

Let G be a group which has no non-trivial normal p-subgroup for any p. If L is a number field, the only Hopf order in LG is $\mathcal{O}_L G$. Let L be a Galois extension of K with Galois group Γ, and let $\beta : \Gamma \to Aut(G)$ be a homomorphism. Then $(LG)^\Gamma = A$ is a K-Hopf algebra.

(20.6) THEOREM. *A has an $\mathcal{O}_K$-Hopf order H, namely $H = (\mathcal{O}_L G)^\Gamma$, if and only if $L^{ker(\beta)}/K$ is unramified.*

PROOF. By replacing L by $L^{ker(\beta)}$ we can assume that β is 1-1.

If L/K is unramified, then $\mathcal{O}_L G$ is an $D(\mathcal{O}_L, \Gamma)$-module, hence $H = (\mathcal{O}_L G)^\Gamma$ is a Hopf order and an $\mathcal{O}_K$-form of $\mathcal{O}_K G$ by Galois descent.

Conversely, suppose H exists.

For each $p \in Spec(\mathcal{O}_K)$, let K_p be the completion of K at p. Then $K_p \otimes L \cong L_1 \times \dots \times L_g$ and L_i is a Galois field extension of K_p with Galois group $\Gamma(i)$, the stabilizer of L_i. If L_i/K is unramified for all i and p, then L/K is unramified.

So fix p and i. Then $H \otimes_{\mathcal{O}_K} \mathcal{O}_{K_p} = (\mathcal{O}_{L_i} G)^{\Gamma(i)}$, so replacing K, L, Γ and H by their local counterparts, we may assume that K is a local field, L is a Galois field extension with Galois group Γ, and H is an $\mathcal{O}_K$-Hopf order in LG.

Let M be the fixed field under the inertia group of Γ. Then M/K is unramified, and L/M is totally ramified. The residue fields of M and L coincide, and since $\mathcal{O}_L \otimes_{\mathcal{O}_K} H \cong \mathcal{O}_L G$, $\mathcal{O}_M \otimes_{\mathcal{O}_K} H \cong \mathcal{O}_M G$ and $M \otimes_K H \cong MG$ by the same argument as in (20.4). But that implies, in the global setting, that G is invariant under the inertia group $I_{\mathcal{P}_i}$ of the prime ideal $\mathcal{P}_i$ lying over p. Since $\beta : \Gamma \to Aut(G)$ is 1-1, that implies that $I_{\mathcal{P}_i}$ is trivial. Hence L/K is unramified. □

§21. One-parameter Larson orders. Now we relate Larson orders to the Hopf orders of rank p obtained in Chapter 4.

(21.1) PROPOSITION. *Let $K \supset \mathbb{Q}_p$ with valuation ring R, and let G be cyclic of order p, generated by σ. Let $A_a = R[\frac{\sigma-1}{a}]$, a Larson order in KG. Then $A_a \cong H_b = R[x]/(x^p - bx)$ where $a^{p-1}bR = w_pR$.*

PROOF. Since $A_a = R[\frac{\sigma-1}{a}]$ is a Hopf order in KG, $A_a = H_b$ for some b by the Tate-Oort classification of the last chapter. We wish to determine b.

Since H_b is an order in KG, $H_b^* \subset RG^*$, the maximal order in KG^*; hence $RG = H_{\omega_p} \subset H_b$; hence by (16.16), w_p/b is a $p-1$st power, i. e. there is some a_0 so that $ba_0^{p-1} = w_p$, and the map from $RG = H_{w_p} = R[t]$, $t^p = w_pt$, to H_b is given by $t \mapsto a_0x$, where a_0 is determined only up to a unit of R.

Similarly, the a in A_a is only determined up to a unit.

Claim: $a_0R = aR$.

To see that $a_0R = aR$ we compute the discriminants of H_b and A_a (see Section 22, below). Now

$$\begin{aligned} disc(H_b) &= disc(\{1, x, \dots, x^{p-1}\}) \\ &= (detM)R \end{aligned}$$

where $M = (m_{i,j})$, a $p \times p$-matrix with $m_{i,j} = tr(T_{x^{i+j}})$ and $T_{x^{i+j}}$ is left multiplication on H_b by x^{i+j}. Since $x^p = bx$, one sees easily that $m_{0,0} = p$, $m_{i,(p-1)-i} = (p-1)b$ for $i = 0, \dots, p-1$, $m_{p-1,p-1} = (p-1)b^2$ and all other $m_{i,j} = 0$. Thus

$$M = \begin{pmatrix} p & 0 & \dots & (p-1)b \\ 0 & 0 & \dots & 0 \\ \vdots & \vdots & & \vdots \\ 0 & (p-1)b & \dots & 0 \\ (p-1)b & 0 & \dots & (p-1)b^2 \end{pmatrix}$$

and so

$$\begin{aligned} disc(H_b) &= (p-1)^{p-2}b^{p-2}(p(p-1)b^2 - (p-1)^2b^2)R \\ &= (p-1)^{p-1}b^pR \\ &= b^pR. \end{aligned}$$

Also

$$\begin{aligned} disc(A_a) &= disc(\{1, \frac{\sigma-1}{a}, (\frac{\sigma-1}{a})^2, \dots, (\frac{\sigma-1}{a})^{p-1}\})R \\ &= \frac{1}{a^{(p-1)p}} disc(\{1, \sigma, \dots, \sigma^{p-1}\})R \\ &= \frac{1}{a^{(p-1)p}} p^p R. \end{aligned}$$

Since $disc(A_a) = disc(H_b)$ we have $p^p R = a^{(p-1)p} b^p R$. Since $b a_0^{p-1} = \omega_p$ and $\omega_p R = pR$, it follows that $a_0 R = aR$. □

COROLLARY 21.2. *Let $K \supset \mathbb{Q}_p[\zeta]$, ζ a primitive pth root of unity. Let $G = \langle\sigma\rangle$ be cyclic of order p, and $\hat{G} = Hom(G, K^\times) = \langle\hat{\sigma}\rangle$ where $\langle\hat{\sigma}, \sigma\rangle = \zeta$. Let $A_j = R[\frac{\sigma-1}{\pi^j}]$ and $\hat{A}_k = R[\frac{\hat{\sigma}-1}{\pi^k}]$. Then $A_j^* = \hat{A}_{e'-j}$.*

PROOF. We have $A_j = H_c$ where $c = \omega_p/\pi^{(p-1)j}$, and so $A_j^* = H_{\omega_p/c} = H_{\pi^{(p-1)j}}$, while $\hat{A}_{e'-j} = H_{\omega_p/\pi^{(p-1)(e'-j)}} = H_{\pi^{(p-1)j}u}$ for some unit u of R since $\omega_p = u\pi^{e'}$.

We show $\hat{A}_{e'-j} \subset A_j^*$. Let $\eta = \frac{\sigma-1}{\pi^j}$ and $\xi = \frac{\hat{\sigma}-1}{\pi^{e'-j}}$. We have

$$\langle\xi, \eta\rangle = \langle\frac{\hat{\sigma}-1}{\pi^{e'-j}}, \frac{\sigma-1}{\pi^j}\rangle = \frac{\zeta-1}{\pi^{e'}}$$

is a unit of R. Then

$$\langle\xi, \eta^{k+1}\rangle = \langle\Delta(\xi), \eta^k \otimes \eta\rangle,$$

and since

$$\Delta(\xi) = \xi \otimes 1 + 1 \otimes \xi + \pi^{e'-j} \xi \otimes \xi$$

and $\langle 1, \eta^k\rangle = 0$ for all k, one sees easily by induction on k that $\langle\xi, \eta^k\rangle \in R$ for all k, and so $\xi \in A_j^*$. Since A_j^* is an R-algebra, $\xi^\ell \in A_j^*$ for all ℓ. Thus

$$\hat{A}_{e'-j} = R[\xi] \subset R[\eta]^* = A_j^*.$$

Now by the proof of (21.1), for any b, $disc(H_b) = b^p R$. Since u is a unit of R,

$$\begin{aligned} disc(A_j^*) &= disc(H_{\pi^{j(p-1)}}) = \pi^{j(p-1)p} R \\ &= disc(H_{\pi^{j(p-1)}u}) = disc(\hat{A}_{e'-j}). \end{aligned}$$

Since $\hat{A}_{e'-j} \subset A_j^*$ and the discriminants are equal, $\hat{A}_{e'-j} = A_j^*$ (see (22.4) below).□

Finally we observe

(21.3) COROLLARY. *Let $A_a = R[\frac{\sigma-1}{a}]$ where $a^{p-1}bR = w_p R$ and b is not a unit of R. Then A_a is a local ring.*

Combined with Waterhouse's result (13.7), this implies that any tame A_a module algebra is Galois, a fact which will be useful in the next chapter.

PROOF. $A_a \cong H_b = R[x]$ with $x^p = bx$. If b is not a unit, then, modulo πR, $x^p = 0$, and so if $k = R/\pi R$, then $k \otimes_R A_a$ is a local ring. Hence A_a is local. $\square$

If b is a unit of R, then $A_a \cong H_b \cong R[x]/(x^p - bx)$ is not local: in fact, H_b is separable over R.

(21.4) REMARKS.

Among Hopf orders in KG, Larson orders have the simplest algebra structure. For an abelian p-group G of order p^s, constructions in [**CS98**] and [**GC98**] suggest that an R-Hopf order in KG should involve s "Larson parameters"and $\frac{s(s-1)}{2}$ "unit"parameters. Larson orders involve only the Larson parameters, namely the valuations $\{v(g_i) | i = 1, \dots s\}$ of the elements $g_1, \dots, g_s$ arising from the filtration of G described above (19.5).

Thus in attempting to generalize results about Hopf orders in KG with G of rank p, researcher often begin by considering Larson orders, and sometimes monogenic Larson orders (for G cyclic, involving only one independent Larson parameter). For example, R. Underwood [**Un96**] obtained all Hopf orders in KC_{p^3} which are extensions of a rank p Larson order by the dual of a monogenic rank p^2 Larson order. Some global results involving Larson orders include [**Ca98a**], which extends [**Ro73**], [**Hu87**] and [**CMS98**] to compute the group of H-principal homogeneous spaces for H a monogenic Larson order, and [**By00**], generalizing [**CP94**], which shows that the Picard invariant map ([**Ch71**], [**CM74**], [**Ch77**]) is onto for H-principal homogeneous spaces when H is a Larson order satisfying a certain compatibility condition at various completions. [**By00**] contains useful information on bases for Larson orders. (See [**Ag99**] for a definitive result on the Picard invariant map). Caenepeel constructs a generalization of monogenic Larson orders over Dedekind rings in [**Ca98b, Section 7.1**].

Finding the largest Larson order inside a given Hopf order (c.f. (19.3)) has been useful for certain constructions: see [**Un96, (3.1.1)**] and [**GC98, section 2**].

CHAPTER 6

Cyclic Extensions of Degree p

Let $K \supset \mathbb{Q}_p$ and $L \supset K$ be a Galois extension of local fields with Galois group G. Let S be the valuation ring of L, and R the valuation ring of K. In this chapter we assume G has order p and determine when S can be a Hopf Galois extension of R for some Hopf algebra $H \subset KG$. We begin this chapter by reviewing some concepts related to ramification. Our basic reference is Serre, *Corps Locaux* [**CL**].

§**22. Different and discriminant.** Let π be a parameter for R, that is, a generator of the maximal ideal of R. Normalize the valuation ord_K on K so that its image is $\mathbb{Z} \cup \{\infty\}$. Then $ord_K(\pi) = 1$. Let k be the residue field of R: $k = R/\pi R$.

If ω is a parameter for S, then $\pi S = \omega^e S$ defines the ramification index $e = e_{L/K}$. Also, $F = S/\omega S$ is a field extension of k: $f = f_{L/K} = [F : k]$ is the residue field degree. Then $ef = [L : K]$. L/K is totally ramified if $e = [L : K]$, and unramified if $e = 1$.

If K is a finite extension of $\mathbb{Q}_p$, then $pR = \pi^{e_0}R$: $e_0 = e_{K/\mathbb{Q}_p}$ is the absolute ramification index of K.

(22.1) PROPOSITION. *If L/K is totally ramified then for any parameter ω of L, $S = R[\omega]$.*

See [**CL, Ch. I, sec. 6, Props. 17 and 18**]. The idea is that ω is a root of an Eisenstein polynomial over R: a polynomial $m(x) = x^n + a_{n-1}x^{n-1} + \ldots + a_1 x + a_0$ with $n = [L : K]$, $ord_K(a_i) > 0$ for all $i < n$ and $ord_K(a_0) = 1$. If $m(x)$ is Eisenstein, then $m(x)$ is irreducible over K and $R[\omega] \cong R[x]/(m(x))$ is a discrete valuation ring, hence is the integral closure of R in L.

Now we examine the discriminant.

Let R be a commutative ring, and let A be a R-algebra which is finitely generated as a R-module. For $x \in A$, let $tr(x)$ be the trace of the R-module homomorphism $T_x : A \to A$ defined by $T_x(y) = xy$: then $tr : A \to R$ is a R-module homomorphism.

If A is an order over R in L, a Galois field extension of K with Galois group G, then the trace of an element of L can be computed as $tr(x) = \sum_{\sigma \in G} \sigma(x)$.

If $\{x_1, \ldots, x_n\}$ is an R-basis of A, then the *discriminant* of $\{x_1, \ldots, x_n\}$ is defined by $disc\{x_1, \ldots, x_n\} = det(tr(x_i x_j))$.

(22.2) LEMMA. *Let A, B be R-algebras, free of rank n over R, with $A \subset B$. Let $\{x_1, ..., x_n\}$ be an R-module basis of A, and $\{y_1, ..., y_n\}$ be an R-module basis of B. Then $disc\{x_1, ..., x_n\} = c^2 disc\{y_1, ..., y_n\}$ for some $c \in R$.*

PROOF. Let $x_i = \sum r_{i,j} y_j$ for some $r_{i,j}$ in R. Then

$$(tr(x_i x_j)) = (r_{i,j}) tr(y_i y_j) (r_{i,j})^{tr}.$$

So $c = det(r_{i,j})$. □

If $\{x_1, ..., x_n\}$ is an R-basis of A, then the ideal of R generated by $det(tr(x_i x_j))$ is the *discriminant* of A/R, denoted by $disc_R(A)$, or, if R is understood, $disc(A)$.

The discriminant of A is well-defined:

(22.3) COROLLARY. *If A is a commutative R-algebra which is a free R-module of finite rank, then the discriminant ideal of A does not depend on the choice of R-basis of A.*

PROOF. Let $A = B$ in the proof of (22.2): $(r_{i,j})$ is invertible, so c is a unit. □

DEFINITION. Let A be a finite R-algebra ("finite" means finitely generated and projective as R-module). The discriminant of A, $disc_R(A)$, is the ideal of R generated by $disc\{y_1, \ldots, y_n\}$ for all subsets $\{y_1, \ldots, y_n\}$ of A.

If A is R-free with basis $\{x_1, \ldots, x_n\}$ then $disc_R(A) = disc\{x_1, \ldots, x_n\}R$.

(22.4) COROLLARY. *Let R be a Noetherian domain with field of fractions K. Let $T_1 \subset T_2$ be finite R-algebras which are orders over R in a finite separable K-algebra A. If $disc_R(T_1) = disc_R(T_2)$, then $T_1 = T_2$.*

PROOF. Since $T_1 \subset T_2$, we have $T_1 = T_2$ iff for every prime ideal p of R,

$$R_p \otimes_R T_1 = R_p \otimes_R T_2.$$

Now the discriminant localizes: for any prime ideal p of R,

$$R_p \otimes_R disc_R(T_i) = disc_{R_p}(R_p \otimes T_i).$$

Since R_p is local, $R_p \otimes_R T_1$ and $R_p \otimes_R T_2$ are free R_p-modules. so to prove (22.4) we can assume $T_1 \subset T_2$ are orders in A which are free of rank n over R.

Thus let $\{x_1, ..., x_n\}$ be an R-module basis of T_1, then $\{x_1, ..., x_n\}$ is a K-basis of A, so $disc\{x_1, \ldots, x_n\} \neq 0$. If $\{y_1, ..., y_n\}$ is an R-module basis of T_2, then $x_i = \sum r_{i,j} y_j$ for $r_{i,j} \in R$, hence if $C = (r_{i,j})$ then $disc(T_1) = det(C)^2 disc(T_2)$. Thus the discriminants are equal iff C is invertible, iff $T_1 = T_2$. □

If K is a field and A is a finite commutative separable K-algebra, then the discriminant of A over K is non-zero. (Showing this reduces, after base change, to showing it when A is a direct product of copies of K, in which case one can compute $disc(A)$ by using a basis of A consisting of pairwise orthogonal idempotents.) It follows that the bilinear map

$$b : A \times A \to K$$

given by $b(x, y) = tr(xy)$ is non-degenerate, that is, $\beta : A \to A^* = Hom_K(A, K)$ by $\beta(x)(y) = b(x, y) = tr(xy)$ is 1-1 and K-linear, hence an isomorphism. This allows one to construct the dual basis with respect to b to any basis $\{x_1, ..., x_n\}$ of A by pulling back the dual basis $\{e_1, ..., e_n\}$ of A^* to a basis $\{y_1, ..., y_n\}$ of A via β.

Now we consider the different.

Let L/K be a Galois extension of local fields containing $\mathbb{Q}_p$ with valuation rings $R \subset S$. Let

$$\mathcal{C} = \{y \in L | tr(xy) \in R \text{ for all } x \in S\},$$

the *complementary module*. Then $\mathcal{C}$ is the preimage in L of $Hom_R(S,R) \subset L^*$ under β: since $Hom_R(S,R)$ is a finitely generated left R-module, so is $\mathcal{C}$, hence $\mathcal{C}$ is a fractional ideal of K. Since $tr(S) \subset R$, $S \subset \mathcal{C}$. Hence $\mathcal{D} = \mathcal{C}^{-1} \subset S$ is an integral ideal: $\mathcal{D}$ is the *different* of S/R.

(22.5) PROPOSITION. *Let T be an order over R in L, a finite field extension of K. Let $T = R[\alpha]$, where α is a root of a monic irreducible polynomial $f(X)$ in $R[X]$. Then $\mathcal{C}(T) = \frac{1}{f'(\alpha)}T$.*

This may be found at [**CL III, Sec. 7, Corollary 6**]. Since this result is needed for results in Chapter 8, we give a proof here for L/K Galois for the convenience of the reader. We need

(22.6) LEMMA (EULER). *Let $L = K[\alpha]$ where the minimal polynomial of α over K is $f(x)$ of degree n. Then $tr(\alpha^i/f'(\alpha)) = 0$ if $i < n-1$; $= 1$ if $i = n-1$.*

PROOF OF (22.6). We have

$$f(x) = \prod_{i=0}^{n-1}(x - \sigma_i(\alpha)).$$

Hence

$$f'(x) = \sum_i \prod_{j \neq i}(x - \sigma_j(\alpha))$$

and

$$f'(\sigma_i(\alpha)) = \prod_{j \neq i}(\sigma_i(\alpha) - \sigma_j(\alpha)).$$

Now apply partial fractions:

$$\frac{1}{f(x)} = \frac{1}{\prod_{i=0}^{n-1}(x - \sigma_i(\alpha))} = \sum_{i=1}^{n} \frac{a_i}{x - \sigma_i(\alpha)}$$

for some $a_i \in K$. To determine a_i, we multiply through by $f(x) = \prod_{j=1}^{n}(x - \sigma_i(\alpha))$ to get

$$1 = \sum a_i \prod_{j \neq i}(x - \sigma_j(\alpha)).$$

Set $x = \sigma_i(\alpha)$ to get a_i:

$$\frac{1}{a_i} = \prod_{j \neq i}(\sigma_i(\alpha) - \sigma_j(\alpha)) = f'(\sigma_i(\alpha)).$$

Now expand each term

$$\frac{a_i}{(x - \sigma_i(\alpha))} = \frac{1}{x f'(\sigma_i(\alpha))(1 - \frac{\sigma_i(\alpha)}{x})}$$

of $1/f(x)$ as a geometric series (i. e. as a power series in $1/x$), and expand $1/f(x)$ as a power series in $1/x$. One sees that

(22.7) $$\frac{1}{f(x)} = \frac{1}{x^n} + \frac{c_{n+1}}{x^{n+1}} + \frac{c_{n+2}}{x^{n+2}} + \dots$$

for some c_{n+j} in K, while

$$\sum_{i=1}^{n} \frac{1}{xf'(\sigma_i(\alpha))} \frac{1}{(1-\frac{\sigma_i(\alpha))}{x})} = \sum_{r=0}^{\infty} tr(\frac{\alpha^r}{f'(\alpha)}) \frac{1}{x^{r+1}}.$$

So

$$tr(\frac{\alpha^r}{f'(\alpha)}) = 0 \text{ if } r \leq n-1; = 1 \text{ if } r = n-1$$

as claimed. □

(22.8) LEMMA. *Let $T = R[\alpha]$, an order over R in a Galois extension L of K with Galois group G, and let $f(x) \in R[x]$ be the minimal polynomial of α. Then $\mathcal{C}(T)$, the complementary module of T with respect to the trace, has basis*

$$\{\frac{1}{f'(\alpha)}, \frac{\alpha}{f'(\alpha)}, \ldots, \frac{\alpha^{n-1}}{f'(\alpha)}\}.$$

PROOF. Let $\{\beta_0, \ldots, \beta_{n-1}\}$ be the dual basis in L of the basis $\{1, \alpha, \ldots, \alpha^{n-1}\}$ of L with respect to the bilinear form $b : L \times L \to K$ given by

$$b(x,y) = tr(xy).$$

Then the matrix $(tr(\alpha^i \beta_j)) = I$, the identity matrix, and $\{\beta_0, \ldots, \beta_{n-1}\}$ is a basis of $\mathcal{C}(T)$. Write

$$\frac{\alpha^j}{f'(\alpha)} = \sum_{i=0}^{n-1} r_{i,j} \beta_j.$$

Then

$$r_{i,j} = tr(\alpha^i(\frac{\alpha^j}{f'(\alpha)})),$$

and $R_{i,j} = 0$ if $i+j < n-1$; $= 1$ if $i+j = n-1$, and $= c_{i+j+1}$ for $i+j \geq n$. Since $f(x) \in R[x]$, the coefficients c_{n+j} in (22.7) are in R, hence the matrix $(r_{i,j})$ is an invertible matrix over R; since $\{\beta_0, \ldots, \beta_{n-1}\}$ is a basis of $\mathcal{C}(T)$, so is $\{\frac{1}{f'(\alpha)}, \frac{\alpha}{f'(\alpha)}, \ldots, \frac{\alpha^{n-1}}{f'(\alpha)}\}$. □

PROOF OF (22.5). Given the order $T = R[\alpha]$, (22.8) shows that its complementary module $\mathcal{C}(T)$ has basis $\{\frac{1}{f'(\alpha)}, \frac{\alpha}{f'(\alpha)}, \ldots, \frac{\alpha^{n-1}}{f'(\alpha)}\}$. Since $\{1, \alpha, \ldots, \alpha^{n-1}\}$ is an R-basis of T, it is clear that $\mathcal{C}(T) = \frac{1}{f'(\alpha)}T$. □

(22.9) COROLLARY. *Let L/K be a Galois extension of fields with valuation rings S/R. If $S = R[\alpha]$ where α has minimal polynomial $f(x)$, then the different $\mathcal{D}(S/R)$ satisfies $\mathcal{D}(S/R) = f'(\alpha)S$.* □

If ω is a parameter for S, the norm $N(\omega^k S)$ of an ideal $\omega^k S$ of S is defined to be $N(\omega^k S) = \pi^{kf} R$ where f is the residue field degree. The relationship between the different and the discriminant is then given by:

(22.10) FACT. $disc_R(S) = N_{L/K}(\mathcal{D}(S/R))$.

See [**CL, III, Sec. 3, Prop. 6**].

In particular, if L/K is totally ramified and $\mathcal{D}(S/R) = \omega^d S$, then $disc_R(S) = \pi^d R$, so

$$(22.11) \qquad ord_K(disc_R(S)) = ord_L(\mathcal{D}(S/R)).$$

If L is a Galois extension of K with Galois group G of order n, then, since $\sigma(wS) = wS$, we have $\prod_{\sigma\in G}\sigma(\omega^k S) = \omega^{kn}S = \omega^{kef}S = \pi^{kf}S = S\cdot N(\omega^k S)$. So $N(\omega^k S) = \prod_{\sigma\in G}\sigma(\omega^k S)$. Thus $disc_R(S) = \prod_{\sigma\in G}\sigma(\mathcal{D}(S/R))$.

(22.12) PROPOSITION. *L/K is unramified iff $\mathcal{D}(S/R) = S$, iff $disc_R(S) = R$.*

This is [**CL, III, Sec. 5, Theorem 1**].

(22.13) PROPOSITION. *Let L be an A-Hopf Galois extension of K, H an R-Hopf order in A and S an R-order in L which is an H-module algebra. Then S is an H-Galois extension of R iff*

$$disc(S) = disc(H^*).$$

PROOF. The coaction $\alpha : S \to S\otimes H^*$ is an R-algebra homomorphism, so

$$\gamma : S\otimes S \to S\otimes H^*$$

is an S-algebra homomorphism, where S embeds in both sides via the left factor. Since $L = K\otimes S$ is an $A = K\otimes H$-Galois extension of K, $K\otimes\gamma$ is an isomorphism, hence γ is 1-1. Then γ is surjective iff S is H-Galois. But γ is surjective iff

$$disc_S(S\otimes S) = disc_S(S\otimes H^*).$$

Since $disc_S(S\otimes S) = S\cdot disc_R(S)$, and $disc_S(S\otimes H^*) = S\cdot disc_R(H^*)$, $disc_S(S\otimes S = disc_S(S\otimes H^*)$ iff $disc_R(S) = disc_R(H^*)$. □

From (22.13) we may prove (22.12) when $L/K$is a Galois extension with Galois group G, as follows: For L/K a Galois extension, L/K is unramified iff S is an RG-Galois extension of R. Now if S is an RG-Galois extension of R, then $disc(S) = disc(RG^*) = R$. (The second equality follows because RG^* is a direct product (as rings) of copies of R, and the R-basis of RG^* consisting of the dual basis to G is a basis of pairwise orthogonal idempotents of RG^*: it is easy to see that the discriminant of that basis is the determinant of the identity matrix, hence $= 1$.) On the other hand, let $S = R[\alpha]$ and let $f(x)$ be the minimal polynomial of α. If $\mathcal{D}(S/R) = R$, then $f'(\alpha) = \prod_{\sigma\neq 1}(\alpha - \sigma(\alpha))$ is a unit of S by (22.9). But then $\sigma(\alpha)\not\equiv\alpha \pmod{\omega S})$ for all $\sigma\neq 1$ in G, and so by condition (f) of the definition of Galois extension (2.5), S is a Galois extension of R, hence L/K is unramified.

We note the transitivity of the discriminant of valuation rings of local fields:

(22.14) PROPOSITION [**CL III, §4, Proposition 8**]. *Let $K\subset M\subset L$ be finite extensions of local fields with valuation rings $R\subset T\subset S$. Then*

$$disc_R(S) = disc_R(T)^{[L:M]}\cdot N_{M/K}(disc_T(S)).$$

This result holds more generally, for R a Dedekind ring, T a finite R-algebra and S a finite T-algebra: see [**Gr92, II, Lemma 1.2**].

The discriminant of a Hopf algebra. Let R be the valuation ring of a local field K, let H be a commutative, cocommutative R-Hopf algebra which is free of rank n over R, and let $L \supset K$ be a field extension containing a primitive nth root of unity and such that $LH = LG \cong LG^*$ for some abelian group G of order $n = rank_R H$. Let $e_1 = \frac{1}{n}\sum_{\sigma \in G} \sigma$ in LG, then e_1 is an integral of LG.

(22.15) PROPOSITION. *With R, H, e_1 as above, if $\theta = me_1$ generates the module of integrals of H, then $disc_R(H) = m^n R$.*

PROOF. Let $\{h_1, \dots, h_n\}$ be an R-basis of H, then $\{h_1, \dots, h_n\}$ is an L-basis of LH. Let $T_{h_i h_j}$ be the R-linear map on H given by left multiplication by $h_i h_j$. Then $tr(T_{h_i h_j})$ is the trace of the matrix of $T_{h_i h_j}$ with respect to the basis $\{h_1, \dots, h_n\}$ over R, or over L: it is the same.

But LG^* contains the group of characters

$$\hat{G} = Hom(G, L^\times) = \{\hat{\sigma}_1, \dots, \hat{\sigma}_n\}.$$

If $e_j = \frac{1}{n}\sum_{\sigma \in G} \langle \hat{\sigma}_i, \sigma \rangle \sigma$, then $\{e_1, \dots, e_n\}$ is a basis of pairwise orthogonal minimal idempotents of LG which is the dual basis in LG to $\{\hat{\sigma}_1, \dots, \hat{\sigma}_n\}$. Over L, $tr(T_{h_i h_j})$ may be computed with respect to the basis $\{e_1, \dots, e_n\}$.

Suppose $h_i h_j = \sum c_{i,j}^k e_k$. Then

$$T_{h_i h_j}(e_\ell) = \sum c_{i,j}^k e_k e_\ell = c_{i,j}^\ell e_\ell,$$

and so

$$tr(T_{h_i h_j}) = \sum_k c_{i,j}^k.$$

We may then compute $tr(T_{h_i h_j})$ as follows: let $\hat{\Sigma} = \sum_{i=1}^n \hat{\sigma}_i$. Then $\langle e_k, \hat{\Sigma} \rangle = 1$ for all k, so

$$tr(T_{h_i h_j}) = \sum_k c_{i,j}^k = \langle \sum c_{i,j}^k e_k, \hat{\Sigma} \rangle = \langle h_i h_j, \hat{\Sigma} \rangle.$$

Now $\hat{\Sigma} = m\hat{\theta}$ for some $m \in R$, where $\hat{\theta}$ is a generator of the module of integrals of H^*. So

$$\langle h_i h_j, \hat{\Sigma} \rangle = m \langle h_i h_j, \hat{\theta} \rangle$$

and so

$$disc_R(H) = m^n det(\langle h_i h_j, \hat{\theta} \rangle) R.$$

But $det(\langle h_i h_j, \hat{\theta} \rangle)$ is a unit of R by (3.6), so $disc_R(H) = m^n R$ where $\hat{\Sigma} = m\hat{\theta}$. But by (5.7), if $\hat{\theta} = \frac{1}{m}\hat{\Sigma}$, then $\theta = me_1$. □

(22.16) EXAMPLE. Let $G = \langle \sigma \rangle$ of order p and let $H = R[\frac{\sigma - 1}{\pi^b}]$. Then $disc(H) = (\frac{p}{\pi^{b(p-1)}})^p R$, by (21.1).

Define

$$rdisc_R(H) = disc_R(H)^{\frac{1}{[H:R]}},$$

the *root discriminant* of H. The next corollary is immediate from (5.6) and (22.15):

(22.17) COROLLARY. *Let R be the valuation ring of a local field K. Let*

$$K \to A_1 \to A \to \overline{A} \to K$$

be a short exact sequence of finite, commutative, cocommutative K-Hopf algebras. Let H be a Hopf order over R in A, and let

$$R \to H_1 \to H \to \overline{H} \to R$$

be the corresponding short exact sequence of R-Hopf algebras. Then

$$disc_R(H) = disc_R(H_1)^{[\overline{H}:R]} disc_R(\overline{H})^{[H_1:R]}$$

hence

$$rdisc_R(H) = rdisc_R(H)_1 \cdot rdisc_R(\overline{H}).$$

(22.18) COROLLARY. *Let L/K be a Galois extension with Galois group G. Let H be an R-Hopf algebra in KG so that the valuation ring S of L is an H-module algebra. Let N be a normal subgroup of G, $M = L^N$, $T =$ valuation ring of M, $H_1 = KN \cap H$ and $\overline{H} = H//H_1$. Then $T = S^{H_1}$ is an $\overline{H}$-module algebra, and S is H-Galois over R iff S is $T \otimes_R H_1$-Galois over T and T is $\overline{H}$-Galois over R.*

PROOF. We have

$$R \to \overline{H}^* \to H^* \to H_1^* \to R$$

an exact sequence of R-Hopf algebras by (4.24). The corollary is true for L/K by the fundamental theorem of Galois theory, so

$$S \otimes_R S \to S \otimes_R H^*,$$

$$S \otimes_T S \to S \otimes_T (T \otimes_R H_1^*)$$

and

$$T \otimes_R T \to T \otimes_R \overline{H}^*$$

are 1-1, hence $disc_R(S) = c\ disc_R(H^*)$ for some c in R, $disc_R(T) = c_1\ disc_R(\overline{H}^*)$ for some c_1 in R, and $disc_T(S) = d\ disc_T(T \otimes_R H_1^*)$ for some d in T; moreover, c, c_1, d is a unit iff S/R is H-Galois, T/R is $\overline{H}$-Galois, S/T is $T \otimes H_1$-Galois, respectively, by (2.13). Now

$$disc_R(H^*) = disc_R(\overline{H}^*)^{[H_1:R]} disc_R(H_1^*)^{[\overline{H}:R]}$$

by (22.17), while

$$disc_R(S) = disc_R(T)^{[S:T]} N_{M/K} disc_T(S)$$

by (22.14), hence

$$c\ disc_R(H^*) = c_1\ disc_R(\overline{H}^*)^{[H_1:R]} N_{M/K}(d\ disc_T(T \otimes H_1^*))\ .$$

Since $N_{M/K} disc_T(T \otimes_R H_1^*) = disc_R(H_1^*)^{[T:R]}$, we have

$$c = c_1^{[H_1:R]} N_{M/K}(d)$$

in R. Thus c is a unit iff c_1 and d are units. □

The result, if S is H-Galois over R then T is $\overline{H}$-Galois over R and S is $T \otimes H_1$-Galois over T, holds in general – see [**CS69, Theorem 7.6**] or [**Gr92, II, Lemma 1.7**].

§**23. Ramification groups.** Let L/K be a Galois extension of local fields, with Galois group G, and with R, S, π, ω as at the beginning of section 22.

Let $G_{-1} = G$, and for $i \geq 0$ define the ith ramification group by

(23.1) $$G_i = \{\sigma \in G | \sigma(x) - x \in \omega^{i+1} S \text{ for all } x \in S\}.$$

If L/K is totally ramified, then the parameter ω of S is a test element:

$$G_i = \{\sigma \in G | \sigma(\omega) - \omega \in \omega^{i+1} S\}.$$

To see this, note that by (22.1), since L/K is totally ramified, any x in S has the form:

$$x = \sum_{j=0}^{n-1} r_j \omega^j$$

with $r_{i,j} \in R$, hence if $\sigma(\omega) - \omega \in \omega^{i+1} S$ then

$$\begin{aligned} \sigma(x) &= \sum_{j=0}^{n-1} r_j \sigma(\omega)^j \\ &= \sum_{j=0}^{n-1} r_j (\omega + a\omega^{i+1})^j \\ &= \sum_{j=0}^{n-1} r_j (\omega^j + ja\omega^{i+j} + \dots) \\ &= \sum_{j=0}^{n-1} r_j \omega^j + \sum_{j=1}^{n-1} r_j j a \omega^{i+j} + \dots \\ &= x + w^{i+1} b \end{aligned}$$

for some b in S, so $\sigma(x) - x \in \omega^{i+1} S$ for all x in S.

The group

$$G_0 = \{\sigma \in G | \sigma(x) - x \in \omega S \text{ for all } x \in S\}$$

is the inertia group. Since $\sigma(\omega) \in \omega S$, we have a natural homomorphism from G onto $Gal(F/k)$, the Galois group of the extension of residue fields, whose kernel is G_0: see [**CL, I, sec. 7, Prop. 20**]).

Suppose G is cyclic of prime order p. Then each ramification group is either G or $\{1\}$.

(23.2). Define the ramification number, or break number t by

$$G_t = G, \ G_{t+1} = \{1\}.$$

If σ is a generator of G, then $\sigma \in G_t$, so $\sigma(\omega) - \omega \in w^{t+1} S$, while $\sigma \notin G_{t+1}$, so $\sigma(\omega) - \omega \notin w^{t+2} S$. Thus the ramification number t may be defined by

$$t + 1 = ord_L(\sigma(\omega) - \omega).$$

There are relations among the ramification number, the image of the trace map $tr : S \to R$ and the valuation of the different. $\lfloor b \rfloor$ denotes the greatest integer $\leq b$.

(23.3) PROPOSITION [**CL, p. 91**]. *Let L/K be totally ramified and cyclic of order p. Let t be the ramification number, and $\mathcal{D}(S/R)$ the different. Then $ord_L(\mathcal{D}(S/R)) = ord_K(disc_R(S)) = (t+1)(p-1)$ and $ord_K(tr(S)) = \lfloor \frac{(t+1)(p-1)}{p} \rfloor$.*

PROOF. The second formula follows from the first, as follows: since

$$\mathcal{D}(S/R)^{-1} = \mathcal{C} = \{y \in L | tr(xy) \in R \text{ for all } x \in S\},$$

we have

$$\begin{aligned} tr(S) \subset \pi^r R &\text{ iff } tr(\pi^{-r}S) \subset R \\ &\text{ iff } \pi^{-r}S \subset \mathcal{C} \\ &\text{ iff } \mathcal{D}(S/R) \subset \pi^r S. \end{aligned}$$

Thus $tr(S) = \pi^d R$ iff d is the maximal r so that $\mathcal{D}(S/R) \subset \pi^r S$, or $pr \leq (t+1)(p-1)$, or $r \leq \frac{(t+1)(p-1)}{p}$. Hence

$$d = \lfloor \frac{(t+1)(p-1)}{p} \rfloor.$$

As for the first formula: if ω is a parameter for S, then $S = R[\omega]$, and if $f(X)$ is the minimal polynomial of ω over K, then $\mathcal{D}(S/R) = Sf'(\omega)$ by (22.9). Now

$$f'(\omega) = \prod_{\sigma \neq 1} (\omega - \sigma(\omega))$$

and $ord_L(\omega - \sigma(\omega)) = t + 1$ for each $\sigma \neq 1$ in G, since for $\sigma \neq 1, \sigma \in G_t \setminus G_{t+1}$. So we have

$$\begin{aligned} ord_L(\mathcal{D}(S/R)) &= \sum_{\sigma \neq 1} ord_L(\omega - \sigma(\omega)) \\ &= (p-1)(t+1). \end{aligned}$$

Since L/K is totally ramified, $ord_L(\mathcal{D}(S/R)) = ord_K(disc_R(S))$ by (22.11). □

§24. Cyclic extensions of degree p. In the rest of this chapter, we illustrate how these concepts relate to the behavior of S as a Galois module when L/K is cyclic of order p. These results are from [**Gr92**] and [**Ch87**].

We characterize those totally ramified L/K for which S is Hopf Galois over R.

(24.1) THEOREM. *Let L/K be a totally ramified Galois extension of local fields containing $\mathbb{Q}_p$, with Galois group G, cyclic of order p and suppose the ramification number $t < pe'$. Then S is Hopf Galois over R with Hopf algebra $A_{\pi^b} = R[\frac{\sigma-1}{\pi^b}]$, $1 \leq b \leq e'$ iff $t = -1 + bp$ with $b > 0$.*

PROOF. By Lemma (22.13), if S is H-Galois, then $disc(S) = disc(H^*)$. If $H = A_a$, $a = \pi^b$, then by the proof of (21.2), $disc(A_a)^* = \pi^{b(p-1)p}$, and so

$$b(p-1)p = ord_K(disc_R(S)) = ord_L(\mathcal{D}(S/R)) = (t+1)(p-1)$$

by (23.3). Thus $t = bp - 1$.

Conversely, suppose $t = -1 + bp$ for some b with $0 < b \leq e'$. Let

$$\mathfrak{A}_S = \{\alpha \in KG | \alpha S \subset S\}$$

be the associated order of S. We show $A_{\pi^b} \subset \mathfrak{A}_S$. Since $\mathfrak{A}_S$ is an R-algebra, it suffices to show that $\frac{\sigma-1}{\pi^b} \in \mathfrak{A}_S$. Let ω be a parameter for S, then by definition of t, we have

$$\frac{\sigma(x) - x}{\omega^{t+1}} \in S \text{ for all } x \in S,$$

or

$$(\frac{\sigma - 1}{\omega^{pb}})x \in S \text{ for all } x \in S,$$

that is,

$$(\frac{\sigma - 1}{\pi^b})x \in S \text{ for all } x \in S$$

where π is a parameter for R. Thus $A_{\pi^b} \subset \mathfrak{A}_S$. Now since $ord_K(disc_R(S)) = ord_L(\mathcal{D}(S/R)) = (t+1)(p-1)$ by (23.3), $disc(S) = \pi^{pb(p-1)}R$, and $disc(A^*_{\pi^b}) = (\pi^b)^{(p-1)p}R$ by the proof of (21.2). Thus S is an A_{π^b}-Galois extension of R by (22.13). □

Kummer extensions of order p. Let L/K be cyclic of order p with Galois group $G = \langle\sigma\rangle$. If $\zeta = \zeta_p$ is a primitive pth root of unity in K, then $L = K[z]$ with $z^p = b \in R$. We can describe when the valuation ring S of L is Hopf Galois over R in terms of z. We first need to normalize z. Set $e = ord_K(p), e' = e/(p-1) = ord_K(\zeta - 1)$.

(24.2) PROPOSITION. *We may choose z so that exactly one of the following cases holds:*

(i) $z^p = u\pi, u$ *a unit of* R

(ii) $z^p = 1 + u\pi^k, u$ *a unit of* $R, k = pq + r \geq 1, k < pe', 0 < r \leq p-1$

(iii) $z^p = 1 + u\pi^{pe'}, u$ *a unit of* R.

If $z^p = 1 + a\pi^k, a \in R, k > pe'$, *then* z *is in* K.

PROOF. Suppose $L = K[y]$ with $y^p = u\pi^d$, u a unit of R. If p does not divide d, find an s so that $sd = 1 + pb$, then $z = y^s/\pi^b$ satisfies $z^p = u^s\pi$ and $L = K[z]$.

If $d = pt$, then setting $z = y/\pi^t$, then $z^p = u$ is a unit of R. If $L = K[y]$ with $y^p = u$, a unit of R, then, since $R/\pi R$ is a finite field of characteristic p, there is some unit v of R so that $v^p \equiv u^{-1} \pmod{\pi R}$. Then $z_1 = yv$ satisfies $z_1^p = uv^p \equiv 1 \pmod{\pi R}$.

Suppose $z_1^p = 1 + u\pi^{pq+r}$ where u is a unit of R, $pq + r > 0$ and $0 \leq r < p$. Suppose $r = 0$. Then we can assume $q > 0$. Set $c = 1 + v\pi^q$ where $v^p \equiv -u \pmod{\pi R}$ and $z_2 = z_1c$, then

$$\begin{aligned} z_2^p &= (z_1c)^p \\ &= (1 + u\pi^{pq})(1 + v\pi^q)^p \\ &= (1 + u\pi^{pq})(1 + \sum_{k=1}^{p-1} \binom{p}{k} v^k\pi^{qk} + v^p\pi^{pq}) \\ &= 1 + (u + v^p)\pi^{pq} + uv^p\pi^{2pq} + p\pi^q w \end{aligned}$$

for some $w \in R$. If $pq < (p-1)e' + q$, that is, $q < e'$, then $ord_K(z_2^p - 1) > ord_K(z_1^p - 1)$. Repeating this construction as needed, we may find z so that $L = K[z]$ and $z^p = u\pi^{pq+r}$ with u a unit and $1 \leq r < p$ or $pq + r \geq pe'$.

If $L = K[z]$ and $z^p = b = 1 + u\pi^k$ with $k > pe'$, then b is a pth power in K [**FV93, I (5.8), Corollary 2**], so since $\zeta \in K$, $z \in K$. $\square$

We now consider each of the three cases of (24.2) in turn, beginning with the extreme cases (i) and (iii). First, case (i). Let $\hat{G} = Hom(G, U(R))$ be the character group of G.

(24.3) PROPOSITION. *If $L = K[z]$ with $z^p = u\pi$, u a unit of R, then $S = R[z]$ is an $R\hat{G}$-comodule and the associated order $\mathfrak{A}(S/R) = \overline{RG}$, the maximal order of RG, and is Hopf. Then L/K is totally ramified and the ramification number $t = pe'$. S is a tame $\overline{RG}$-extension, but not an $\overline{RG}$-Hopf Galois extension of R.*

PROOF. Since z satisfies an Eisenstein polynomial over R, S is totally ramified over R and z is a parameter for S, hence $S = R[z]$.

Now $\hat{G} = \langle \hat{\sigma} \rangle$ with $\langle \hat{\sigma}, \sigma \rangle = \zeta$. Since $S = R[z]$ with $z^p \in R$, S is $\hat{G}$-graded, hence an $R\hat{G}$-comodule algebra via

$$\alpha : S \to S \otimes R\hat{G},$$

$\alpha(z^i) = z^i \otimes \hat{\sigma}^i$. Thus (c.f. (2.1)), S is an $(R\hat{G})^*$-module algebra and $(R\hat{G})^* \subset \mathfrak{A}$. If $e_j = \frac{1}{p}\sum \zeta^{-jk}\sigma^k$, then $\{e_0, \ldots, e_{p-1}\}$ is the set of minimal pairwise orthogonal idempotents of KG and is the dual basis to $\hat{G}$, so $(R\hat{G})^* = \sum Re_j = \overline{RG}$, the maximal order in KG. Hence $\overline{RG} = \mathfrak{A}$ is a Hopf order in KG.

Finally, S is a tame $\mathfrak{A}$-module algebra, since the integral of $\mathfrak{A} = \overline{RG}$ is $e_0 = \frac{1}{p}\sum_{i=0}^{p-1} \sigma^i$ and $e_0 \cdot 1 = 1$, hence $e_0 S = R$. However S is not $\mathfrak{A}$-Galois. To see this, observe that the map

$$\gamma : S \otimes S \to S \otimes R\hat{G}$$

by $\gamma(s \otimes t) = (s \otimes 1)\alpha(t)$ sends $z^i \otimes z^j$ to $(z^i \otimes 1)(z^j \otimes \hat{\sigma}^j) = z^{i+j} \otimes \hat{\sigma}^j$. Hence, over K, $1 \otimes \hat{\sigma} = \gamma(\frac{z^{p-1}}{u\pi} \otimes z)$; since $K \otimes \gamma$ is an isomorphism, $\frac{z^{p-1}}{u\pi} \otimes z$ is the unique element of $K \otimes (S \otimes S) \cong L \otimes_K L$ which maps to $1 \otimes \hat{\sigma}$ in $K \otimes (S \otimes R\hat{G}) \cong L \otimes K\hat{G}$. Thus $1 \otimes \hat{\sigma}$ in $S \otimes R\hat{G}$ is not in the image of $S \otimes S$ under γ. $\square$

Now for case (iii) of (24.2).

(24.4) PROPOSITION. *Let $L = K[z]$ with $z^p = 1 + u\pi^{pe'}$, u a unit of R. Then L/K is unramified, and $S = R[\frac{z-1}{\lambda}]$, where $\lambda = \zeta - 1$. The ramification number $t = -1$.*

PROOF. Let $x = \frac{z-1}{\lambda}$, then $z = 1 + \lambda x$, so

$$1 + u\pi^{pe'} = z^p = \sum_{k=1}^{p-1} \binom{p}{k} \lambda^k x^k + \lambda^p x^p.$$

Divide by λ^p, noting that $\lambda^p R = \pi^{pe'} R = p\lambda R$: we obtain

$$u(\frac{\pi^{pe'}}{\lambda^p}) = x^p + \sum_{k=1}^{p-1} (\frac{1}{p}) \binom{p}{k} \frac{p\lambda^k x^k}{\lambda^p}$$

and so x satisfies a monic polynomial in $R[X]$. Hence $R[x] \subset S$.

Now $R[x]$ is an RG-Galois extension of S. For

$$\sigma(\frac{z-1}{\lambda}) = \frac{\zeta z - 1}{\lambda} = \frac{(\zeta - 1)z}{\lambda} + \frac{z-1}{\lambda},$$

so $\sigma(x) = z + x$. Since z is a unit of $R[x]$, the inertia group

$$G_0 = \{\tau \in G | \tau(x) \equiv x \pmod{m} \text{ for all maximal ideals } m \text{ of } R[t]\}$$

is trivial. Thus $t_{L/K} = -1$ and $R[x]$ is a Galois extension of R with group G. Hence $disc(R[x]) = disc(RG^*) = R$. Since $R[x] \subset S$, $disc(R[x]) \subset disc(S) \subset R$, and hence $disc(S) = R$. But then by (22.4), $R[x] = S$, S is Galois over R with group G, $\mathfrak{A} = RG$ and L/K is unramified. □

Now we consider the remaining case (ii) of (24.2):

$$L = K[z], z^p = 1 + u\pi^k$$

where $1 \leq k < pe'$ and p does not divide k. Here and below, u is a unit of R. Write $k = pq + r$ where $1 \leq r < p$.

(24.5) LEMMA. *L/K is totally ramified. If $x = \frac{z-1}{\pi^q}$ then $ord_L(x) = r$.*

PROOF. We have $z = 1 + \pi^q x$, so $1 + u\pi^{pq+r} = z^p = (1 + \pi^q x)^p$, hence

$$1 + u\pi^{pq+r} = 1 + \sum_{j=1}^{p-1} \binom{p}{j} \pi^{qj} x^j + \pi^{pq} x^p;$$

dividing by π^{pq} gives

$$0 = -u\pi^r + \sum_{j=1}^{p-1} \binom{p}{j} \pi^{qj-qp} x^j + x^p.$$

Two terms in the right side must have equal minimal valuations. Let $ord_L(x) = v$, then $ord_L(x^p) = pv$. Let h be the ramification index of L/K ($h = 1$ or p). Then $ord_L(-u\pi^r) = hr$, and the middle terms have valuation

$$ord_L(\binom{p}{j} \pi^{qj-qp} x^j) = he + h(qj - qp) + vj$$

for $j = 1, \ldots p-1$. Now the middle terms cannot have minimal valuation. To see this, we have $pq < pe'$, so $q < e'$, and

$$\begin{aligned} he + h(qj - qp) + vj &= he - hq(p-j) + vj \\ &\geq he'(p-1) - h(e'-1)(p-1) + vj \\ &= h(p-1) + vj \\ &> hr \end{aligned}$$

unless $v = 0$ and $r = p - 1$. But if $ord_L(x) = v = 0$, then x^p is the unique term with minimal valuation, which is impossible. Thus the two terms with minimal valuation must be x^p and $u\pi^r$, hence $pv = hr$. Since $r < p$, $h = p$, hence L/K is totally ramified and $v = r$. □

(24.6) PROPOSITION. *If $L = K[z]$ with $z^p = 1 + u\pi^k$ with u a unit of R and $0 < k = pq + r < pe'$ with $1 \leq r < p$, then the ramification number $t = pe' - k$.*

PROOF. If $x = \frac{z-1}{\pi^q}$, then $ord(x) = r$ by (24.5), so if s, n are integers with $rs = 1 + pn$, then $\omega = \frac{x^s}{\pi^n}$ is a parameter for S. Recall (24.1) that the ramification number t for L/K satisfies $t + 1 = ord_L(\sigma(\omega) - \omega)$. Choose $\sigma \in G$ with $\sigma(z) = \zeta z$. Then

$$\begin{aligned}
\sigma(\frac{x^s}{\pi^n}) &= \frac{1}{\pi^n}(\frac{\zeta z - 1}{\pi^q})^s \\
&= \frac{1}{\pi^n}(\frac{\lambda z}{\pi^q} + x)^s \\
&= \frac{x^s}{\pi^n} + \frac{1}{\pi^n}\sum_{j=1}^{s}\binom{s}{j}x^{s-j}(\frac{\lambda z}{\pi^q})^j.
\end{aligned}$$

The valuation of the terms in the sum are

$$\begin{aligned}
ord_L(\frac{1}{\pi^n}\binom{s}{j}x^{s-j}(\frac{\lambda z}{\pi^q})^j) &= -pn + r(s-j) + e'pj - pqj \\
&= 1 + j(p(e'-q) - r) \\
&= 1 + j(pe' - k),
\end{aligned}$$

whose minimum is for $j = 1$. Thus $t = pe' - k$. □

Since p does not divide k, we have

(24.7) COROLLARY. *If L/K is a Kummer extension of order p and $t < pe'$ then p does not divide t.*

Applying (24.1) to the Kummer extensions of (24.6) gives

(24.8) COROLLARY. *Let $L = K[z]$ with $z^p = 1 + u\pi^k$, where u is a unit of R and $k = pq + 1 \leq pe'$. Then $S = R[x]$ with $x = \frac{z-1}{\pi^q}$ is H-Hopf Galois over R with $H = A_{\pi^{e'-q}} = R[\frac{\sigma-1}{\pi^{e'-q}}]$.*

Finally, we note the structure of H:

(24.9) PROPOSITION. *Suppose K contains a primitive pth root of unity ζ, and L/K is as in (24.8). If $t = -1 + bp$ with $0 < b \leq e'$, then S is H-Galois for $H = A_{\pi^b}$. If $b < e'$ then H is a local ring. If $b = e'$, then $H \cong RG^*$ is separable.*

PROOF. The case $b < e'$ follows from (21.3). For the case $b = e'$, $L = K[z]$ with $z^p = 1 + u\pi$, u a unit of R. Hence $z - 1$ satisfies an Eisenstein polynomial, so $S = R[z-1] = R[z]$. Thus S is a G-graded R-algebra, and the identification of H with RG^* follows from (2.4). □

(24.10) REMARK. The question of whether the valuation ring S of a totally ramified Galois extension L/K of degree p is free over its associated order $\mathfrak{A}$ was completely resolved by M. J. Ferton [**Fe74**], as follows. Let the ramification number $t = qp + r$ with $0 \leq r < p$.

If $t < pe' - 1$ then S is $\mathfrak{A}$-free iff r divides $p - 1$. (Note: $\mathfrak{A}$ is Hopf only for $r = p - 1$.)

If $pe' - 1 \leq t \leq pe'$ then S is $\mathfrak{A}$-free iff the period of the continued fraction expansion of $\frac{t}{p}$ is ≤ 4. (When $pe' - 1 \leq t \leq pe'$, L/K is called *almost maximally ramified.*)

Criteria for S to be Hopf Galois. We showed for L/K a totally ramified Galois extension of order p that S is Hopf Galois over R, hence the associated order of S in KG is a Hopf order, iff the ramification number $t \equiv -1 \pmod{p}$. The following result, a complement to (23.3), gives alternate criteria.

(24.11) THEOREM. *Let L/K be cyclic of order p, totally ramified. The following are equivalent:*
(i) $t \equiv -1 \pmod{p}$;
(ii) *$disc(S)$ is the pth power of an ideal of R;*
(iii) *$tr(S)$ is the $p-1$st power of an ideal of R and $t \neq \frac{pe}{p-1}$.*

PROOF. The only part which is not immediate from (23.3) is (iii) implies (i). For this we need to extend (24.7) to extensions L/K when $\zeta \notin K$. We begin with the following lemma, from [**Wy69**]. $U(-)$ denotes the group of units.

(24.12) PROPOSITION. *Let K be an extension of $\mathbb{Q}_p$, and $\hat{K}$ a totally ramified extension of K of degree g with g not divisible by p. Let $R, \hat{R}$ be the valuation rings of $K, \hat{K}$, respectively. Then $U(\hat{R}) = U(R)U(\hat{R})^g$.*

PROOF. Let π, ρ be the parameters of $R, \hat{R}$, respectively, and let $\mathcal{P} = \rho\hat{R}$. Then we have a filtration of $U(\hat{R})$:

$$U(\hat{R}) = U_0 \supset U_1 \supset U_2 \supset \dots$$

where $U_i = 1 + \mathcal{P}^i$ for $i \geq 1$. Also (c.f [**CL. p. 74**]),

$$U_0/U_1 \cong k^*$$

by $u \mapsto u \pmod{\mathcal{P}}$, and

$$U_i/U_{i+1} \cong k^+$$

by $1 + a\rho^i \mapsto a \pmod{\mathcal{P}}$ for $i \geq 1$.

Let $u \in U(\hat{R})$. Then $u \equiv r \pmod{\mathcal{P}}$ for some $r \in U(R)$ since $\hat{K}/K$ is totally ramified. So $ur^{-1} \in U_1$.

Now the gth power map on U_i/U_{i+1} corresponds to multiplication by g on k^+. Since p does not divide g, this is onto. Thus $ur^{-1} = v_2^g \pmod{U_2}$, hence $ur^{-1}v_2^{-g} \in U_2$. Proceeding inductively, suppose we have found v_i in $U(\hat{R})$ so that $ur^{-1}v_i^{-g} \in U_i$. Then $ur^{-1}v_i^{-g} = 1 + a_i\rho^i$ for some $a \in \hat{R}$, and there is some $s_i = 1 + b_i\rho^i \in U_i$ so that

$$(1 + b_iu^i)^g \equiv 1 + gb_i\rho^i \pmod{\rho^{i+1}}$$

with $gb_i \equiv -a_i \pmod{\rho}$. Then

$$ur^{-1}v_i^{-g}s_i^g \equiv (1 + a_i\rho^i)(1 + gb_i\rho^i) \equiv 1 \pmod{\rho^{i+1}}.$$

So setting $v_{i+1} = s_i^{-1}v_i$, we have $ur^{-1}v_{i+1}^{-g} \in U_{i+1}$. Since $\hat{R}$ is complete and $v_i \equiv v_{i+1} \pmod{\mathcal{P}^i}$ for all i, the v_i converge to a unit v of $\hat{R}$ so that $u = rv^g$. Thus

$$U(\hat{R}) = U(R) \cdot U(\hat{R})^g. \qquad \square$$

(24.13) PROPOSITION. *Let $K \supset \mathbb{Q}_p$ with absolute ramification index e. Let L/K be cyclic of order p, totally ramified with valuation rings S, R, respectively, and let ω be a parameter for S. Let $\hat{K} = K[\zeta_p]$ and $g = [\hat{K} : K]$. Let $\hat{L} = \hat{K} \cdot L$. Let $\hat{S}, \hat{R}$ be the valuation rings of $\hat{L}, \hat{K}$, respectively, and $\hat{\omega}$ a parameter for $\hat{S}$. If $t, \hat{t}$ are the ramification numbers for $L, \hat{L}$, respectively, then $\hat{t} = gt$.*

PROOF. Since $\mathbb{Q}_p[\zeta]/\mathbb{Q}_p$ is totally ramified and Galois of degree $p-1$, $\hat{K}/K$ is totally ramified of degree dividing $p-1$. Since L/K has order p, $\hat{L}/\hat{K}$ has order p and the restriction map from $Gal(\hat{L}/\hat{K})$ to $Gal(L/K)$ is an isomorphism. Since $\hat{K}/K$ and L/K are totally ramified of relatively prime degrees, $\hat{L}/K$ is totally ramified. Then $ord_{\hat{L}} = g(ord_L)$ and $\hat{\omega}^g \hat{S} = \omega \hat{S}$, so $\hat{\omega}^g = \hat{u}\omega$ for some unit $\hat{u}$ of $\hat{S}$. By multiplying $\hat{\omega}$ by a unit, we are free to alter $\hat{u}$ by the gth power of a unit of $\hat{S}$.

Now $\hat{L}/L$ is totally ramified of degree g coprime to p, so (24.12) yields that $\hat{u} = \hat{v}^g u_0$ for some units $\hat{v}$ of $\hat{S}$ and u_0 of S. Replacing $\hat{\omega}$ by $\hat{\omega}\hat{v}$, we may assume that $\hat{\omega}^g = u_0\omega$ where $u_0 \in S$. Since $u_0 \in S$, we may replace ω by $u_0\omega$ and assume $\hat{\omega}^g = w$.

Now the ramification number $\hat{t}$ of $\hat{L}/L$ is defined by

$$\hat{t} + 1 = ord_{\hat{L}}(\sigma(\hat{\omega}) - \hat{\omega})$$

and t is defined by

$$\begin{aligned} t + 1 &= ord_L(\sigma(w) - w) \\ &= \frac{1}{g} v_{\hat{L}}(\sigma(\hat{\omega}^g) - \hat{\omega}^g)) \\ &= \frac{1}{g}(v_{\hat{L}}(\sigma(\hat{\omega}) - \hat{\omega}) + \sum_{i=1}^{g-1} ord_{\hat{L}}(\sigma(\hat{\omega}) - \xi^i\hat{\omega})) \end{aligned}$$

where ξ is a primitive gth root of unity: since g divides $p-1$, $\xi \in \mathbb{Q}_p \subset K$.

Since $\hat{L}/\hat{K}$ is totally ramified, $\hat{t} > 0$, so $ord_L(\sigma(\hat{\omega}) - \hat{\omega}) \geq 2$. Also, since p and g are coprime, $1 - \xi^i$ is a unit of $\hat{S}$ for $i \neq 0$, , so for $i \neq 0$, $\sigma(\hat{\omega}) - \xi^i\hat{\omega} = (\sigma(\hat{\omega}) - \hat{\omega}) + (1 - \xi^i)\hat{\omega}$ satisfies

$$ord_{\hat{L}}(\sigma(\hat{\omega}) - \xi^i\hat{\omega}) = ord_{\hat{L}}(\hat{\omega}) = 1.$$

Thus

$$t + 1 = \frac{1}{g}(\hat{t} + 1 + (g-1)) = \frac{\hat{t}}{g} + 1.$$

Hence $\hat{t} = gt$. □

(24.14) COROLLARY. *With $\hat{L}, L$ as in the proposition,*
(i) *p and t are coprime iff p and $\hat{t}$ are coprime;*
(ii) *p divides t iff $\hat{t} = \frac{pe_{\hat{K}}}{p-1}$, iff $t = \frac{pe_K}{p-1}$.*

PROOF. (i) is immediate from (24.12). For (ii): $\hat{L}/\hat{K}$ is Kummer, so p divides t iff p divides $\hat{t}$ iff $\hat{L} = \hat{K}[z]$ with z a parameter, iff $\hat{t} = \frac{pe_{\hat{K}}}{p-1}$, iff $t = \frac{1}{g}\frac{pe_{\hat{K}}}{p-1}$. Since $e_{\hat{K}} = ge_K$, (ii) follows. □

(24.15) COROLLARY. *Let L/K be a cyclic extension of degree p with Galois group $G = \langle\sigma\rangle$. Then the ramification number can take on the following values:*

1. *If L/K is unramified then $t = -1$;*
2. *If L/K is totally ramified then $t < \frac{pe}{p-1}$ and p does not divide t, or $t = \frac{pe}{p-1}$.*

If the associated order $\mathfrak{A}$ of the valuation ring S of L is a Hopf order in KG, then $t = -1 + bp$ with $b \leq e'$, or $t = pe'$. If $t = -1 + bp$ with $b \leq e'$, then $\mathfrak{A} = A_b = R[\frac{\sigma-1}{\pi^b}]$. If $t = pe'$, then $\mathfrak{A} = A_{e'} = R[\frac{\sigma-1}{\pi^{e'}}]$.

PROOF. The general assertions about t follow from (24.7) and (24.14). If $\mathfrak{A}$ is Hopf and $t < pe'$ the result follows from (24.1). If $t = pe'$ then as in the proof of (24.1) one sees that $A_{e'} \subset \mathfrak{A}$, and $disc(A_{e'}) = R$ by the proof of (21.1); hence $A_{e'} = \mathfrak{A}$. □

Now we can prove (iii) implies (i) of Theorem (24.11), namely, that if $tr(S)$ is the $p-1$st power of an ideal of R and $t \neq pe'$, then $t \equiv -1 \pmod p$.

PROOF OF (III) IMPLIES (I). Let $t + 1 = qp + r$ with $0 \leq r \leq p - 1$. Then by (23.3),

$$ord_K(tr(S)) = \lfloor\frac{(t+1)(p-1)}{p}\rfloor = q(p-1) + \lfloor\frac{r(p-1)}{p}\rfloor$$

is a multiple of $p - 1$ iff $r = 0$ or 1. But by (24.15), since $t \neq \frac{pe}{p-1}, t \not\equiv 0 \pmod p$, so $r \neq 1$. Hence $r = 0$ and $t \equiv -1 \pmod p$. □

In this chapter we have shown that for L/K cyclic of order p, then excepting the case t divisible by p, S is Hopf Galois over R iff $t \equiv -1 \pmod p$. We shall see that analogous ramification conditions are necessary (Chapter 8) but not sufficient (Chapter 10) for the valuation ring of a cyclic extension of prime power order to be Hopf Galois.

CHAPTER 7

Non-maximal Orders

Let K be a local number field with valuation ring R. When studying subrings of an extension L of K we tend to focus on the integral closure S of R in L, that is, the maximal order of R in L, because S is a Dedekind ring, hence has an attractive ideal theory. However, in Galois module theory certain non-maximal orders over R in L may also be of interest. One example, studied by Frohlich [**Fr62**], is the Kummer order $\hat{S}$ associated to a Kummer extension L/K with Galois group G, defined as follows. If $\chi \in \hat{G} = Hom(G, K)$ and

$$S_\chi = \{s \in S | \sigma(s) = \chi(\sigma)s \text{ for all } \sigma \in G\},$$

then S_χ is a rank one projective R-module and $\hat{S} = \sum_{\chi \in \hat{G}} S_\chi$ is the Kummer order of S. The Kummer order $\hat{S}$ is maximal only under special circumstances; for G cyclic of order p, $\hat{S} = S$ iff the ramification number $t \geq pe' - 1$ (see (24.3), (24.9)).

To illustrate the possibilities, we begin with Kummer extensions of order p, prime.

§25. Hopf Galois orders in Kummer extensions of order p. Let K be a local field with valuation ring R and parameter π with $\mathrm{char}(R/\pi) = p$. Assume K contains a primitive pth root of unity ζ.

Let L be a field which is a Galois extension of K of order p. Then $L = K[z]$ with $z^p \in K$ and $G = \langle\sigma\rangle$, cyclic of order p, is the Galois group of L/K with action by $\sigma(z) = \zeta z$. Recall that $e' = e/(p-1)$ where e is the ramification index of K over $\mathbb{Q}_p$.

From (24.2) we may normalize z so that $z \in R$ and z^p satisfies one of the following conditions:

(1). $z^p = u\pi$ for some unit u of R;

(2). $z^p = 1 + u\pi^k$ for some unit u of R, where $0 < k < pe'$ and p does not divide k;

(3). $z^p = u\pi^{pe'}$ for some unit u of R.

In case (1) the Kummer order $\hat{S}$ is the maximal order: $\hat{S} = S = R[z]$.

Suppose case (2) or (3) holds: $z^p = 1 + u\pi^k, 0 < k \leq pe'$. For $0 \leq i \leq \frac{k}{p}$ let $w = \frac{z-1}{\pi^i}$ and let $S_i = R[w]$. Then

$$\pi^{pi} w^p = z^p - 1 - \sum_{j=1}^{p-1} \binom{p}{j} \pi^{ij} w^j,$$

so if $i \leq \frac{k}{p}$, w is the root of a monic polynomial of degree p with coefficients in R, hence is integral over R. Thus S_i is an order over R in L, that is, an R-subalgebra of L which is finitely generated as an R-module and spans L over K.

Evidently

$$S_0 \subset S_1 \subset \ldots \subset S_b \subset S$$

where $b = \lfloor \frac{k}{p} \rfloor$ and S is the maximal order of R in L.

Recall that KG contains $e' + 1$ Larson orders $L_b = R[\frac{\sigma - 1}{\pi^b}]$ for $0 \leq b \leq e'$. Evidently,

$$RG = L_0 \subset L_1 \subset \ldots \subset L_{e'}$$

and since R contains ζ, $L_{e'} = \overline{RG}$, the maximal order over R in KG, and $\overline{RG} \cong (RG)^*$.

Set $i' = e' - i$ for $0 < i < e'$.

(25.1) PROPOSITION. *For* $0 \leq i \leq \lfloor \frac{k}{p} \rfloor$, S_i *is an* $L_{i'}$*-Galois extension.*

PROOF. We first show that $L_{i'}$ acts on S_i. To do so, it suffices to show that $L_{i'}$ is contained in $\mathfrak{A}_i$, the associated order of S_i,

$$\mathfrak{A}_i = \{\alpha \in KG | \alpha S_i \subset S_i\}.$$

Since $\mathfrak{A}_i$ is an R-algebra, it suffices to show that the R-algebra generator $\xi = \frac{\sigma - 1}{\pi^{i'}}$ of $L_{i'}$ is in $\mathfrak{A}_i$. Since $S_i = R[w]$ with $w = \frac{z-1}{\pi^i}$, we need to check that $\xi w^k \in S_i$ for all k, which we do by induction on k.

For $k = 1$ we have

$$\xi w = (\frac{\sigma - 1}{\pi^{i'}})(\frac{z - 1}{\pi^i}) = \frac{1}{\pi^{e'}}(\zeta - 1)z \in S_i$$

since $(\zeta - 1)R = \pi^{e'} R$. We have

$$\Delta(\xi) = \pi^{i'}(\xi \otimes \xi) + \xi \otimes 1 + 1 \otimes \xi \ .$$

So

$$\begin{aligned} \xi w^k &= \Delta(\xi)(w \otimes w^{k-1}) \\ &= \pi^{i'}(\xi w \cdot \xi w^{k-1}) + \xi w \cdot w^{k-1} + w \cdot \xi w^{k-1}. \end{aligned}$$

Thus if ξw and $\xi w^{k-1} \in S_i$, then $\xi w^k \in S_i$. Hence $L_{i'}$ acts on S_i.

Since $L_{i'}$ acts on S_i, we obtain the coaction

$$\alpha : S_i \to S_i \otimes L_{i'}^*$$

and the S_i-algebra homomorphism

$$\gamma : S_i \otimes S_i \to S_i \otimes L_{i'}^*.$$

Since L is a KG-Galois extension of K, the map $K \otimes \gamma : L \otimes L \to L \otimes KG^*$ is an isomorphism, and hence γ is 1-1.

To show that γ is an isomorphism, it suffices by (22.4) to show that

$$disc_{S_i}(S_i \otimes S_i) = disc_{S_i}(S_i \otimes L_{i'}^*).$$

To see this, we first note that if

$$\hat{G} = Hom(G, K^\times) = \langle \hat{\sigma} \rangle$$

then $L^*_{i'} \cong R[\frac{\hat{\sigma}-1}{\pi^i}] \cong L_i$. Now if $T = R[t]$ is an R-algebra isomorphic to $R[T]/(T^p - a)$ for some a in R, then $\{1, t, \ldots, t^{p-1}\}$ is an R-basis of T and

$$tr(t^i t^j) = \begin{cases} p \text{ if } i = j = 0 \\ ap \text{ if } i + j = p \\ 0 \text{ otherwise.} \end{cases}$$

Thus the discriminant of T is $p^p a^{p-1} R = p^p R$ if a is a unit of R. Applying this to $S_i[z]$ (recall: $z^p = 1 + u\pi^k$) and $S_i[\hat{\sigma}]$ shows that

$$disc_{S_i}(S_i[z]) = disc_{S_i}(S_i \otimes R[\hat{G}]).$$

Now the change of basis matrix C in $M_p(S_i)$ from $\{1, \frac{z-1}{\pi^{i'}}, \ldots, (\frac{z-1}{\pi^{i'}})^{p-1})\}$ to $\{1, z, \ldots, z^{p-1}\}$ is the same as that from $\{1, \frac{\hat{\sigma}-1}{\pi^{i'}}, \ldots, (\frac{\hat{\sigma}-1}{\pi^{i'}})^{p-1})\}$ to $\{1, \hat{\sigma}, \ldots, \hat{\sigma}^{p-1}\}$. Hence by (22.2),

$$disc_{S_i}(S_i \otimes L_i) \cdot (det C)^2 = disc_{S_i}(S_i[\hat{\sigma}]),$$

$$disc_{S_i}(S_i \otimes S_i) \cdot (det C)^2 = disc_{S_i}(S_i[z])$$

and so $disc_{S_i}(S_i \otimes S_i) = disc_{S_i}(S_i \otimes L_i)$, and γ is onto. □

Thus we have the chain of orders in S:

$$S_0 \subset S_1 \subset \ldots \subset S_b \subset S$$

where $b = \lfloor \frac{k}{p} \rfloor \leq e'$, and the chain of Hopf algebras

$$L_{e'} \supset L_{e'-1} \supset \ldots \supset L_{e'-b} \supset \ldots L_0$$

such that for $0 \leq i \leq b$, S_i is an $L_{e'-i}$-Hopf Galois extension of R. The maximal order S is $L_{e'-b}$-Galois iff $k = pb + 1$ (in which case $w = \frac{z-1}{\pi^b}$ is a parameter of L) or $k = pe'$ (in which case L/K is unramified).

The minimal order S_0 in the chain is the Kummer order $\hat{S} = \sum S_\chi$, an $L_{e'}$-Galois extension where $L_{e'} = \overline{RG} = \sum_{i=0}^{p-1} Re_i \cong (RG)^*$.

The situation becomes more natural if we view all this from the point of view of principal homogeneous spaces (or Galois objects). Then the chain

$$S_0 \subset S_1 \subset \ldots \subset S_b$$

corresponds to

$$L_0 \subset L_1 \subset \ldots \subset L_b,$$

where S_i is an L_i-principal homogeneous space, or L_i-Galois object. This point of view arises quite naturally when we look at Hopf algebras arising from isogenies of formal groups, as we shall see in Chapter 12.

Kummer orders are used in the classification of unramified p-power extensions of fields containing $\mathbb{Q}_p$: see [**Gr89**], generalizing [**Ha36**], and [**Gr92b**].

§**26. Orders and associated orders.** The idea of constructing orders over R in L associated to a Hopf order over R in KG is quite general.

Let L be an A-Hopf Galois extension of K, a local number field, for some cocommutative K-Hopf algebra A. Let R be the valuation ring of K, let S be the integral closure of R in L (we do not assume L is a field) and let H be a Hopf order over R in A. Let

$$\hat{\mathcal{O}}(H) = \{s \in L | hs \in S \text{ for all } h \in H\}.$$

Taylor has observed:

(26.1) Proposition. *$\hat{\mathcal{O}}(H)$ is an order over R in L.*

Proof ([**Ta87, Lemma 3.1)**]. First, for any R-algebra H, $\hat{\mathcal{O}}(H)$ is an R-lattice in L, that is, an R-finitely generated submodule of L which contains a K-basis of L. To see this, observe that since 1 is in H, $\hat{\mathcal{O}}(H) \subset S$; on the other hand, if $\{h_i\}$ is an R-basis of H and $\{s_j\}$ is an R-basis of S, then there is some $r \in R$ so that r multiplies the product $h_i s_j$ into S for all i and j. So $rS \subset \hat{\mathcal{O}}(H)$ and so $\hat{\mathcal{O}}(H)$ is a lattice.

If H is an R-Hopf algebra, then $\hat{\mathcal{O}}(H)$ is an R-algebra. For if s, t are in $\hat{\mathcal{O}}(H)$, then, for all $h \in H$, $h(st) = \sum_{(h)} h_{(1)}(s) \cdot h_{(2)}(t)$ is in S. So st is in $\hat{\mathcal{O}}(H)$. Also, 1 is in $\hat{\mathcal{O}}(H)$ because for all $h \in H$, $h \cdot 1 = \varepsilon(h) \cdot 1$ and $\varepsilon(h)$ is in R, hence $h \cdot 1$ is in S for all $h \in H$. Thus $\hat{\mathcal{O}}(H)$ is an R-algebra, and hence is an order in L. □

(26.2) Examples. If $A = KG$, then $\hat{\mathcal{O}}(RG) = S$. If G is abelian of order n and R contains a primitive nth root of unity, then $\overline{RG} \cong RG^*$ is a Hopf order and $\hat{\mathcal{O}}(\overline{RG}) = \hat{S}$, the Kummer order.

Given a A-Hopf Galois extension L/K of number fields for A some K-Hopf algebra, we have the associated order map $\mathfrak{A}$, from orders over R in L to orders over R in A, and the map $\hat{\mathcal{O}}$, from orders over R in A to lattices over R in L.

If H is a Hopf order over R in A, $\hat{\mathcal{O}}(H)$ is always an order over R in L. On the other hand, for an order S over R in L, often $\mathfrak{A}(S)$ is not a Hopf order in A. Thus it is not the case that $\hat{\mathcal{O}}$ and $\mathfrak{A}$ are inverse maps. The simplest example is to take a wildly ramified extension L/K with valuation ring S and Galois group G; then RG acts on S, so, since S is the maximal order of L, $\hat{\mathcal{O}}(RG) = S$. But $\mathfrak{A}(S)$ is necessarily larger than RG, for by the extension of Noether's theorem (Chapter 3), if $\mathfrak{A}(S) = RG$, then $\mathfrak{A}(S)$ is a Hopf order and so S is RG-tame, i.e. L/K is tamely ramified.

The following results (from [**Ch88**]) bear on the question of when $\mathfrak{A}$ and $\hat{\mathcal{O}}$ are inverses of each other.

(26.3) Proposition. *Let K be a local field with valuation ring R. Let H be a cocommutative R-Hopf algebra, finitely generated and free as R-module, and $A = K \otimes_R H$. Let L be an A-Hopf Galois extension of K. Let S be an order over R in L.*

(1) *If S is a tame H-extension of R, then S is a free rank one H-module and $H = \mathfrak{A}(S)$.*

(2) *If S is an H-Galois extension of R, then $S = \hat{\mathcal{O}}(H)$, and S is the unique order over R in L which is H-Galois.*

PROOF. (1). Since S/R is H-tame, S is free of rank one. To show $H = \mathfrak{A}(S)$, first note that since H acts on S and

$$\mathfrak{A}(S) = \{a \in A | aS \subset S\},$$

we have $H \subset \mathfrak{A}(S)$. Let $S = Hw$, the free rank one H-module with basis w. Then $L = Aw$. If a is in $\mathfrak{A}(S)$, then $aw \in S$, so $aw = hw$ for some $h \in H$. But since L is A-free on w, $a = h \in H$. Hence $\mathfrak{A}(S) = H$.

For part (2): we first show $S = \hat{\mathcal{O}}(H)$, where

$$\hat{\mathcal{O}}(H) = \{s \in L | Hs \subset O_L\}$$

where O_L is the integral closure of R in L. Since H acts on $S \subset O_L$, $S \subset \hat{\mathcal{O}}(H)$. To show equality, observe that the inclusion $S \subset \hat{\mathcal{O}}(H)$ is an R-algebra, H-module homomorphism. Also $\hat{\mathcal{O}}(H)^H = R$: for evidently $R \subset \hat{\mathcal{O}}(H)^H$, while $\hat{\mathcal{O}}(H)^H \subset \hat{\mathcal{O}}(H) \cap L^A \subset O_L \cap K = R$. Thus $S = \hat{\mathcal{O}}(H)$ by (14.4). $\hat{\mathcal{O}}(H)$ is therefore the unique order over R in L which is H-Galois. □

(26.4) THEOREM. *Let L/K be an A-Galois extension of local fields, and R be the valuation ring of K. Let H_0 be a Hopf order in A so that $\hat{\mathcal{O}}(H_0)$ is H_0-tame. Then $H_0 = \mathfrak{A}\hat{\mathcal{O}}(H_0)$. If H is any Hopf order in A containing H_0, then $\hat{\mathcal{O}}(H)$ is tame over H and $\mathfrak{A}\hat{\mathcal{O}}(H) = H$.*

PROOF. The first part of the last proposition shows that $H_0 = \mathfrak{A}\hat{\mathcal{O}}(H)$. Let θ generate the module of left integrals of H_0. Since $S_0 = \hat{\mathcal{O}}(H_0)$ is H_0-tame, there is some z_0 in S_0 so that $\theta_0 z_0 = 1$. Let θ generate the module of left integrals of H, then $\theta_0 = r\theta$ for some r in R, since $H_0 \subset H$. Let $z = rz_0$. We show: $z \in \hat{\mathcal{O}}(H)$. To prove this, first note that since $H_0 \subset H$, their duals satisfy $H^* \subset H_0^*$. We have $H = H^*\theta$, so for any $\xi \in H$, there exists f in H^* with $\xi = f \cdot \theta$. To show z is in $\hat{\mathcal{O}}(H)$, we need to show that for any ξ in H, ξz is in O_L, the valuation ring of L. But $\xi z = (f \cdot \theta)z = (f \cdot (\theta_0/r))(rz_0) = (f \cdot \theta_0)z_0$. Now since f is in $H^* \subset H_0^*$, $f \cdot \theta_0$ is in H_0, and since z_0 is in $\hat{\mathcal{O}}(H_0)$, $(f \cdot \theta_0)z_0$ is in O_L. Thus ξz is in O_L, and z is therefore in $\hat{\mathcal{O}}(H)$. Now $\theta z = (\theta_0/r)(rz_0) = \theta_0 z_0 = 1$, so S is H-tame. Thus S is H-free of rank one, and $H = \mathfrak{A}(S) = \mathfrak{A}\hat{\mathcal{O}}(H)$. □

(26.5) COROLLARY. *If L/K is a Galois extension of local fields with Galois group G and L/K is tamely ramified, then for every Hopf order H in KG, $\hat{\mathcal{O}}(H)$ is tame, hence free of rank one over H, and $H = \mathfrak{A}\hat{\mathcal{O}}(H)$.*

This follows immediately from the theorem and the fact that any Hopf order in KG contains RG (because the dual of any Hopf order in KG is contained in the maximal order of KG^*, namely RG^*).

(26.6). The Kummer theory of formal groups (Section 39) provides many examples of Hopf orders H such that $\hat{\mathcal{O}}(H^*)$ is a Galois extension of R: in general these will be non-maximal orders (see (39.8). Interesting global examples may be found associated to elliptic curves or more general abelian varieties: see [**Ta88**], [**ST90**] and its predecessors, and [**TB92**].

CHAPTER 8

Ramification Restrictions

Let $L \supset K$ be a Galois extension of local fields containing $\mathbb{Q}_p$. Let $G = Gal(L/K)$ be the Galois group of L/K. Let S be the valuation ring of L, and R the valuation ring of K.

In Chapter 6 we assumed G has order p and obtained necessary and sufficient conditions for the associated order $\mathfrak{A}_S$ of S,

$$\mathfrak{A}_S = \{\alpha \in KG | \alpha S \subset S\}$$

to be a Hopf order, namely, that the ramification number, or break number $t(G) \equiv -1$ or $0 \pmod p$.

In this chapter we present two theorems of Byott which extend this result.

Let L/K be totally ramified and Galois with group G of order p^n, not necessarily abelian, and assume that the associated order $\mathfrak{A}$ of S is Hopf. There is a Galois extension $K \subset L_1 \subset L$ with $Gal(L_1/K) = G'$ cyclic of order p such that $t_1(G) = t(G')$ and the associated order $\mathfrak{A}'$ of the valuation ring S_1 of L_1 is Hopf. It follows by (24.10) that $t(G') \equiv -1$ or $0 \pmod p$. The case $t(G') \equiv 0 \pmod p$ is exceptional, as will be noted at the end of the chapter. Suppose, then, that $t(G') \equiv -1 \pmod p$. Under that assumption, we show that S/R is $\mathfrak{A}$-Hopf Galois (not just $\mathfrak{A}$-tame), and all ramification numbers $t_i(G) \equiv -1 \pmod{p^n}$. The first result is Theorem 5 of [**By95c**], which we prove using Waterhouse's theorem (13.7); the second is Theorem 4.4 of [**By97c**] .

§**27. Ramification numbers.** We begin by defining the ramification numbers, or break numbers, for G or L/K, when L/K is Galois and totally ramified of degree p^n.

Let ω be a parameter for L. Since L/K is totally ramified, $S = R[\omega]$, and ω is a test element for membership in the ith ramification group

$$G_i = \{\sigma \in G | \sigma(x) - x \in \omega^{i+1}S \text{ for all } x \in S\},$$

namely,

$$G_i = \{\sigma \in G | \sigma(\omega) - \omega \in \omega^{i+1}S\}.$$

(27.1) Definition. The ramification numbers, for G are defined by

$$t_j(G) = max\{i | |G/G_i| < p^j\}.$$

(27.2). Suppose G is cyclic of order p. Then

$$G = G_{-1} = \ldots = G_s \supset G_{s+1} = \{1\}.$$

So $|G/G_i| = 1$ for $i = -1, \dots, s$, while $|G/G_i| = p$ for $i > s$. So $t_1 = s$, and (27.1) extends the definition of ramification number for groups of order p of (23.2).

(27.3). Suppose G is cyclic of order p^n. By [**CL IV, Cor. 3 to Prop. 7**], G_i/G_{i+1} has exponent p. Thus

$$G = G_{-1} = \dots = G_{s_1} \supset G_{s_1+1} = \dots = G_{s_2} \supset G_{s_2+1} = \dots$$
$$\dots = G_{s_{n-1}} \supset G_{s_{n-1}+1} = \dots G_{s_n} \supset \{1\}$$

and $G_{s_j}/G_{s_{j+1}}$ has order p. Then, for all j, $|G/G_{s_j}| = p^{j-1}$, while $|G/G_{s_{j+1}}| = p^j$, and so $t_j = s_j$.

In general, consider the set of natural numbers

$$\{ord_L(\sigma(\omega) - \omega) | \sigma \neq 1 \in G\}.$$

Denote the numbers in this set, in increasing order, by

$$\nu_1 < \nu_2 < \dots < \nu_{r+1}.$$

Then the ramification groups satisfy

$$\begin{aligned} G = G_0 = G_1 = \dots &= G_{\nu_1 - 1} \\ \supset G_{\nu_1} = \dots &= G_{\nu_2 - 1} \\ \supset G_{\nu_2} = \dots &= G_{\nu_3 - 1} \\ &\dots \\ \supset G_{\nu_k} = \dots &= G_{\nu_{k+1} - 1} \\ &\dots \\ \supset G_{\nu_{r+1}} &= 1 , \end{aligned}$$

and

$$\begin{aligned} G/G_0 = G/G_1 = \dots = G/G_{\nu_1 - 1} &\text{ has order } 1 \\ G/G_{\nu_1} = \dots = G/G_{\nu_2 - 1} &\text{ has order } p^{i_1} \\ G/G_{\nu_2} = \dots = G/G_{\nu_3 - 1} &\text{ has order } p^{i_2} \\ &\dots \\ G/G_{\nu_k} = \dots = G/G_{\nu_{k+1} - 1} &\text{ has order } p^{i_k} \\ &\dots \\ G/G_{\nu_{r+1}} &\text{ has order } p^{i_{r+1}} = p^n \end{aligned}$$

for some $i_1 < i_2 < \dots < i_{r+1} = n$. Then, since $t_j(G) = max\{i | |G/G_i| < p^j\}$, we have

$$\begin{aligned} t_1(G) = \dots = t_{i_1}(G) &= \nu_1 - 1; \\ t_{i_1+1}(G) = \dots = t_{i_2}(G) &= \nu_2 - 1; \\ &etc. \end{aligned}$$

If $i_{k-1} < j \leq i_k$, then $t_j(G) = \nu_k - 1$.

We note that $G = G_1$ since L/K is totally ramified and G is a p-group [**CL IV, Corollary 1 to Proposition 7**].

If H is a subgroup of G, then $H = Gal(L/L^H)$, so the ramification groups for H are computed using the parameter for L, just as with G. Thus $H_i = G_i \cap H$. The analogous result for quotient groups is more subtle, since it involves a change in parameter.

(27.4) LEMMA **[CL IV, Cor. to Prop. 3]**. *Let L/K be totally ramified and Galois with group G. Let $G_j \supset H \supseteq G_{j+1}$. Then $(G/H)_i = G_i/H$ for $i \leq j$, $(G/H)_i = 1$ for $i > j$.*

PROOF. Let $\overline{\sigma} \neq \overline{1} \in G/H$, then for some $i \leq j$, $\overline{\sigma} \in G_i/H, \overline{\sigma} \notin G_{i+1}/H$. Let $\sigma \in G$ map to $\overline{\sigma}$. Then $\sigma \in G_i$, $\sigma \notin G_{i+1}$. Hence $ord_L(\sigma(\omega) - \omega) = i + 1$. By **[CL IV, Prop. 3]**, if $M = L^H$ with parameter ρ, then

$$ord_M(\overline{\sigma}(\rho) - \rho) = \frac{1}{e_{G/H}} \sum_{\sigma \mapsto \overline{\sigma}} ord_L(\sigma(\omega) - \omega).$$

Since L/K is totally ramified, $e_{G/H} = [L : M]$, so

$$ord_M(\overline{\sigma}(\rho) - \rho) = ord_L(\sigma(\omega) - \omega) = i + 1.$$

So $\overline{\sigma} \in (G/H)_i \setminus (G/H)_{i+1}$. Hence $(G/H)_i = G_i/H$ for $i \leq j$. For $\overline{\sigma} \in G_j/H$, $\overline{\sigma} \neq \overline{1}$, we have $ord_M(\overline{\sigma}(\rho) - \rho) = j + 1$. Hence $\overline{\sigma} \notin (G/H)_{j+1}$, and so $(G/H)_{j+1} = \{\overline{1}\}$. $\square$

The ramification numbers $t_j(G)$ may be identified as the ramification numbers of order p subquotients of G. To see this, we refine the chain

$$G = G_{\nu_1 - 1} \supset G_{\nu_1} = G_{\nu_2 - 1} \supset G_{\nu_2} = \ldots G_{\nu_k - 1} \supset G_{\nu_k} = \ldots = G_{\nu_{k+1} - 1} \cdots$$

into a normal series

$$G = N_0 \supset N_1 \supset N_2 \supset \ldots \supset N_n = 1$$

("normal series" means N_k is a normal subgroup of G and N_k/N_{k+1} has order p for all k). Then N_k has order p^{n-k} for all k. Thus

$$\begin{aligned}
G = G_0 = G_1 = \ldots = G_{\nu_1 - 1} &= N_0 \text{ has order } p^n, \\
G_{\nu_1} = \ldots = G_{\nu_2 - 1} &= N_{i_1} \text{ has order } p^{n-i_1}, \\
&\ldots \\
G_{\nu_{k-1}} = \ldots = G_{\nu_k - 1} &= N_{i_{k-1}} \text{ has order } p^{n-i_{k-1}}, \\
G_{\nu_k} = \ldots = G_{\nu_{k+1} - 1} &= N_{i_k} \text{ has order } p^{n-i_k}, \\
&\ldots \\
G_{\nu_{r+1}} = N_{i_{r+1}} &= N_n = 1 \text{ has order } p^{n-i_{r+1}} = 1,
\end{aligned}$$

so $i_{r+1} = n$.

(27.5) PROPOSITION. $t_j(G) = t_j(G/N_j) = t_1(N_{j-1}/N_j)$.

PROOF. Suppose $i_{k-1} < j \leq i_k$, so that $G_{\nu_k} \subseteq N_j \subset N_{j-1} \subseteq G_{\nu_k - 1}$. Consider G/N_j with $i_{k-1} < j \leq i_k$. Then the ramification groups $(G/N_j)_i$ of G/N_j are G_i/N_j for $1 \leq i \leq \nu_k - 1$ by (27.4), while $(G/N_j)_i = \{1\}$ for $i \geq \nu_k$. Thus, looking at the ramification groups modulo $N_j \subset G_{\nu_k - 1}$ and applying (27.4), we have

$$\begin{aligned} G/N_j = G_0/N_j = G_1/N_j = \ldots = G_{\nu_1 - 1}/N_j &\text{ has order } p^j, \\ G_{\nu_1}/N_j = \ldots = G_{\nu_2 - 1}/N_j &\text{ has order } p^{j-i_1}, \\ \ldots \\ G_{\nu_{k-1}}/N_j = \ldots = G_{\nu_k - 1}/N_j &\text{ has order } p^{j-i_{k-1}}, \end{aligned}$$

while $(G/N_j)_{\nu_k} = 0$. Thus the ramification number

$$\begin{aligned} t_j(G/N_j) &= max\{i|(G/N_j)/(G/N_j)_i \text{ has order } < p^j\} \\ &= max\{i|(G/N_j)_i \neq \{1\}\} = \nu_k - 1. \end{aligned}$$

We show that $t_1(N_{j-1}/N_j) = t_j(G/N_j)$. The ramification groups of G/N_j are $G_i/N_j, 1 \leq i \leq \nu_k - 1$, and $(G/N_j)_i = 0$ for $i \geq \nu_k$. The ramification groups of N_{j-1}/N_j are $(N_{j-1}/N_j)_i = (N_{j-1}/N_j) \cap (G_i/N_j)$; thus if $i \leq \nu_k - 1$, then $(N_{j-1}/N_j)_i = N_{j-1}/N_j$ has order p, while $(N_{j-1}/N_j)_i = 0$ if $i \geq \nu_k$. So

$$t_1(N_{j-1}/N_j) = \nu_k - 1 = t_j(G/N_j).$$

□

We note [**CL IV, Proposition 11**]:

$$\nu_1 - 1 \equiv \nu_2 - 1 \equiv \ldots \equiv \nu_{r+1} - 1 \pmod{p},$$

hence

$$t_1(G) \equiv \ldots \equiv t_n(G) \pmod{p}.$$

The main result of this chapter strengthens this congruence considerably when L/K is totally ramified of degree p^n and the associated order of S is Hopf.

We will need, in this chapter and later,

(27.6) PROPOSITION [**Wy69, Theorem 22**] = [**CL IV Section 2, Ex. 3a**]. *Let L/K be a totally ramified Galois extension of local fields containing $\mathbb{Q}_p$ with Galois group G. Let $\sigma \in G$ and suppose $\sigma \in G_b$, $\sigma \notin G_{b+1}$. Let*

$$U_i = \{s \in S | s \equiv 1 \pmod{\omega^i S}\}$$

and let $f : L \setminus \{0\} \to L$ by $f(x) = \sigma(x)/x$. Then $f(U_i) \subseteq U_{i+b}$. If p does not divide i, then f induces an isomorphism from U_i/U_{i+1} onto U_{i+b}/U_{i+1+b}.

The group U_1 is the group of principal units of S, and the U_i form a filtration of U_1, such that for all $i \geq 1$, U_i/U_{i+1} is isomorphic to the additive group $S/\omega S$ via the map $1 + a\omega^i \mapsto a + \omega S$.

PROOF. Recall that if $b \geq 1$ and $\sigma \in G_b, \notin G_{b+1}$, then $\sigma(\omega) = \omega + u\omega^{b+1}$ for some unit u of S, and for all $s \in S$, $\sigma(s) - s \in \omega^{b+1}S$.

Let $x \in U_i \setminus U_{i+1}$, then $x = 1 + v\omega^i$ for v a unit of S. We consider $f(x) = \frac{\sigma(x)}{x}$ modulo $\omega^{i+b+1}S$. We have

$$\sigma(x) = 1 + \sigma(v)\sigma(\omega)^i$$

where

$$\sigma(v) = v + t\omega^{b+1}$$

for some $t \in S$, and

$$\sigma(\omega) = \omega + u\omega^{b+1}$$

for u a unit of S. Hence

$$\begin{aligned}\sigma(x) &= 1 + (v + t\omega^{b+1})(\omega + u\omega^{b+1})^i \\ &= 1 + v(\omega + u\omega^{b+1})^i + t\omega^{b+1}(\omega + u\omega^{b+1})^i.\end{aligned}$$

The last term is in $\omega^{i+b+1}S$. Hence, modulo $\omega^{i+b+1}S$,

$$\sigma(x) = 1 + v\omega^i + vui\omega^{b+i}$$

and so

$$\begin{aligned}f(x) = \frac{\sigma(x)}{x} &\equiv (1 + v\omega^i)(1 + v\omega^i)^{-1} + vui\omega^{b+i}(1 + v\omega^i)^{-1} \\ &\equiv 1 + vui\omega^{b+i} \pmod{\omega^{i+b+1}}\end{aligned}$$

hence $f(x) \in U_{i+b}$.

If p does not divide i, then since u and v are units, f induces a 1-1 map $\overline{f}$ from U_i/U_{i+1} to U_{i+b}/U_{i+b+1}. If $L/\mathbb{Q}_p$ is finite, then U_i/U_{i+1} and U_{i+b}/U_{i+b+1} are isomorphic finite sets, so $\overline{f}$ is onto. More directly, let $y = 1 + u_1\omega^{i+b}$ be in the range of $\overline{f}$, where u_1 is a unit of S. We have $\sigma(\omega) = \omega + u\omega^{b+1}$. Since p does not divide i, we can find a unit v of S so that $vui \equiv u_1 \pmod{\omega S}$. Then, setting $x = 1 + v\omega^i$ and repeating the calculation above shows that

$$f(x) \equiv y \pmod{\omega^{i+b+1}S}$$

and so the induced map from U_i/U_{i+1} to U_{i+b}/U_{i+b+1} is onto. (Note that if p divides i then $f(U_i) \subset U_{i+b+1}$.) □

(27.7) COROLLARY. *If $ord_L(x) = i$ and $\sigma \in G_b \setminus G_{b+1}$, then $ord_L(\sigma(x) - x) \geq i + b$. If p does not divide i, then $ord_L(\sigma(x) - x) = i + b$.*

PROOF. Let $v = 1 + x$. Then $v \in U_i \setminus U_{i+1}$, so $\frac{\sigma(v)}{v} \in U_{i+b}$, and $\notin U_{i+b+1}$ if p does not divide i. Now

$$\begin{aligned}\frac{\sigma(v)}{v} &= \frac{1 + \sigma(x)}{1 + x} \\ &= 1 + \frac{\sigma(x) - x}{1 + x} \\ &= 1 + (\sigma(x) - x)(1 - x + x^2 - \dots) \\ &\equiv 1 + (\sigma(x) - x) \pmod{x(\sigma(x) - x)}.\end{aligned}$$

So $ord_L(\sigma(x) - x) \geq i + b$, and $= i + b$ if p does not divide i. □

We need in (32.11) the following refinement of (27.6), due to Greither [**Gr92, II, 3.6**]:

(27.8) PROPOSITION. *Let L/K be a totally ramified Galois extension of order p, with Galois group $G = \langle\sigma\rangle$ and ramification number t. If $N_{L/K}(\alpha) = 1$ and $\alpha \in U_b(S) \setminus U_{b+1}(S)$ where $b > t$ and p does not divide $b - t$, then there exists $\gamma \in U_{b-t}(S)$ with $\frac{\sigma(\gamma)}{\gamma} = \alpha$.*

PROOF. Set $D : L^* \to L^*$ by $D(\gamma) = \frac{\sigma(\gamma)}{\gamma}$. If π is a parameter for R and ω is a parameter for S, let $\gamma_0 \in U(S)$ so that $\omega^p = \gamma_0\pi$.

By Hilbert's Theorem 90, if $N_{L/K}(\alpha) = 1$, then $\alpha = D(\delta)$ for some $\delta \in L^*$. We use an argument similar to the normalization lemma (24.2) for generators of Kummer extensions to show that if $\alpha = D(\delta)$ for some $\delta \in L^*$ and $\alpha \in U_b(S) \setminus U_{b+1}(S)$ where $b > t$, then $\alpha = D(\gamma)$ with $\gamma \in U_{b-t}(S)$. The idea is that since $D(a) = 1$ for any $a \in K$, we are free to multiply δ by any element of K.

First, write $\delta = \gamma_1\omega^\ell$ for some $\gamma_1 \in U(S)$. If $\ell = pq + r$ with $0 \leq r < p$, then $\delta = \gamma_1\gamma_0^q\pi^q\omega^r$. So $\alpha = D(\gamma\omega^r)$ where $\gamma = \gamma_1\gamma_0^q \in U(S)$ and $0 \leq r < p$.

Now we show that in fact r must equal 0. To see this, note that

$$\sigma(\omega) - \omega = \gamma_2\omega^{t+1}$$

for some $\gamma_2 \in U(S)$, and so

$$ord_S(\frac{\sigma(\omega)}{\omega} - 1) = t$$

and also

$$ord_S(\frac{\sigma(\omega^r)}{\omega^r} - 1) = t$$

for $1 \leq r < p$. On the other hand, for $\gamma \in U(S)$,

$$ord_S(\sigma(\gamma) - \gamma) \geq t + 1,$$

so

$$ord_S(\frac{\sigma(\gamma)}{\gamma} - 1) \geq t + 1.$$

If $ord_S(\alpha - 1) = b > t$, then

$$D(\omega^r) = \alpha D(\gamma^{-1}) \in U_{t+1}(S),$$

impossible unless $r = 0$. Thus $\alpha = D(\gamma)$ for some $\gamma \in U(S)$.

Now $\gamma \equiv c \pmod{\omega S}$ for some $c \in U(R)$ since L/K is totally ramified. So replacing γ by $c^{-1}\gamma$, we may assume $\gamma \in U_1(S)$.

Suppose $ord_S(\gamma - 1) = pk$ for some k, so that

$$\gamma = 1 + \gamma_1\omega^{pk} = 1 + \gamma_1\gamma_0^{pk}\pi^k$$

with $\gamma_1 \in U(S)$. Letting

$$\gamma_1\gamma_0^{pk} \equiv d \pmod{\omega S}$$

with $d \in U(R)$, then $a = 1 - d\pi^k \in U(R)$ and

$$\begin{aligned}\gamma a &= (1 + \gamma_1 \gamma_0^{pk} \pi^k)(1 - d\pi^k) \\ &= 1 + \gamma_2 \omega \pi^k,\end{aligned}$$

hence $ord_S(\gamma a - 1) > ord_S(\gamma - 1)$. Repeat this argument if p divides $ord_S(\gamma a - 1)$. Note that for any unit $\delta \in U(S)$, if $D(\delta) = \alpha$, then $ord_S(\delta - 1) \leq b$. For $ord_S(\delta - 1) = ord_S(\sigma(\delta) - 1)$: apply the isosceles triangle inequality to

$$(\sigma(\delta) - 1) + (1 - \delta) = \delta(\alpha - 1).$$

Thus eventually one obtains some $\gamma \in U(S)$ with $ord_S(\gamma - 1)$ prime to p and $D(\gamma) = \alpha$.

Now repeat the proof of (27.6): if $ord_S(\gamma - 1) = i$ for i prime to p, then

$$ord_S(\frac{\sigma(\gamma)}{\gamma} - 1) = i + t.$$

Since $\frac{\sigma(\gamma)}{\gamma} = \alpha$, $ord_S(\alpha - 1) = i + t = b$, hence $i = b - t$, as claimed. □

§**28. Tameness.** Suppose $L \supset K$ is Galois with group G, N is a normal subgroup of G and $L_1 = L^N$. In this section we prove that if S is H-tame over R for some Hopf order over R in KG, $H_1 = H \cap KN$ and $\overline{H} = H//H_1$, then S is $S_1 \otimes_R H_1$-tame over S_1 and S_1 is $\overline{H}$-tame over R. This will yield congruence conditions on the ramification numbers of a totally ramified p-extension.

(28.1) THEOREM. *Let S be the valuation ring of a Galois extension L/K of local fields with Galois group G. Let H be a Hopf order over R in KG such that S is H-tame over R, the valuation ring of K. Let N be a normal subgroup of G, let $H_1 = H \cap K[N]$, $\overline{H} = H//H_1$, and $T = S^N$. Then*

1. *S is $T \otimes_R H_1$-tame over T*
2. *T is $\overline{H}$-tame over R.*

PROOF. $T = I(H_1)S = I(T \otimes_R H_1)S$ by (5.8). Since S has rank $|N|$ as a projective T- module and S is a faithful $T \otimes_R H_1$-module, S is $T \otimes_R H_1$-tame.

Let $(\frac{1}{n})\Sigma_N$, $\frac{1}{b}\Sigma_G$ and $\frac{1}{m}\Sigma_{\overline{G}}$ generate the integrals of H_1, H and $\overline{H}$, respectively. We may assume $mn = b$ by (5.6). Since S is H-tame over R, there is some $z \in S$ so that $\frac{1}{b}\Sigma_G z = 1$ in R. Let $w = \frac{1}{n}\Sigma_N z$, then, since $I(H_1) = (\frac{1}{n})\Sigma_N R$, $w \in T$; also since $\Sigma_{\overline{G}}\Sigma_N(z) = \sum_\tau \tau(\sum_\eta \eta(z)) = \sum_{\sigma \in G} \sigma(z) = \Sigma_G z$, we have

$$\frac{1}{m}\Sigma_{\overline{G}}(w) = \frac{1}{m}\Sigma_{\overline{G}}\frac{1}{n}\Sigma_N z = \frac{1}{mn}\Sigma_G z = \frac{1}{b}\Sigma_G z = 1.$$

Hence T is $\overline{H}$-tame over R. □

The following corollary is immediate:

(28.2) COROLLARY. *Suppose L/K is Galois and totally ramified with Galois group G of order p^n. Let*

$$L = L_n \supset L_{n-1} \supset \ldots \supset L_1 \supset L_0 = K$$

correspond to a normal series

$$G = N_0 \supset N_1 \supset \ldots \supset N_{n-1} \supset N_n = (0)$$

for G which refines the chain of ramification groups of G. Suppose $S_n/S_0 = S/R$ is H-tame for some Hopf algebra order H in KG. Let $H_r = H \cap KN_r$. Then S_{r+1}/S_r is tame for $S_r \otimes H_r//S_r \otimes H_{r+1} \cong S_r \otimes (H_r//H_{r+1})$.

(28.3) COROLLARY. *With L/K as in (28.2), for each r, the rth ramification number $t_r(G)$ satisfies either $t_r(G) = -1 + b_r p^r$ with $1 \leq b_r \leq e'$, or $t_r(G) = p^r e'$.*

PROOF. The ramification numbers of G satisfy $t_r(G) = t_r(L/K) = t(L_r/L_{r-1}) = t(N_{r-1}/N_r)$. Since L_r/L_{r-1} is tame and totally ramified, $S_{r-1} \otimes (H_{r-1}//H_r)$ is isomorphic to the Larson order

$$S_{r-1} \otimes A_{\pi^{b_r}} = S_{r-1}[\frac{\sigma_r - 1}{a}],$$

where $a = \omega_{r-1}^{b_r p^{r-1}}$, σ_r generates N_r/N_{r+1} and $b_r \leq e'$. Then (24.15) applied to L_r/L_{r-1} yields $t_r(L/K) = t(N_{r-1}/N_r) = -1 + b_r p^r$ or $= p(p^{r-1}e')$. □

The last possibility of (28.3) either occurs for no r or for all r: by [**CL, IV, Prop. 11**], all ramification numbers of G are congruent modulo p. Thus

COROLLARY (28.4). *With L/K as in (28.2), one of the following cases holds:*
1. *$t_r = p^r e'$ for all r*
2. *$t_1(L/K) = -1 + b_1 p$ with $1 \leq b_1 \leq e'$, in which case $t_r(L/K) = -1 + b_r p^r$ with $1 \leq b_r \leq e'$ for all r.*

We will discuss the second case in detail below.

(28.5). If p is odd and $t_n(G) = -1 + e'p^n$, then $t_r(G) = -1 + e'p^r$ for all r, by a result of Vostokov [**V078, Proposition 1**].

§**29. When does tame imply Galois.** In this section we apply Waterhouse's theorem (13.7) to give an alternate proof of a theorem of Byott which answers the question. Waterhouse's theorem applies when the Hopf algebra is local, so we begin by studying localness.

A not necessarily commutative ring H is *local* if it has a unique maximal left ideal. Then H local means that the radical $rad(H)$ is the maximal left ideal, and, since $rad(H)$ is two-sided, also a maximal right ideal.

Continue to assume that R is the valuation ring of a local field K.

(29.1) PROPOSITION. *Let $R \to H_1 \to H \to \overline{H} \to R$ be a short exact sequence of finite R-Hopf algebras. Then H_1 and $\overline{H}$ are local rings iff H is a local ring.*

PROOF. It suffices to prove it modulo the maximal ideal of R, and so we can assume that R is a field. Then the radicals of H, H_1 and $\overline{H}$ are nilpotent ideals.

Suppose H_1 and $\overline{H}$ are local. Then $rad(H_1) = ker(\varepsilon_1)$ and $rad(\overline{H}) = ker(\overline{\varepsilon})$. Let $x \in ker(\varepsilon)$. Then $\pi(x) \in ker(\overline{\varepsilon})$, so $\pi(x^n) = 0$ for some n, hence $x^n \in H_1^+ H$, a two-sided ideal of H. Since $H_1^+ = ker(\varepsilon_1)$ is nilpotent, x^n is nilpotent in H, so $x \in rad(H)$. Thus $ker(\varepsilon) \subset rad(H)$. But $ker(\varepsilon)$ is a maximal left ideal since $H/ker(\varepsilon) = R$, a field, so $ker(\varepsilon) = rad(H)$ and H is local.

Conversely, if $ker(\varepsilon) \subset H$ is nilpotent, then $ker(\overline{\varepsilon})$ and $ker(\varepsilon_1)$ are nilpotent, so if H is local, so are $\overline{H}$ and H_1. □

We now have the following result of Byott ([**By95c, Theorem 5**]) for the last case of (28.6) :

(29.2) THEOREM. *Let L/K be a totally ramified Galois extension of odd order p^n with Galois group G. Suppose the valuation ring S of L is a tame H-extension of R for some Hopf order $H \subset KG$. Suppose the first ramification number $t_1(L/K) = -1 + b_1 p$ with $b_1 < e'$. Then S is an H-Galois extension of R.*

PROOF. By Waterhouse's theorem, it suffices to show that H is local. Let

$$G = N_0 \supset N_1 \supset \ldots \supset N_n = 1$$

be a normal series for G which refines the chain of ramification groups of G, and let

$$K = L_0 \subset L_1 \subset \ldots \subset L_n = L$$

be the corresponding chain of fields. Then for all r,

$$t_r(G) = t(N_{r-1}/N_r) = -1 + b_r p^r$$

with $b_r < e'$. Hence S_r/S_{r-1} is Hopf Galois for the Hopf algebra $S_{r-1} \otimes (H_{r-1}//H_r)$ which is a local Larson order by (21.3).

Since each layer $H_{r-1}//H_r$ of H is local, it follows easily by (29.1) that H is local. □

Note that for $n = 1$, if $t = -1 + pe'$ then H is not local, as noted in (21.3).

§ **30. Ramification conditions.** Now we show Byott's result [**By97c, Theorem 4.4**] that if L/K is a totally ramified Galois extension of local fields containing $\mathbb{Q}_p$, with Galois group G of order p^n, the first ramification number $t_1(L/K) = -1 + b_1 p$ with $b_1 < e'$, and the associated order $\mathfrak{A} = \mathfrak{A}_{S/R}$ is a Hopf order in KG, then all ramification numbers $t_r = t_r(G)$ are congruent to -1 modulo p^n. We assume throughout this section that p is odd. For the case $p = 2$ we refer to [**By97c**].

If $\mathfrak{A}$ is Hopf and $b_1 < e'$, then $\mathfrak{A}$ is a local ring, and so S/R is $\mathfrak{A}$-Galois, by (29.2).

Here is an outline of the proof that all t_i are congruent to -1 modulo p^n, not just p^i.

Suppose $\mathfrak{A}_S$ is Hopf and S/R is $\mathfrak{A}_S$-Hopf Galois. Then S is $\mathfrak{A}_S$-free. Let $\mathcal{C}$ be the inverse different of S/R, $\mathcal{C} = \{y \in L | tr(yS) \subset R\}$, and let $\mathcal{D} = \mathcal{C}^{-1}$, the different of S/R. Then $\mathcal{C} = \omega^\delta S$ where δ is given in terms of $t_1, \ldots, t_n$. All $t_r = -1 + b_r p^r$ with $b_r < e'$. Then $\delta = p^n d$ for some d, so $\mathcal{C} = \pi^d S$, hence $\mathcal{C} \cong S$ as RG-modules. Then $\mathfrak{A}_S = \mathfrak{A}_\mathcal{C}$ and $\mathcal{C} \cong S$ as $\mathfrak{A}_S$-modules. Hence $\mathcal{C}$ is $\mathfrak{A}_\mathcal{C}$-free. The argument then concludes by showing that if $\mathcal{C}$ is $\mathfrak{A}_\mathcal{C}$-free, then $t_i \equiv -1 \pmod{p^n}$ for all i.

We begin.

(30.1) PROPOSITION. *Suppose $\mathfrak{A}_S$ is Hopf and $t_1(G) \equiv -1 \pmod{p}$. Then $\mathcal{D} = \pi^c S$ for some integer c.*

PROOF. From [**CL, p. 72**]. Let L/K be Galois with Galois group G. Then the different $\mathcal{D}$ satisfies the formula

$$\mathcal{D} = \omega^{\delta} S$$

where ω is a parameter of S and

$$\delta = \sum_{j=0}^{\infty} (|G_j| - 1).$$

Recall that the ramification groups are defined in terms of the numbers ν_i as follows:

$$\begin{aligned} G = G_0 = G_1 = \ldots = G_{\nu_1 - 1} \\ \supset G_{\nu_1} = \ldots = G_{\nu_2 - 1} \\ \supset G_{\nu_2} = \ldots = G_{\nu_3 - 1} \\ \ldots \\ \supset G_{\nu_k} = \ldots = G_{\nu_{k+1} - 1} \\ \ldots \\ \supset G_{\nu_{r+1}} = 1 \ . \end{aligned}$$

Thus

$$\begin{aligned} w &= \nu_1(p^n - 1) + (\nu_2 - \nu_1)(p^{n-i_1} - 1) + \ldots + (\nu_{r+1} - \nu_r)(p^{n-i_r} - 1) \\ &= \nu_1 p^n + (\nu_2 - \nu_1)p^{n-i_1} + \ldots + (\nu_{r+1} - \nu_r)p^{n-i_r} - \nu_{r+1}. \end{aligned}$$

Also, $t_j(G) = \nu_k - 1$ if $i_{k-1} < j \leq i_k$, and $t_j(G) \equiv -1 \pmod{p^j}$ by (28.4). So

$$t_{i_k}(G) = \nu_k - 1 \equiv -1 \pmod{p^{i_k}}.$$

and so p^{i_k} divides ν_k, hence $\nu_{k+1} - \nu_k$; also $p^n = p^{i_{r+1}}$ divides ν_{r+1}. Hence δ is divisible by p^n, $\delta = p^n c$ for some $c > 0$. Now $\pi S = \omega^{p^n} S$, hence $\mathcal{D} = \omega^{p^n c} S = \pi^c S$. □

(30.2) COROLLARY. *If $\mathfrak{A}_S$ is Hopf and $t_1(G) \equiv -1 \pmod{p}$, then $\mathfrak{A}_S = \mathfrak{A}_{\mathcal{C}}$ and $\mathcal{C}$ is $\mathfrak{A}_{\mathcal{C}}$-free.*

PROOF. Since $\mathcal{D} = \pi^c S$, $\mathcal{C} = \pi^{-c} S$. Then $\mathfrak{A}_S = \mathfrak{A}_{\mathcal{C}}$, for if $\alpha \in KG$, then $\alpha \in \mathfrak{A}_S$ iff for all s in S, $\alpha s \in S$, iff $\alpha(\pi^{-c}s) = \pi^{-c}\alpha s \in \pi^{-c}S = \mathcal{C}$ iff $\alpha \in \mathfrak{A}_{\mathcal{C}}$. Setting $\mathfrak{A}_S = \mathfrak{A}_{\mathcal{C}} = \mathfrak{A}$, if $S = \mathfrak{A}x$ is $\mathfrak{A}$-free, then $\mathcal{C} = \mathfrak{A}\pi^{-c}x$ is $\mathfrak{A}$-free. □

We are reduced to showing:

If $t_1(G) = -1 + b_1 p$ with $b_1 < e'$ and $\mathcal{C}$ is $\mathfrak{A}$-free, then $t_i \equiv -1 \pmod{p^n}$ for $i = 1, \ldots, n$.

Byott's approach to this result is to obtain and apply a criterion for a module (like $\mathcal{C}$) over a local ring (like $\mathfrak{A}$) to not be free.

(30.3) PROPOSITION. *Let A be a local ring, M a finitely generated A-module. If there is a generating set $\{m_1, \ldots, m_n\}$ for M as $\mathfrak{A}$-module so that $\mathfrak{A}m_i \neq M$ for all i, then M is not $\mathfrak{A}$-free.*

PROOF. If M is $\mathfrak{A}$-free, then $M = \mathfrak{A}m$ for some m in M. Then $m_i = \alpha_i m$ for some $\alpha_i \in \mathfrak{A}$. For all i, since $\mathfrak{A}m_i \neq M$, α_i is not a unit of $\mathfrak{A}$, hence (since $\mathfrak{A}$ is local) α_i is in J, the radical of $\mathfrak{A}$ [**CR81, (5.21)**]. But then

$$M = \sum \mathfrak{A}m_i = \sum \mathfrak{A}\alpha_i m \subset JM$$

so by Nakayama's Lemma [**CR81, (5.7)**] $M = 0$, a contradiction. □

Let A be a K-algebra (e.g. KG) which is finite dimensional as a K-module, let V be a free left A-module of rank one, let M be a lattice in V (i. e. a finitely generated R-submodule of V such that $KM = V$) and $\mathfrak{A} = \{\alpha \in A | \alpha M \subset M\}$. (The application is to $M = \mathcal{C}, V = L$). Suppose $\mathfrak{A}$ is local. Recall that π is a parameter for R, the valuation ring of $K \supset \mathbb{Q}_p$. To use (30.3) we need a sufficient condition for an element of M to not be a generator of M as a free $\mathfrak{A}$-module:

(30.4) PROPOSITION. *With $A, V, M, \mathfrak{A}$ as above, given $n \in M$, suppose there exists some $\alpha \in A$ so that $\alpha n \in \pi M$ and $\alpha n' \notin \pi M$ for some $n' \in M$. Then $\mathfrak{A}n \neq M$.*

PROOF. Consider the map $\phi : \mathfrak{A} \to M$ by $\phi(\gamma) = \gamma n$. The canonical quotient maps yield the commutative diagram:

$$\begin{array}{ccc} \mathfrak{A} & \xrightarrow{\phi} & M \\ \downarrow & & \downarrow \\ \mathfrak{A}/\pi\mathfrak{A} & \xrightarrow[\overline{\phi}]{} & M/\pi M \end{array}$$

where $\overline{\phi}$ is the induced map on $\mathfrak{A}/\pi\mathfrak{A}$.

If ϕ is onto, then $\overline{\phi}$ is onto. Since $\mathfrak{A}$ and M are free of equal ranks over R, $\mathfrak{A}/\pi\mathfrak{A}$ and $M/\pi M$ are vector spaces over $R/\pi R$ of equal dimensions. Thus $\overline{\phi}$ is onto iff $\overline{\phi}$ is 1-1. If there is some α in A so that $\alpha n \in \pi M$ but $\alpha M \not\subset \pi M$, then $\alpha \notin \pi\mathfrak{A}$. Let $j \geq 0$ be minimal so that $\beta = \pi^j \alpha \in \mathfrak{A}$. Then $\overline{\beta} \neq \overline{0}$ in $\mathfrak{A}/\pi\mathfrak{A}$, but $\beta n \in \pi M$ so $\overline{\phi}(\overline{\beta}) = \overline{0}$. Hence $\overline{\phi}$ is not 1-1 and ϕ is not onto. □

Using these ideas we prove

(30.5) THEOREM. *Let L/K be a totally ramified Galois extension of order p^n with Galois group G and ramification numbers $t_1, \ldots, t_n$ such that $t_1(L/K) = -1 + b_1 p$ with $b_1 < e'$. If $t_i \not\equiv 1 \pmod{p^n}$ for some i, then $\mathcal{C}$ is not free over $\mathfrak{A}$, hence $\mathfrak{A}$ is not a Hopf order.*

PROOF. Assume, to the contrary, that $\mathfrak{A}$ is Hopf. Then, since $t_1 = -1 + b_1 p$ with $b_1 < e'$, $\mathfrak{A}$ is local. Suppose $t_i \not\equiv -1 \pmod{p^n}$ for some i: write $t_i = p^n b + r$. Then $p - 1 \leq r \leq p^n - 2$.

We shall prove that $\mathcal{C}$ is not free over the local ring $\mathfrak{A}$ using (30.3): we'll find a generating set $\{y_1, \ldots, y_n\}$ for $\mathcal{C}$ as an $\mathfrak{A}$-module with the property that $\mathfrak{A}y_i \neq \mathcal{C}$ for all i. By (30.3), that will show $\mathcal{C}$ is not $\mathfrak{A}$-free.

We choose the generating set to be an R-basis of $\mathcal{C}$. The first element of the basis, x, is any element of $\pi\omega^{-1}\mathcal{C}$ so that $tr(x) = z$ is a unit of R, where ω is a parameter for S. To see that such an x exists, recall that

$$\mathcal{C} = \{y \in L | tr(yS) \subset R\},$$

and so for all $y \in \mathcal{C}$, $tr(y) = tr(y \cdot 1) \in R$, i. e. $tr(\mathcal{C}) \subset R$. On the other hand, $tr(\omega^{-1}\mathcal{C}) \not\subset R$. For suppose for all $y \in \omega^{-1}\mathcal{C}$, $tr(y) \in R$. Then for any $s \in S$, $sy \in \omega^{-1}\mathcal{C}$ for all $y \in \omega^{-1}\mathcal{C}$, so $tr(sy) \in R$, hence $y \in \mathcal{C}$. But then $\omega^{-1}\mathcal{C} = \mathcal{C}$, impossible.

Now since $tr(\mathcal{C}) \subset R, tr(\pi\mathcal{C}) \subset \pi R$, while since $tr(\omega^{-1}\mathcal{C}) \not\subseteq R$, $tr(\pi\omega^{-1}\mathcal{C}) \not\subseteq \pi R$. Let $x \in \pi\omega^{-1}\mathcal{C}$ with $tr(x) = z \notin \pi R$. Since $\pi\omega^{-1}\mathcal{C} \subset \mathcal{C}, z \in R$, so z is a unit of R.

Since $x \notin \pi\mathcal{C}, \overline{x} \neq \overline{0}$ in $\mathcal{C}/\pi\mathcal{C}$. Therefore we can use $x = x_1$ as part of an R-basis $\{x_1, \ldots, x_n\}$ of $\mathcal{C}$.

Now choose a new R-basis $\{y_1, \ldots, y_n\}$ of $\mathcal{C}$ as follows: $y_1 = x_1 = x$ and for $2 \leq j \leq n$, $y_j = x_j - z^{-1}tr(x_j)x_1$. This is obviously an R-basis of $\mathcal{C}$, hence a set of generators for $\mathcal{C}$ as an $\mathfrak{A}$-module. Note that $tr(y_j) = 0$ for $j \geq 2$.

We show that $\mathcal{C} \neq \mathfrak{A}y_j$ for all j.

The choice of $y_2, \ldots, y_n$ makes showing $\mathcal{C} \neq \mathfrak{A}y_j$ easy. For if $\mathcal{C} = \mathfrak{A}y_j$ then L is free over KG with generator y_j. But

$$(\sum_{\sigma \in G} \sigma)y_j = tr(y_j) = 0,$$

hence $L \neq KGy_j$ for $j = 2, \ldots, n$.

We're left with showing that $\mathfrak{A}y_1 = \mathfrak{A}x \neq \mathcal{C}$.

To do this, we apply (30.4): we find $\alpha \in A$ so that $\alpha x \in \pi\mathcal{C}$ but there is some $y \in \mathcal{C}$ with $\alpha y \notin \pi\mathcal{C}$.

We have $t_i = p^n b + r$ with $p-1 \leq r \leq p^n - 2$. We set $\alpha = \frac{\rho-1}{\pi^b}$ where $\rho \in G_{t_i}, \rho \notin G_{t_i+1}$. By (27.7), $ord_L((\rho-1)x) \geq ord_L(x) + t_i$, and $ord_L((\rho-1)x) = ord_L(x) + t_i$, if p does not divide $ord_L(x)$. Using this, we show that $\alpha x \in \pi\mathcal{C}$.

We have x in $\pi\omega^{-1}\mathcal{C}$ but not in $\pi\mathcal{C}$. If $\mathcal{C} = \omega^{-\delta}S$ then $ord_L(x) = p^n - 1 - \delta$. So

$$\begin{aligned}
ord_L(\alpha x) &= ord_L((\frac{\rho - 1}{\pi^b})x) \\
&\geq ord_L(x) + t_i - p^n b \\
&= p^n - 1 - \delta - r \\
&\geq p^n - \delta = ord_L(\pi\mathcal{C})
\end{aligned}$$

since $r \geq 1$. Hence $\alpha x \in \pi\mathcal{C}$.

On the other hand, choose $y \in \mathcal{C}$ with $ord_L(y) = -\delta + 1$. Then, since $\delta = p^n c$ by (30.1), p does not divide $ord_L(y)$, so by (27.7),

$$\begin{aligned}
ord_L(\alpha y) &= ord_L((\frac{\rho - 1}{\pi^b})y) \\
&= ord_L(y) + t_i - p^n b \\
&= ord_L(y) + r
\end{aligned}$$

$$\leq (1-\delta) + p^n - 2$$
$$= p^n - \delta - 1$$
$$< p^n - \delta = ord_L(\pi\mathcal{C}).$$

So $\alpha y \notin \pi\mathcal{C}$. By Proposition (30.4), $\mathfrak{A}y = \mathfrak{A}y_1 \neq \mathcal{C}$. Hence $\mathcal{C}$ is not a free $\mathfrak{A}$-module. □

(30.6) COROLLARY. *If $\mathfrak{A}_S$ is an Hopf order in KG, G totally ramified of order p^n, and $t_1(G) = -1 + b_1 p$ with $b_1 < e'$, then S is $\mathfrak{A}_S$-Galois and the ramification numbers t_i are all $\equiv -1 \pmod{p^n}$.*

PROOF. If $t_1(G) \equiv -1 \pmod p$, then $\mathfrak{A} = \mathfrak{A}_S$ is local. If $\mathfrak{A}$ is also Hopf, then S is $\mathfrak{A}$-Galois, so $\mathcal{C} \cong S$ as RG-modules, hence $\mathfrak{A} = \mathfrak{A}_\mathcal{C}$ and, since S is $\mathfrak{A}$-free, $\mathcal{C}$ is $\mathfrak{A}$-free. Since $\mathcal{C}$ is $\mathfrak{A}$-free, all $t_i \equiv -1 \pmod{p^n}$. □

We conclude this chapter by noting the "exceptional" cases: L/K totally ramified and Galois with group G of order p^n, $\mathfrak{A}(S/R)$ a Hopf order in KG and $t_1(G)$ either $= -1 + e'p$ or $= e'p$.

(30.7) PROPOSITION [**By97, Theorem 4.4**]. *Let L/K be totally ramified and Galois with group G of order p^n. If $t_1(G) \equiv 0 \pmod p$, then G is cyclic, S is not an $\mathfrak{A}$-Galois extension of R, and $\mathfrak{A}$ is the maximal order in KG. If $t_1 = -1 + e'p$ then S is an $\mathfrak{A}$-Galois extension of R and $t_r \equiv -1 \pmod{p^n}$, but $\mathfrak{A}$ is not local.*

PROOF. Let $G = N_0 \supset N_1 \supset \ldots \supset N_n = (1)$ be a normal series which refines the series of ramification groups of G. Let $L_1 = L^{N_1}$, S_1 the valuation ring of L_1. Then $t_1(G) = t(G/N_1)$. Since L/K is $\mathfrak{A}$-tame, L_1/K is $\overline{\mathfrak{A}}$- tame where $\overline{\mathfrak{A}}$ is the image of $\mathfrak{A}$ in $K[G/N_1]$. If $t_1(G) \equiv 0 \pmod p$, then S_1 is not an $\overline{\mathfrak{A}}$-Galois extension of R, and hence S is not an $\mathfrak{A}$-Galois extension of R. In that case, by [**CL IV, Sec. 2, Ex. 3e**], G is cyclic of order p^n. If $t_1 = -1 + e'p$ then $\overline{\mathfrak{A}}$ is not local by the remark following (21.3), hence A is not local by (29.1). However, each layer S_i/S_{i-1} is tame and has $t \equiv -1 \pmod p$, hence is Galois. By (22.18), S is an $\mathfrak{A}$-Galois extension of R. □

Byott shows in this case that $\mathfrak{A}$ is the maximal order in KG.

CHAPTER 9

Hopf Algebras of Rank p^2

In Chapter 6 we classified R-Hopf algebras of rank p. In this chapter we construct a collection of Hopf orders inside KC_{p^2} and in Chapter 10 we determine which of them are realizable, that is, arise as the associated orders of valuations rings of local number fields.

§**31. Hopf orders in** KC_{p^2}**.** Let K be a finite field extension of $\mathbb{Q}_p$; let e be the absolute ramification index of $K/\mathbb{Q}_p$. Assume that K contains a primitive pth root of unity ζ, from which it follows that $e' = e/(p-1)$ is an integer. For $0 < i < e'$ let $i' = e' - i$. Let R be the valuation ring of K, and π a parameter for R.

We begin the construction of Hopf orders with an adaptation of [**GC, Lemma 2.1**].

Let G be an abelian p-group. Let G' be a subgroup of index p, so that $G = G' \cdot \langle\sigma\rangle$, where $\sigma^p \in G'$.

(31.1) PROPOSITION. *Let A be an order over R in KG'. Let $y = \frac{u\sigma - 1}{\pi^i} \in KG'\langle\sigma\rangle$, where $0 \leq i \leq e'$ and u is some element of KG', and let $B = A[y]$. Then B is A-free on 1, $y, y^2, \dots, y^{p-1}$ iff $1 - u^p\sigma^p \in \pi^{pi}A$*

PROOF. Evidently $1, y, y^2, \dots, y^{p-1}$ are linearly independent over KG', so form a basis over A iff y^p is an A-linear combination of $1, y, \dots, y^{p-1}$.

Now $u\sigma = 1 + \pi^i y$, so y is a root of

$$\begin{aligned} q(Y) &= \frac{1}{\pi^{pi}}((\pi^i Y + 1)^p - u^p\sigma^p) \\ &= Y^p + \sum_{j=1}^{p-1} \binom{p}{j} \frac{\pi^{ji}Y^j}{\pi^{pi}} + \frac{1 - u^p\sigma^p}{\pi^{pi}} ; \end{aligned}$$

setting $Y = y$ gives y^p uniquely as a KG'-linear combination of $1, y, \dots, y^{p-1}$. So B is A-free on $\{1, y, \dots, y^{p-1}\}$ iff all coefficients of $q(Y)$ are in A. But $i \leq e'$, so for $j = 1, \dots, p-1$, $(p-j)i \leq e$, hence $pi \leq e + ji$. Thus $q(Y)$ has coefficients in A iff $\frac{1-u^p\sigma^p}{\pi^{pi}} \in A$. □

(31.2) PROPOSITION. *Let A be a Hopf order over R in KG', $y = \frac{u\sigma-1}{\pi^i}$ and $B = A[y]$. If u is a unit of A, then B is a subcoalgebra of KG iff*

$$\Delta(u) \equiv u \otimes u \pmod{\pi^i(A \otimes A)}.$$

Thus for $u \in KG'$, B is a Hopf order in KG iff $u^p\sigma^p - 1 \in \pi^{pi}A$ and $\Delta(u) \equiv u \otimes u$ $\pmod{\pi^i(A \otimes A)}$.

PROOF. If B is a subcoalgebra then $\Delta(y) \in B \otimes B$. Conversely, if we show that $\Delta(y) \in B \otimes B$, then, since B is generated as an A-algebra by y and Δ is an R- algebra homomorphism, it will follow that $\Delta(B) \subset B \otimes B$.

Now

$$\begin{aligned}
\Delta(y) &= \Delta(\frac{u\sigma - 1}{\pi^i}) \\
&= \frac{1}{\pi^i}(\Delta(u)(\sigma \otimes \sigma) - 1 \otimes 1) \\
&= \frac{1}{\pi^i}[(\Delta(u) - u \otimes u)(\sigma \otimes \sigma) + u\sigma \otimes u\sigma - 1 \otimes 1] \\
&= (\frac{\Delta(u) - u \otimes u}{\pi^i})(\sigma \otimes \sigma) + \frac{u\sigma - 1}{\pi^i} \otimes u\sigma + 1 \otimes \frac{u\sigma - 1}{\pi^i} \\
&= (\frac{\Delta(u) - u \otimes u}{\pi^i})(\sigma \otimes \sigma) + y \otimes u\sigma + 1 \otimes y \ .
\end{aligned}$$

Since u is a unit of A and $\sigma \otimes \sigma$ is a unit of $B \otimes B$, $\Delta(y) \in B \otimes B$ iff

$$\frac{\Delta(u) - u \otimes u}{\pi^i} \in B \otimes B.$$

If $u^p\sigma^p - 1 \in \pi^{pi}A$, then u is a unit of A, and the last statement follows from (31.1). □

Under the canonical map from KG to KG/G', B maps onto $H_j = R[\frac{\overline{\sigma}-1}{\pi^j}]$. Thus u satisfies the conditions of (31.2) iff B lies in a short exact sequence of R-Hopf algebras (4.12)

$$R \to A \to B \to H_j \to R.$$

In that case we call B an extension of H_j by A.

We apply (31.2) to Hopf orders inside KG, $G = \langle\sigma\rangle$ cyclic of order p^2. Let $\tau = \sigma^p$, $G' = \langle\tau\rangle$.

Since K contains ζ, a primitive pth root of unity, $K\langle\tau\rangle \cong K \times \cdots \times K$ via the map $\gamma \to (\varepsilon(\gamma), \hat{\tau}(\gamma), \hat{\tau}^2(\gamma), \ldots, \hat{\tau}^{p-1}(\gamma))$ where ε is the counit map and $\hat{\tau}$ is the map from $K[\tau]$ to K induced by the character $\hat{\tau} : G' \to K$ by $\hat{\tau}(\tau^i) = \zeta^i$. Thus

$$\tau \mapsto (1, \zeta, \zeta^2, \ldots, \zeta^{p-1}) \ .$$

For $v \in K$ let a_v be the preimage in $K[\tau]$ of $(1, v, v^2, \ldots, v^{p-1})$ in $K \times \ldots \times K$. Then the map from K to $K[\tau]$ given by

$$v \mapsto (1, v, \ldots, v^{p-1}) \mapsto a_v$$

yields a homomorphism of multiplicative groups

$$a : K^\times \to (KG')^\times \ .$$

Notice that $\tau = a_\zeta$.

For the rest of this chapter, assume that $0 \leq i, j \leq e'$ and $j < pi$. We will need this last inequality in (31.10). As we shall see in Chapter 10, if an R-Hopf algebra extension E of H_j by H_i is the associated order of a cyclic extension L/K of local fields of order p^2, then the ramification numbers of L/K are $t_1 = pj - 1$ and $t_2 = p^2 i - 1$, so necessarily $j < pi$. Thus that inequality is a reasonable assumption.

Let $A = H_i = R[\frac{\sigma^p - 1}{\pi^i}]$ and $B = A[\frac{a_v\sigma - 1}{\pi^j}]$. Let $y = \frac{a_v\sigma-1}{\pi^j}$. Then from (31.1) B is a Hopf algebra extension of H_j by $A = H_i$ iff

$$1 - a_v^p\sigma^p = 1 - a_v^p\tau = 1 - a_{v^p\zeta} \in \pi^{pj}A$$

and

$$\Delta(a_v) \equiv a_v \otimes a_v \pmod{\pi^j(A \otimes A)}.$$

These are equivalent to the two conditions:

$$\zeta v^p \equiv 1 \pmod{\pi^{i'+pj}R}$$

by Proposition (31.8) below (where $i' = e' - i$), and

$$v^p \equiv 1 \pmod{\pi^{pi'+j}R}$$

by Proposition (31.10) below. That is, assuming Propositions (31.8) and (31.10), we have:

(31.3) PROPOSITION. *Let $0 \leq i, j \leq e'$ and $j < pi$. For $v \in R$,*

$$E_v = H_i[\frac{a_v\sigma - 1}{\pi^j}]$$

is a Hopf order extension of H_j by H_i in KC_{p^2} iff

$$v^p \equiv 1 \pmod{\pi^{pi'+j}R}$$

and

$$\zeta v^p \equiv 1 \pmod{\pi^{i'+pj}R}\ .$$

Note: the case for $G = C_p \times C_p$ is almost identical: the congruences become

$$v^p \equiv 1 \pmod{\pi^{pi'+j}R}$$
$$v^p \equiv 1 \pmod{\pi^{i'+pj}R}\ .$$

See [**GC98**] or [**Sm97**].

Continuing with the cyclic case, we have:

(31.4) COROLLARY. *If $0 < i, j < e'$ and v in K satisfies*

$$v^p \equiv 1 \pmod{\pi^{pi'+j}},$$

$$\zeta v^p \equiv 1 \pmod{\pi^{i'+pj}}\ ,$$

then $i \geq pj$ or $j' \geq pi'$.

PROOF. We apply the isosceles triangle inequality to the equation

$$\zeta v^p - 1 = (\zeta - 1)v^p + (v^p - 1):$$

then two of $ord_R(\zeta v^p - 1), ord_R(v^p - 1)$ and $e' = ord_R((\zeta - 1)v^p)$ must be equal, and both $\leq$ the third.

Since

$$ord_R(\zeta v^p - 1) \geq i' + pj$$

and

$$ord_R(v^p - 1) \geq pi' + j\ ,$$

we must have $e' \geq i' + pj$ or $e' \geq pi' + j$. The first is equivalent to $i \geq pj$, the second to $j' \geq pi'$. □

Corollary (31.4) is the "valuative condition for n = 2"of [**Un96**], c.f. [**Un94**]. The first inequality $i \geq pj$ goes back to Larson [**La76**], c.f. (17.1)(5).

Note that if $i \geq pj$ or $j' \geq pi'$, then $j \leq pi$.

For $0 \leq i, j \leq e'$, a Hopf order $E_v = R[\frac{\sigma^p - 1}{\pi^i}, \frac{a_v\sigma - 1}{\pi^j}]$ satisfying (31.3) with $i \geq pj$ is called a *Greither order*. A Hopf order E_v satisfying (31.3) with $j' \geq pi'$ is called a *dual Greither order*, because of

(31.5) PROPOSITION. *Let ζ_2 be a primitive p^2 root of unity with $\zeta_2^p = \zeta$. Let $\hat{G} = Hom(G, U(R)) = \langle\hat{\sigma}\rangle$ with $\langle\hat{\sigma}, \sigma\rangle = \zeta_2$. Let $E_v = R[\frac{\sigma^p - 1}{\pi^i}, \frac{a_v\sigma - 1}{\pi^j}]$ with $ord_R(v^p - 1) \geq pi' + j$ and $ord_R(\zeta v^p - 1) \geq i' + pj$. Then $E_v^* = \hat{E}_w = R[\frac{\hat{\sigma}^p - 1}{\pi^{j'}}, \frac{a_w\hat{\sigma} - 1}{\pi^{i'}}]$ with $w = (\zeta_2 v)^{-1}$. If E_v is a dual Greither order then $\hat{E}_w$ is a Greither order in $K\hat{G}$.*

PROOF. $\hat{E}_w$ is a Hopf order extension of $H_{i'}$ by $H_{j'}$ because

$$ord_R(w^p - 1) = ord_R(\zeta^{-1}v^{-p} - 1) = ord_R(\zeta v^p - 1) \geq i' + pj$$

and

$$ord_R(\zeta w^p - 1) = ord_R(v^{-p} - 1) = ord_R(v^p - 1) \geq pi' + j\ .$$

So $\hat{E}_w$ is a Hopf order in $K\hat{G}$ which is an extension of $H_{i'}$ by $H_{j'}$.

Since

$$rdisc_R\hat{E}_w = rdisc_R(H_{j'}) \cdot rdisc_R(H_{i'}) = rdisc_R(E_v^*)$$

by (4.24), (21.2) and (22.17), it suffices to show that $\hat{E}_w \subset E_v^*$. By an argument similar to that in (32.9) below, it suffices to show that the generators of $\hat{E}_w$ map the generators of E_v to R.

Now a computation shows that since $\langle\hat{\sigma}, \sigma\rangle = \zeta_2$, if $k = qp + r$ with $0 \leq r < p$, then $\langle a_w\hat{\sigma}, \sigma^k\rangle = w^r\zeta_2^r\zeta^q$. Hence

$$\langle\frac{\hat{\sigma}^p - 1}{\pi^{j'}}, \frac{\sigma^p - 1}{\pi^i}\rangle = 0;$$

$$\langle\frac{\hat{\sigma}^p - 1}{\pi^{j'}}, \frac{a_v\sigma - 1}{\pi^j}\rangle = \frac{\zeta - 1}{\pi^{e'}} \in R;$$

$$\langle\frac{a_w\hat{\sigma} - 1}{\pi^{i'}}, \frac{\sigma^p - 1}{\pi^i}\rangle = \frac{\zeta - 1}{\pi^{e'}} \in R;$$

and

$$\langle\frac{a_w\hat{\sigma} - 1}{\pi^{i'}}, \frac{a_v\sigma - 1}{\pi^j}\rangle = 0$$

by the choice of w. Thus $\hat{E}_w \subset E_v^*$, hence $= E_v^*$. The final statement that if one of E_v, $\hat{E}_w$ is a Greither order, then the other is a dual Greither order, is clear. □

For a Greither order we can simplify the conditions on v if $i + j \leq e'$:

(31.6) PROPOSITION. *Let i, j be such that*

$$0 < i, j \leq e'$$

and

$$i \geq pj.$$

If also $i + j \leq e'$, then E_v is a Hopf algebra free of rank p over H_i iff

$$ord_R(v^p - 1) \geq pi' + j \ .$$

PROOF. If $i \geq pj$, then we have

$$e' \geq i' + pj.$$

If also $i + j \leq e'$, then $j \leq i'$, so $(p-1)j \leq (p-1)i'$, hence

$$i' + pj \leq pi' + j \ .$$

Therefore, if $ord_R(v^p - 1) \geq pi' + j$, then

$$\begin{aligned} ord_R(\zeta v^p - 1) &\geq min\{e', ord_R(v^p - 1)\} \\ &\geq min\{e', pi' + j\} \\ &\geq i' + pj \end{aligned}$$

and both conditions of Proposition (31.3) follow from the single condition

$$ord_R(v^p - 1) \geq pi' + j.$$

□

To complete Proposition (31.3) we need the next four results of Greither [**Gr92**]:

(31.7) PROPOSITION. *Let $x = (x_0, x_1, \ldots, x_{p-1}) \in R^p$. Then $x \in H_i$ iff $d^k(x)$ is divisible by $\pi^{ki'}$ for $k = 1, \ldots, p-1$.*

Here $d^k(x)$, the k-th difference function of the sequence x at x_0, is defined by

$$d^0(x) = x_0 \ ,$$

$$d^1(x) = x_0 - x_1 \ ,$$

$$d^2(x) = d^1(x_0) - d^1(x_1) = (x_0 - x_1) + (x_1 - x_2) = x_0 - 2x_1 + x_2,$$

and for all $k \geq 1$,

$$d^k(x) = \sum_{j=0}^{k} (-1)^j \binom{k}{j} x_j \ .$$

PROOF. Let $\langle -, - \rangle : KG \otimes K\hat{G} \to K$ be the evaluation map, and let $\hat{\sigma}$ be the character of G such that

$$\langle \sigma, \hat{\sigma} \rangle = \zeta.$$

Then $\langle \sigma^i, \hat{\sigma}^j \rangle = \zeta^{ij}$, hence

$$\begin{aligned} \langle e_i, \hat{\sigma}^j \rangle &= \langle \frac{1}{p} \sum \zeta^{-ik} \sigma^k, \hat{\sigma}^j \rangle \\ &= \frac{1}{p} \sum \zeta^{-ik} \zeta^{jk} = \delta_{i,j} \end{aligned}$$

so, identifying $K \times \ldots \times K$ with KG via $x \to \sum x_i e_i$, we have

$$\langle x, \hat{\sigma}^j \rangle = \langle \sum x_i e_i, \hat{\sigma}^j \rangle = x_j \ .$$

Thus

$$\begin{aligned} \langle x, (1 - \hat{\sigma})^k \rangle &= \langle x, \sum_{j=0}^{k} (-1)^j \binom{k}{j} \hat{\sigma}^j \rangle \\ &= \sum_{j=0}^{k} (-1)^j \binom{k}{j} x_j \\ &= d^k(x) \ . \end{aligned}$$

Since $H_i = H_{i'}^*$, we have: $x \in H_i$ iff $\langle x, (\frac{1-\hat{\sigma}}{\pi^{i'}})^k \rangle \in R$ for $k = 0, \ldots, p-1$, iff $d^k(x) \in \pi^{i'k} R$ for $k = 0, \ldots, p-1$. □

From Proposition (31.7) we obtain

(31.8) PROPOSITION. *Let* $a_v = (1, v, v^2, ..., v^{p-1})$ *for* v *in* R. *Then* $a_v - 1 \in \pi^\ell H_i$ *iff* $v - 1 \in \pi^{i'+\ell} R$.

PROOF. For $x = a_v$, the criterion of Proposition (31.7) assumes a very nice form:

$$d^k(a_v) = \sum_{j=0}^{k} (-1)^k \binom{k}{j} v^j = (1 - v)^k$$

for $k = 1, \ldots, p-1$. Hence $a_v - 1 \in \pi^\ell H_i$ iff $\frac{a_v - 1}{\pi^\ell} \in H_i$, iff $d^k(\frac{a_v-1}{\pi^\ell}) \in \pi^{i'k} R$, iff $d^k(a_v - 1) \in \pi^{i'k+\ell} R$ for $k = 1 \ldots p-1$.

Now $d^k(a_v - 1) = d^k(a_v) - d^k(1) = d^k(a_v)$. So $a_v - 1 \in \pi^\ell H_i$ iff $(1-v)^k \in \pi^{i'k+\ell} R$ for $k = 1, \ldots, p-1$. This implies that $1 - v \in \pi^{i'+\ell} R$, and conversely, if $1 - v \in \pi^{i'+\ell} R$, then $(1-v)^k \in \pi^{i'k+\ell} R$ for all $k = 1, \ldots, p-1$. □

To verify the congruence condition on comultiplication:

$$\Delta(a_v) \equiv a_v \otimes a_v \pmod{\pi^j H_i \otimes H_i} \text{ iff } v^p \equiv 1 \pmod{\pi^{pi'+j} R},$$

we view $H_i \otimes H_i \subset R^p \otimes R^p$ and identify $R^p \otimes R^p$ as $p \times p$ matrices with pointwise operations, as follows:

For $a = (a_0, \ldots, a_{p-1})$ and $b = (b_0, \ldots, b_{p-1})$ in R^p, set

$$a \otimes b = \begin{pmatrix} a_0 b_0 & \ldots & a_0 b_{p-1} \\ \vdots & & \vdots \\ a_{p-1} b_0 & \ldots & a_{p-1} b_{p-1} \end{pmatrix}.$$

With this notation, we have

(31.9) PROPOSITION. *The $p \times p$ matrix*

$$M = \begin{pmatrix} m_{0,0} & \ldots & m_{0,p-1} \\ \vdots & & \vdots \\ m_{p-1,0} & \ldots & m_{p-1,p-1} \end{pmatrix}$$

is in $H_i \otimes H_i$ iff for all k_1, k_2,

$$d^{k_1,k_2}(M) = \sum_{j_1=0}^{k_1} \sum_{j_2=0}^{k_2} \binom{k_1}{j_1} \binom{k_2}{j_2} (-1)^{j_1+j_2} m_{j_1,j_2}$$

is in $\pi^{i'(k_1+k_2)} R$.

PROOF. Let $M = \sum a^\ell \otimes b^\ell$ with a^ℓ, b^ℓ in R^p. Then $M \in H_i \otimes H_i$ iff M is in the dual of $H_i^* \otimes H_i^*$, that is,

$$\langle M, \frac{(1-\hat{\sigma})^{k_1}}{\pi^{i'k_1}} \otimes \frac{(1-\hat{\sigma})^{k_2}}{\pi^{i'k_2}} \rangle \in R$$

for $k_1, k_2 = 0, \ldots, p-1$. Now

$$\langle M, \frac{(1-\hat{\sigma})^{k_1}}{\pi^{i'k_1}} \otimes \frac{(1-\hat{\sigma})^{k_2}}{\pi^{i'k_2}} \rangle = \sum_{j_1=0}^{k_1} \sum_{j_2=0}^{k_2} (-1)^{j_1+j_2} \binom{k_1}{j_1} \binom{k_2}{j_2} \langle M, \hat{\sigma}^{j_1} \otimes \hat{\sigma}^{j_2} \rangle.$$

From the proof of Proposition (31.7), we have

$$\langle M, \hat{\sigma}^{j_1} \otimes \hat{\sigma}^{j_2} \rangle = \sum_\ell \langle a^\ell \otimes b^\ell, \hat{\sigma}^{j_1} \otimes \hat{\sigma}^{j_2} \rangle = \sum_\ell a^\ell_{j_1} b^\ell_{j_2} = m_{j_1,j_2}.$$

So $M \in H_i \otimes H_i$ iff for all k_1, k_2,

$$\sum_{j_1=0}^{k_1} \sum_{j_2=0}^{k_2} (-1)^{j_1+j_2} \binom{k_1}{j_1} \binom{k_2}{j_2} m_{j_1,j_2} \in \pi^{i'k_1+i'k_2} R.$$

□

Using Proposition (31.9) we have

PROPOSITION (31.10). *Suppose $j \leq pi$. Then*

$$\Delta(a_v) \equiv a_v \otimes a_v \pmod{\pi^j H_i \otimes H_i}$$

iff $v^p \equiv 1 \pmod{\pi^{pi'+j} R}$.

PROOF. We have

$$\begin{aligned}
\frac{\Delta(a_v)-a_v\otimes a_v}{\pi^j} &= \frac{1}{\pi^j}\left[\sum_{s=0}^{p-1}\Delta(v^s e_s)-\sum_{r=0}^{p-1}v^r e_r\otimes\sum_{t=0}^{p-1}v^t e_t\right]\\
&= \frac{1}{\pi^j}\sum_{s=0}^{p-1}\left[v^s\Delta(e_s)-\sum_{\substack{0\le r,t\le p-1\\ r+t\equiv s \pmod p}} v^r e_r\otimes v^t e_t\right]\\
&= \frac{1}{\pi^j}\sum_{s=0}^{p-1}\sum_{\substack{0\le r,t\le p-1\\ r+t\equiv s \pmod p}}(v^s-v^{r+t})(e_r\otimes e_t).
\end{aligned}$$

The coefficient $v^s - v^{r+t} = 0$ if $r+t<p$, and $= v^s(1-v^p)$ if $r+t = p+s$, $s\ge 0$.
If M is the matrix corresponding to

$$\frac{\Delta(a_v)-a_v\otimes a_v}{\pi^j},$$

then $m_{r,t} = \frac{v^s - v^{r+t}}{\pi^j}$ with $s\equiv r+t \pmod p$ and $s<p$. Thus, setting $w = \frac{1-v^p}{\pi^j}$, we have $m_{r,t} = v^s w$ if $r+t = p+s$, and $=0$ if $r+t<p$. Hence

$$M = \begin{pmatrix}
0 & 0 & 0 & \dots & 0 & 0\\
0 & 0 & 0 & \dots & 0 & w\\
0 & 0 & 0 & \dots & w & vw\\
0 & 0 & 0 & \dots & vw & v^2w\\
 & & \dots & & & \\
0 & w & vw & \dots & v^{p-3}w & v^{p-2}w
\end{pmatrix}.$$

We apply Proposition (31.9): the matrix M is in $H_i\otimes H_i$ iff for all k_1, k_2,

$$d^{k_1,k_2}(M) = \sum_{r=0}^{k_1}\sum_{t=0}^{k_2}\binom{k_1}{r}\binom{k_2}{t}(-1)^{r+t}m_{r,t}\in \pi^{i'(k_1+k_2)}R.$$

For $k_1+k_2<p$, $d^{k_1,k_2}(M)=0$. For $k_1+k_2\ge p$, set $k_1+k_2 = p+\ell$. Then

$$d^{k_1,k_2}(M) = \sum_{s=0}^{\ell}\sum_{\substack{r+t=p+s,\\ 0\le r,t\le p-1}}(-1)^{p+s}\binom{k_1}{r}\binom{k_2}{t}v^s w$$

Some manipulations with binomial coefficients show that

$$\sum_{\substack{0\le r,t\le p-1\\ r+t=p+s}}\binom{k_1}{r}\binom{k_2}{t} = \binom{k_1+k_2}{p+s}$$

and if $k_1+k_2 = p+\ell$, then

$$\binom{k_1+k_2}{p+s} = \binom{p+\ell}{p+s}\equiv\binom{\ell}{s}\pmod p\,.$$

Hence

$$d^{k_1,k_2}(M) \equiv \sum_{s=0}^{\ell}(-1)^{p+s}\binom{\ell}{s}v^s w \pmod p$$
$$= -w(1-v)^\ell \pmod p \ .$$

We therefore wish to show:

if $v^p \equiv 1 \pmod{\pi^{pi'+j}R}$, then $w(1-v)^\ell \in \pi^{i'(p+\ell)}R$, for $\ell = 1, \ldots, p-2$.

We pause for

(31.11) Lemma. *Suppose $j \leq pi$ and $ord_R(v^p - 1) \geq pi' + j$. Then $ord_R(v - 1) \geq i' + j/p$.*

Proof. If $j \leq pi$ then $i' + j/p \leq e'$. Let $ord_R(v-1) = c$. If $c \geq e'$ then $c \geq i' + j/p$. Otherwise we have

$$(v-1)^p = v^p - 1 + p(v-1)y$$

for some $y \in R$. If $c < e'$ then $e + c = (p-1)e' + c > pc$ and so

$$ord_R(v-1)^p = pc < e + c \leq ord_R p(v-1)y,$$

so $ord(v^p - 1) = pc$; since $pc \geq i'p + j$, we have $c \geq i' + j/p$. □

Note that if $i \geq pj$ or $j' \geq pi'$ then $j \leq pi$, so (31.11) applies.

Continuing with the proof of Proposition (31.10), we assume

$$v^p \equiv 1 \pmod{\pi^{pi'+j}R}$$

and wish to show $w(1-v)^\ell \in \pi^{i(p+\ell)}R$ for $\ell = 1, \ldots, p-2$, where $w = \frac{1-v^p}{\pi^j}$. But since $ord_R(v-1) \geq i' + j/p \geq i'$, we have

$$ord(w(1-v)^\ell) = ord(w) + \ell\, ord(1-v)$$
$$\geq i'p + i'\ell \ .$$

Thus

$$M \in H_i \otimes H_i \text{ iff } v^p \equiv 1 \pmod{\pi^{i'p+j}R} \ ,$$

as we wished to show. □

For Greither orders ($i \geq pj$) we obtained in Proposition (31.5) a criterion for E_v to be a Hopf order in KC_{p^2} free of rank p over H_i, namely, $ord_R(v^p - 1) \geq pi' + j$, or

$$v^p \in U_{pi'+j}(R)$$

where $U_n(R) = \{v \in R \big| v \equiv 1 \pmod{\pi^n R}\}$. We have the following criterion for different v to yield the same Hopf order.

(31.12) Proposition. *Let i, j satisfy $0 \leq i, j \leq e'$ and $i \geq pj$. Then for $v^p, v_1^p \in U_{pi'+j}$, $E_v = E_{v_1}$ iff $v_1 = cv$ with $c \in U_{i'+j}(R)$.*

PROOF. Lemma 31.11 implies that

$$ord_R(v-1) \geq i' + j/p$$

and also

$$ord_R(v_1 - 1) \geq i' + j/p\ .$$

Now $E_v = H_i[\frac{a_v\sigma-1}{\pi^j}]$ and $E_{v_1} = H_i[\frac{a_{v_1}\sigma-1}{\pi^j}]$. So $E_v = E_{v_1}$ iff

$$\frac{a_{v_1}\sigma - 1}{\pi^j} \in E_v$$

and

$$\frac{a_v\sigma - 1}{\pi^j} \in E_{v_1},$$

iff

$$\sigma^{-1}\left(\frac{a_v\sigma-1}{\pi^j} - \frac{a_{v_1}\sigma-1}{\pi^j}\right) = \frac{a_v - a_{v_1}}{\pi^j} \in H_i\ .$$

Set $x = a_{v_1} - a_v = a_{vc} - a_v$ for some c in R. Then

$$x = (x_0, x_1, \ldots, x_{p-1})$$

where $x_i = (c^i - 1)v^i$. Applying (31.7), $\pi^{-j}x \in H_i$ iff $\pi^{-j}d^k(x)$ is divisible by $\pi^{ki'}$ for $k = 0, \ldots, p-1$, iff $d^k(x)$ is divisible by $\pi^{ki'+j}$ for $k \geq 0$.

If $E_v = E_{vc}$ then $\pi^{-j}x \in H_i$, so $d^1(x) = (c-1)v - 0$ is divisible by $\pi^{i'+j}$, hence $c \in U_{i'+j}(R)$.

Conversely, if $c \in U_{i'+j}(R)$, we show that $d^k(x)$ is divisible by $\pi^{ki'+j}R$ for $k \geq 0$. This is clear for $k = 0$. For $k > 0$,

$$\begin{aligned}
d^k(x) &= \sum_{r=0}^{k}(-1)^r\binom{k}{r}x_r \\
&= \sum_{r=-}^{k}(-1)^r\binom{k}{r}(c^r - 1)v^r \\
&= \sum_{r=0}^{k}(-1)^r\binom{k}{r}c^rv^r - \sum_{j=0}^{k}(-1)^r\binom{k}{r}v^r \\
&= (1-cv)^k - (1-v)^k \\
&= ((1-v) + (1-c)v)^k - (1-v)^k \\
&= \sum_{r=1}^{k}\binom{k}{r}(1-v)^{k-r}(1-c)^rv^r\ .
\end{aligned}$$

Set $v - 1 = \pi^{i'+j/p}s$, and $c - 1 = \pi^{i'+j}d$. Then

$$d^k(x) = \sum_{r=1}^{k}\binom{k}{r}(\pi^{i'+j/p}s)^{k-r}(\pi^{i'+j}d)^rv^r\ .$$

But

$$(i' + j/p)(k-r) + (i'+j)r \geq ki' + j\ .$$

for $1 \leq r \leq k$. Hence $\pi^{ki'+j}$ divides $d^k(x)$ for $k \geq 1$.

That completes the proof. $\square$

(31.13) Notes. If $0 < i, j < i + j \leq e'$ and $j \leq pi$, then we have shown that $E_v = H_i[\frac{a_v\sigma - 1}{\pi^j}]$ is a Hopf order in KC_{p^2} which is an extension of H_j by H_i, that is, E_v fits into a short exact sequence

$$R \to H_i \to E_v \to H_j \to R$$

of R-Hopf algebras, iff $ord_R(v^p - 1) \geq pi' + j$ and $ord_R(\zeta v^p - 1) \geq i' + pj$. We have not shown that any R-Hopf order E in KC_{p^2} with $E \cap K[\sigma^p] = H_i$ and $\bar{E} = H_j$ is of the form E_v for some $v \in R$. That is true, however: Greither, in part I of [**Gr92**], showed that if $i \geq pj$ then any such E is a Greither order E_v for some v. His argument involves a cohomological classification of extensions of $Spec(H_i)$ by $Spec(H_j)$. We refer to [**Gr92**] for the proof. The classification was also obtained in [**By93**].

Underwood [**Un94**] subsequently showed that any Hopf order in KC_{p^2} is either a Greither order ($i \geq pj$) or a dual Greither order ($j' \geq pi'$), and is an E_v for some v.

Hopf orders in group rings where the group has p-power rank p^n, $n > 2$, are somewhat understood when the group is elementary abelian (see [**CGMSZ98**] and [**Sm98**]), but much less so when the group is cyclic (see [**Un96**] when $n = 3$).

The question of whether a given Hopf order in KG is realizable, that is, is the associated order of the valuation ring of some Galois extension L/K with Galois group G, is understood well only for G of order p^2. See Chapter 10.

CHAPTER 10

Cyclic Hopf Galois extensions of degree p^2

Let L/K be a totally ramified Galois extension with Galois group G, cyclic of order p^n. Let S, R be the valuation rings of L, K, respectively.

In Chapter 8 we showed that if the associated order $\mathfrak{A} = \mathfrak{A}_{L/K}$ of L/K is a Hopf order such that S is an $\mathfrak{A}$-Hopf Galois extension of R, then the ramification numbers $t_1, \ldots, t_n$ of L/K satisfy the congruence conditions

$$t_i \equiv -1 \pmod{p^n} .$$

In this chapter, we examine for L/K cyclic of order p^2 exactly when its associated order is Hopf. The congruence conditions on t_1, t_2 is not sufficient. We also determine which Hopf orders E_v constructed in Chapter 9 are realizable, that is, may be identified as the associated order $\mathfrak{A}_{L/K}$ for some cyclic extension L/K.

§32. Valuation rings of cyclic extensions of degree p^2. We begin by recalling the situation for Kummer extensions of prime order p. Assume throughout that K contains ζ, a primitive pth-root of unity.

(32.1) PROPOSITION. *Let M be a totally ramified Kummer extension of K of order p, $M = K[z]$ with Galois group $G = \langle\sigma\rangle$ acting by $\sigma(z) = \zeta z$. Let T be the valuation ring of M. Then the following are equivalent*

i) T/R is H_j-Hopf Galois for $H_j = R[\frac{\sigma-1}{\pi^j}]$,

ii) The ramification number $t^G = pj - 1$

iii) z may be chosen so that $ord_R(z^p - 1) = pj' + 1$ and $t = \frac{z-1}{\pi^{j'}}$ is a parameter for T, where $j + j' = e'$.

The proof is in Chapter 6.

Now let L be a Galois extension of K with Galois group $G = C_{p^2} = \langle\sigma\rangle$. If M is the intermediate field, $M = L^{G'}$, then both L/M and M/K are Kummer extensions of order p.

Assume L/K is totally ramified. Let $S \supset T \supset R$ be the valuation rings of $L \supset M \supset K$, respectively.

Suppose S/R is E_v-Hopf Galois, where

$$E_v = R[\frac{\sigma^p - 1}{\pi^i}, \frac{\sigma a_v - 1}{\pi^j}]$$

for some v as in Proposition (31.3). Then T/R is H_j-Hopf Galois, where H_j is the image of E_v under the map from KG to $K(G/G')$ given by $\sigma \mapsto \bar{\sigma}$, so by (32.1) we can write $M = K[z]$ with $z^p = 1+\pi^{pj'+1}u$, u a unit of R, and $T = R[t]$ with $t = \frac{z-1}{\pi^{j'}}$ a parameter for T. Then $t^{G/G'} = pj - 1$, and hence by (27.5), $t_1^G = t^{G/G'} = pj - 1$.

Also, S/T is $T\otimes_R H_i$-Hopf Galois for $H_i = E_v \cap KG'$. So we can write $L = M[x]$ with $ord_T(x^p - 1) = pi' + 1$, so $x^p = 1 + \pi^{p^2 i'} tu$, u a unit of T, and $w = \frac{x-1}{\pi^{i'}}$ is a parameter for S. Hence $S = T[w]$, and $t^{G'} = p^2 i - 1$, and so $t_2^G = p^2 i - 1$.

Conversely, if $t_1^G = pj - 1$ and $t_2^G = p^2 i - 1$ then T/R and S/T are Hopf Galois extensions for H_j and $T \otimes H_i$, respectively. Thus the condition that $t_i \equiv -1 \pmod{p^i}$, $i = 1, 2$ is equivalent to the layers T/R and S/T being Hopf Galois.

Thus we are interested in asking: if the ramification numbers of an extension L/K satisfy

$$t_1 = pj - 1, t_2 = p^2 i - 1$$

for some i, j, is E_v the associated order of L/K for some v?

In view of the assumptions made in the construction of E_v in Chapter 9, we assume

$$0 < i, j < i + j \leq e' \text{ and } i \geq pj. \tag{32.2}$$

Then by (31.6), E_v is Hopf iff $ord_R(v^p - 1) \geq pi' + j$. We recall (31.11): If $ord_R(v^p - 1) \geq pi' + j$ then $ord_R(v - 1) \geq i' + j/p$.

The key to understanding the Galois structure of L/K, cyclic of order p^2, when K contains a primitive pth root of unity ζ, is how the Galois group of M/K acts on L.

Let $\bar{G} = \langle \bar{\sigma} \rangle$ be the Galois group of M/K and let $M = K[z]$ with $\bar{\sigma}(z) = \zeta z$ and $z^p = 1 + u\pi^{pj'+1}$, u a unit of R. Then $t = \frac{z-1}{\pi^{j'}}$ is a parameter for T. Let $G' = \langle \tau \rangle$ be the Galois group of L/M and let $L = M[x]$ with $\tau(x) = \zeta x$ and $x^p = 1 + u_1 t^{p^2 i' + 1}$, u_1 a unit of T. Then $w = \frac{x-1}{\pi^{i'}}$ is a parameter for S. Let $G = \langle \sigma \rangle$ be the Galois group of L/K, with $\sigma_{|M} = \bar{\sigma}$ and $\sigma^p = \tau$. Then the action of G on L is determined by the way σ acts on x. Hence define $\beta \in L$ by

$$\beta = \frac{\sigma(x)}{x}, \tag{32.3}$$

then $\sigma(x) = \beta x$, so β describes the action of G on L, given the Galois action of the layers L/M and M/K.

The next three results describe β.

(32.4) LEMMA. *$\beta \in M$ and $N(\beta) = \zeta \in K$.*

PROOF. If $\sigma(x) = \beta x$ then

$$\sigma^2(x) = \sigma(\beta x) = \sigma(\beta)\beta x$$
$$\sigma^3(x) = \sigma(\sigma(\beta)\beta x) = \sigma^2(\beta)\sigma(\beta)\beta x, \text{ etc.,}$$

So

$$\zeta x = \sigma^p(x) = \sigma^{p-1}(\beta)\sigma^{p-2}(\beta)\ldots\sigma(\beta)\beta x\ .$$

But then

$$\zeta = \sigma^{p-1}(\beta)\sigma^{p-2}(\beta)\cdot\ldots\cdot\sigma(\beta)\beta\ ; \tag{*}$$

apply σ to both sides and note that $\sigma(\zeta) = \zeta$, then

$$\sigma^p(\beta)\sigma^{p-1}(\beta)\cdot\ldots\cdot\sigma(\beta) = \sigma^{p-1}(\beta)\cdot\ldots\cdot\sigma(\beta)\beta\ ,$$

hence $\sigma^p(\beta) = \beta$ and $\beta \in M$. Since $\beta \in M$, (*) says that $N_{M/K}(\beta) = \zeta$. □

(32.5) PROPOSITION. $ord_T(\beta - 1) = pi' + j$.

PROOF. The valuation of β is obtained from the ramification number $t_1^G = pj - 1$, as follows: since $\sigma(x) = \beta x$,

$$\sigma(w) = \sigma(\frac{x-1}{\pi^{i'}}) = \frac{\beta x - 1}{\pi^{i'}} = \frac{\beta x - x}{\pi^{i'}} + \frac{x-1}{\pi^{i'}}$$
$$= \frac{\beta - 1}{\pi^{i'}} x + w ,$$

hence $\sigma(w) - w = \frac{\beta-1}{\pi^{i'}}x$. Since w is a parameter for S, $t_1 = t_1^G$ is determined by $ord_S(\sigma(w) - w) = t_1 + 1$.

Since x is a unit of T, $ord_S(\frac{\beta-1}{\pi^{i'}}) = t_1 + 1 = pj$, and so

$$p^2 i' + pj = ord_S(\beta - 1) = p(ord_T(\beta - 1)).$$

□

The element β of T is determined by the action of σ on x. We want to see how β varies if we replace x by another element x_1 of L such that $\frac{\sigma^p(x_1)}{x_1} = \frac{\sigma^p(x)}{x} = \zeta$. Since $t_2^G = t_1^{G'} = p^2 i - 1$, if $x_1 \in L$ so that $L = M[x_1]$ where $\sigma^p(x_1) = \zeta x_1$ and p does not divide $ord_T(x_1^p - 1)$, then x satisfies $ord_T(x_1^p - 1) = p^2 i' + 1$ by (24.6).

(32.6) LEMMA. *β is unique modulo $\pi^{i'+j}T$.*

PROOF. If we replace x by x_1 with $\frac{\sigma^p(x_1)}{x_1} = \zeta$, then, setting $x_1 = x\alpha$, we have $\frac{\sigma^p(\alpha)}{\alpha} = 1$, so $\alpha \in M$. If $x^p \in U_{p^2 i'+1}(T)$ and $x^p \alpha^p \in u_{p^2 i'+1}(T)$, then $\alpha^p \in U_{p^2 i'+1}(T)$. Since $\beta = \frac{\sigma(x)}{x}$, then $\frac{\sigma(\alpha x)}{\alpha x} = \beta \frac{\sigma(\alpha)}{\alpha}$. We determine the order of $\frac{\sigma(\alpha)}{\alpha} - 1$.

If $ord_T(\alpha - 1) = s$, then

$$ps = ord_T(\alpha - 1)^p \geq \min\{ord_T(\alpha^p - 1), ord_T(p(\alpha - 1))\}$$

since $(\alpha - 1)^p = \alpha^p - 1 + p(\alpha - 1)\gamma$ for some $\gamma \in T$. Since $ord_T(\alpha^p - 1) \geq p^2 i' + 1$, we have

$$ps \geq \min\{p^2 i' + 1, pe + s\} .$$

If $ps \geq p^2 i' + 1$, then $s \geq pi' + 1$. If $ps \geq pe + s$, then, since $i > 0$, $s \geq pe' \geq pi' + 1$. Since $s = ord_T(\alpha - 1) \geq pi' + 1$,

$$ord_T(\frac{\sigma(\alpha)}{\alpha} - 1) \geq ord_T(\alpha - 1) + t_1^G \text{ by (27.6)} ,$$
$$\geq pi' + 1 + pj - 1 = p(i' + j) .$$

Thus $\beta \equiv \frac{\sigma(\alpha)}{\alpha}\beta \pmod{t^{p(i'+j)}T}$ and so β is unique modulo $t^{p(i'+j)}T = \pi^{i'+j}T$, as claimed. □

Here is the basic result describing when L/K is E_v-Galois:

(32.7) THEOREM. *Let L/K be totally ramified with valuation ring S and ramification numbers $t_1 = pj - 1, t_2 = p^2 i - 1$ where $0 < i, j < i + j \leq e'$ and $i \geq pj$. Let $E_v = H[\frac{\sigma^p - 1}{\pi^i}, \frac{\sigma a_v - 1}{\pi^j}]$. Then S/R is E-Hopf Galois for some Hopf order E in KG iff $E = E_v$ with $\beta v \equiv 1 \pmod{\pi^{i'+j}T}$.*

PROOF. If S is an E-Hopf Galois extension of R then the values of the ramification numbers imply that S/T is $T \otimes H_i$- Hopf Galois and T/R is H_j-Hopf Galois. Hence E is a Hopf algebra extension of H_j by H_i. By Greither's classification of such extensions, valid for $i \geq pj$, $E = E_v$ for some v.

Now we show:

(32.8) PROPOSITION. *For any v, $disc_R(S) = disc_R(E_v^*)$.*

PROOF OF (32.8). We have

$$disc_R(S) = disc_R(T)^p \cdot N_{M/K}(disc_T(S))$$

by (22.14). Now T/R is H_j-Galois, so $disc_R(T) = disc_R(H_j^*)$. Also, S/T is $T \otimes H_i$-Galois, and

$$disc_T(S) = disc_T(T \otimes H_i^*) = disc_R(H_i^*)T.$$

So

$$N_{M/K}(disc_T(S)) = disc_R(H_i^*)^p.$$

Thus

$$disc_R(S) = (disc_R(H_j^*))^p(disc_R(H_i^*))^p.$$

But we have

$$disc_R(E_v^*) = (disc_R(H_i^*))^p(disc_R(H_j^*))^p.$$

by (22.17). Hence $disc_R(S) = disc_R(E_v^*)$. □

PROOF OF (32.7). Suppose E_v acts on S. Then S will be E_v-Hopf Galois, since the discriminant of E_v^* equals the discriminant of S. So we ask: when does E_v act on S for some v?

We have $S = T[w], w = \frac{x-1}{\pi^{i'}}$ and $T = R[t], t = \frac{z-1}{\pi^{j'}}$. Let $\xi = \frac{\sigma^p - 1}{\pi^i}$, $\eta = \frac{\sigma a_v - 1}{\pi^j}$, the algebra generators of E_v.

(32.9) LEMMA. *If ξw, ξt, ηw and $\eta t \in S$, then $E_v \cdot S \subset S$.*

PROOF. Let $\mathfrak{A}$ be the associated order of S. Then $\mathfrak{A}$ is an R-algebra, so since E_v is generated over R by ξ and η, we have $E_v \subset \mathfrak{A}$, and hence E_v acts on S, if ξ and $\eta \in \mathfrak{A}$.

We first show $\xi \in \mathfrak{A}$, assuming ξw and $\xi t \in S$.

We have that

$$\begin{aligned}\Delta(\xi) &= \frac{\sigma^p \otimes \sigma^p - 1 \otimes 1}{\pi^i} \\ &= \frac{(\sigma^p - 1) \otimes \sigma^p + 1 \otimes (\sigma^p - 1)}{\pi^i} \\ &= \xi \otimes (1 + \pi^i \xi) + 1 \otimes \xi \ .\end{aligned}$$

Thus the R-submodule C of E_v generated by 1 and ξ is a subcoalgebra of E_v: that is,

$$\Delta(C) \subseteq C \otimes C \ .$$

From this we show that if $s \in S$ and $Cs \subseteq S$, then $Cs^r \subseteq S$ for all $r \geq 1$. For if $Cs^{r-1} \subseteq S$, then $\xi s^r = \Delta(\xi)(s^{r-1} \otimes s)$ since S is an E_v-module algebra. Hence

$$\xi s^r \in Cs^{r-1} \cdot Cs \subset S \ .$$

Applying this for $s = t$ shows that $\xi T \subset S$; then applying this for $s = w$ shows that $\xi S \subset S$. Hence $C \subset \mathfrak{A}$. Since C generates H_i as an R-algebra, therefore, $H_i \subset \mathfrak{A}$.

Now we repeat this argument for η, assuming ηt and $\eta w \in S$. We have

$$\begin{aligned}\Delta(\eta) &= \frac{(\sigma\otimes\sigma)\Delta(a_v) - 1}{\pi^j} \\ &= \frac{1}{\pi^j}((\sigma\otimes\sigma)(\Delta(a_v) - a_v\otimes a_v) + (\sigma a_v \otimes \sigma a_v - 1\otimes 1)) \\ &= \frac{1}{\pi^j}(\sigma\otimes\sigma)(\Delta(a_v) - a_v\otimes a_v) \\ &\qquad + \frac{1}{\pi^j}(\sigma a_v\otimes\sigma a_v - \sigma a_v\otimes 1 + \sigma a_v \otimes 1 - 1\otimes 1) \\ &= \frac{1}{\pi^j}(\sigma\otimes\sigma)(\Delta(a_v) - a_v\otimes a_v) + \sigma a_v \otimes \frac{\sigma a_v - 1}{\pi^j} + \frac{\sigma a_v - 1}{\pi^j}\otimes 1)\ .\end{aligned}$$

Since $v^p \equiv 1 \pmod{\pi^{pi'+j}R}$, we have by (31.9),

$$a_v \otimes a_v \equiv \Delta(a_v) \pmod{\pi^j(H_i\otimes H_i)}.$$

Hence the H_i-module C' generated by 1 and η is a subcoalgebra of E_v. Since $H_i S \subset S$, the same argument which showed that $C \subset \mathfrak{A}$ shows that $C' \subseteq \mathfrak{A}$. Since C' generates E_v as an H_i-algebra, therefore, $E_v \subset \mathfrak{A}$, proving (32.9). □

Continuing with the proof of (32.7), we have that E_v acts on S iff ξw, ξt, ηw and $\eta t \in S$. So we compute these actions. First

$$\xi t = (\frac{\sigma^p - 1}{\pi^i})(\frac{z-1}{\pi^{j'}}) = 0$$

since $\sigma^p z = z$; also

$$\xi w = (\frac{\sigma^p - 1}{\pi^i})(\frac{x-1}{\pi^{i'}}) = \frac{(\zeta - 1)x}{\pi^{i+i'}} = \frac{\zeta - 1}{\pi^{e'}} x \in S\ .$$

Now

$$\eta t = (\frac{a_v\sigma - 1}{\pi^j})(\frac{z-1}{\pi^{j'}}) = \frac{(\zeta-1)z}{\pi^{e'}} \in T\ :$$

to see this, we have $a_v = \sum_{i=0}^{p-1} v^i e_i$ and for $b \in T$,

$$e_i b = \frac{1}{p}\sum_j \zeta^{ij}\sigma^{jp}(b) = \frac{1}{p}\sum_j \zeta^{ij} b = \begin{cases} 0 & \text{if } i \neq 0 \\ b & \text{if } i = 0 \end{cases},$$

hence $\sigma a_v(b) = \sigma(b)$; in particular, $a_v\sigma(z) = \sigma(z) = \zeta z$. Also,

$$e_i x = \frac{1}{p}\sum_j \zeta^{ij}\sigma^{jp}(x) = \frac{1}{p}\sum_j \zeta^{ij}\zeta^j x = \begin{cases} 0 & \text{if } i \not\equiv 1 \pmod p \\ x & \text{if } i \equiv 1 \pmod p \end{cases}\ .$$

Hence

$$a_v\sigma(x) = (\sum_i \sigma v^i e_i)(x) = v\sigma(x) = v\beta x.$$

Thus

$$\eta w = \left(\frac{a_v\sigma - 1}{\pi^j}\right)\left(\frac{x-1}{\pi^{i'}}\right) = \frac{(v\beta - 1)x}{\pi^{i'+j}}.$$

Since x is a unit of S, $\eta w \in S$ iff $\beta \equiv v^{-1} \pmod{\pi^{i'+j}T}$. □

By (32.6), β is unique modulo $\pi^{i'+j}T$. This puts a restrictive condition on v:

(32.10) PROPOSITION. *If $0 < i, j < i + j \leq e'$ and $i \geq pj$, and S is E_v-Hopf Galois for some v, then*

$$ord_R(v - 1) = i' + j/p\ ,$$

and p divides j. Thus E_v is not realizable unless $ord(v - 1) = i' + j/p$.

PROOF. We showed $ord_T(\beta - 1) = pi' + j$, hence if $\beta \equiv v^{-1} \pmod{\pi^{i'+j}}$, then $ord_R(v - 1) = i' + j/p$, and p divides j. □

In (32.3) we determined, assuming $0 < i, j < i + j \leq e'$ and $i \geq pj$, that for E_v to be a Hopf order free of rank p over H_i, we must have $ord_R(v - 1) \geq i' + j/p$. Proposition (32.10) means that for some S to be E_v-Hopf Galois, i.e. for E_v to be realizable, $ord_R(v - 1)$ must take on that minimal value $i' + j/p$.

In the remainder of this chapter we obtain two related results:

1. If $ord_R(v - 1) = i' + j/p$ then E_v is realizable and
2. An L/K with correct ramification numbers need not have an associated order which is Hopf.

To obtain both results we begin with an extension M/K with ramification number $pj - 1$ and construct a Kummer extension L/M with ramification number $p^2i - 1$ and a prescribed Galois structure.

(32.11) PROPOSITION. *Let i, j satisfy: $0 < i, j \leq i + j \leq e'$, $i \geq pj$ and p divides j. Let M/K be a Galois extension with Galois group $\bar{G} = \langle\bar{\sigma}\rangle$ of order p and valuation ring T, with ramification number $t = pj - 1$. Let $\beta \in T$ with $ord_T(\beta - 1) = pi' + j$ and $N_{T/R}(\beta) = \zeta$. Then there exists $\gamma \in T$ so that $L = M[y]$, $y^p = \gamma$ is a Galois extension of K with group $G = \langle\sigma\rangle$ cyclic of order p^2, where $\sigma(y) = \beta y$, σ on M acts by $\bar{\sigma}$, and L/K has ramification numbers $t_1 = pj - 1$, $t_2 = p^2i - 1$.*

PROOF. Since $t_1^G > 0$, M/K is totally ramified, so $ord_T(p) = pe$. Since $N_{T/R}(\beta) = \zeta$, we have $N_{T/R}(\beta^p) = 1$.

We first show $ord_T(\beta^p - 1) = p^2i' + pj$. Since $\beta^p - 1 = (\beta - 1)^p + p(\beta - 1)u$ for some $u \in T$, $ord_T(\beta^p - 1) = p^2i' + pj$ provided

$$(*) \qquad p^2i' + pj < pe + pi' + j\ .$$

But $pj \leq i$, so $j < pi$, so $pi' + j \leq pe'$, so $(p - 1)(pi' + j) \leq pe$ and (*) holds.

Now $N(\beta^p) = 1$ and $\beta^p \in U_{p^2i'+pj} \setminus U_{p^2i'+pj+1}$. Since $p^2i' + pj - t = p^2i' + 1$ is coprime to p, (27.8) shows that there exists γ in T with $ord_T(\gamma - 1) = p^2i' + 1$ so that $\frac{\bar{\sigma}(\gamma)}{\gamma} = \beta^p$.

Let $L = M[y]$ with $y^p = \gamma$. Then L/M is a Kummer extension, $S = T[\frac{y-1}{\pi^{i'}}]$ is the valuation ring of L, S/T is $T \otimes_R H_i$-Hopf Galois, and the ramification number $t_{L/M} = p^2i - 1$. Let $G = \langle\sigma\rangle$ be cyclic of order p^2. Define an action of G on L by $\sigma(y) = \beta y$, and $\sigma(m) = \bar{\sigma}(m)$ for $m \in M$, then $\sigma(y^p) = \beta^p y^p$, or $\sigma(\gamma) = \beta^p\gamma$, so

$\sigma(\gamma) = \bar{\sigma}(\gamma)$ and the action of G on L is well-defined. Also $\sigma^p(y) = N_{T/R}(\beta)y = \zeta y$, so σ^p generates the Galois group of L/M. Thus L/K is Galois with group G, and the ramification numbers of L/K are $t_2^G = p^2 i - 1$ and $t_1^G = pj - 1$. □

(32.12) THEOREM. *If $v \in R$ with $ord_R(v-1) = i' + j/p$, then E_v is realizable.*

PROOF. Given M/K, we need to construct $L = M[y]$, $y^p = \gamma$, with valuation ring S in such a way that E_v acts on S. For this we need to choose $\beta \in T$ so that

1. $\beta \equiv v^{-1} \pmod{\pi^{i'+j}T}$;
2. $ord_T(\beta - 1) = pi' + j$, and
3. $N_{T/R}(\beta) = \zeta$.

We need

(32.13) LEMMA. *Let $t_{M/K} = pj - 1$ and $pj \leq i < e'$. Then there exists $\delta \in U_{pi+pj}(T)$ with $N(\delta) = \zeta$.*

PROOF OF (32.13). We apply [**CL V, p. 93, Corollary 3**]: for $n > t = t_{M/K}$,

$$N(U_{t+p(n-t)}(T)) = U_n(R).$$

Let $n = e'$, then

$$N(U_\ell(T)) = U_{e'}(R)$$

for $\ell = pj - 1 + p(e' - (pj - 1))$. Now $\zeta \in U_{e'}(R)$, so there is some δ in $U_\ell(T)$ with $N(\delta) = \zeta \in U_{e'}(R)$. Since $i \geq pj$ we have

$$\begin{aligned} pi &> p^2 j - (p-1) \\ pe' - pi' &\geq p^2 j - (p-1), \end{aligned}$$

and so

$$\begin{aligned} \ell &= pe' - (p-1)(pj-1) \\ &= pe' - (p^2 j - (p-1)) + pj \geq pi' + pj \end{aligned}$$

hence $U_\ell(T) \subset U_{pi'+pj}(T)$. □

PROOF OF (32.12). First, given E_v with v in R with $ord_T(v-1) = pi' + j$, there exists $\beta' \in T$ with $\beta' \equiv v^{-1} \pmod{t^{pi+pj}T}$ and $N(\beta') = 1$.

To see this, we observe that $N_{M/K}(v) = v^p$ and $v^p \in U_{pi'+j}(R)$. Since $i+j \leq e'$, we have $(p-1)j - 1 + pi < pe'$, or $(p-1)j - 1 < pi'$, hence $t = pj - 1 < pi' + j$. Thus by [**CL p.93, Cor.3**] (see (32.13)),

$$N(U_{t+p(n-t)}(T)) = U_n(R)$$

for $n = pi' + j$. Here

$$t + p(n-t) = pj - 1 + p(pi' + j - (pj - 1).$$

Since $i + j \leq e'$,

$$(p^2 - p)j - (p-1) + (p^2 - p)i \leq (p^2 - p)e'$$

which is equivalent to

$$t + p(n-t) = p^2 i' + 2pj - 1 - p^2 j + p \geq pi' + pj\ .$$

So there exists α in $U_{pi'+pj}(T)$ with $N(\alpha) = v^p$, hence $N(v^{-1}\alpha) = 1$. Setting $\beta' = v^{-1}\alpha$, we have $N(\beta') = 1$ and

$$\beta' \equiv v^{-1} \pmod{t^{pi'+pj}T}.$$

Now from (32.13), we find $\delta \in U_{pi'+pj}(T)$ with $N(\delta) = \zeta$; setting $\beta = \beta'\delta$ we have $N(\beta) = \zeta$ and $\beta \equiv v^{-1} \pmod{t^{pi'+pj}T}$. Since $ord_T(\beta - 1) = ord_T(\beta' - 1) = pi' + j$, that completes the proof that E_v is realizable. □

Now we show that that if p is odd and L/K is cyclic of order p^2, then having prescribed ramification numbers $\equiv 1 \pmod{p^2}$ is not sufficient for the associated order of S/R to be Hopf. We first show:

(32.14) THEOREM. *Each Galois extension L/K cyclic of order p^2 with ramification numbers $t_1 = pj - 1$, $t_2 = p^2 i - 1$ determines a unique class $[\beta]$ in*

$$V = U_{pi'+j}(T)/U_{pi'+j+p-1}(T)$$

with $ord_T(\beta - 1) = pi' + j$. For each class $[\beta]$ in V with $ord_T(\beta - 1) = pi' + j$, there exists a corresponding Galois extension L/K, cyclic of order p^2, with ramification numbers $t_1 = pj - 1$ and $t_2 = p^2 i - 1$.

PROOF. The first statement follows from (32.5), in which we showed that given L/K with prescribed ramification, the β which describes the Galois action has $ord_T(\beta - 1) = pi' + j$ and (32.6), where we showed that β is unique modulo $t^{pi'+pj}T$. Hence each such L/K determines a unique class $[\beta]$ in V since $pi' + j + p - 1 \leq pi' + pj$.

For the second statement, we first show that each class in V contains an element β of norm 1. To do this, we use (27.3): whenever $(k, p) = 1$, the map $\alpha \mapsto \bar{\sigma}(\alpha)/\alpha$ yields an isomorphism

$$U_k(T)/U_{k+1}(T) \to U_{k+t_1}(T)/U_{k+1+t_1}(T)$$

where t_1 is the ramification number of $\bar{G} = \langle \bar{\sigma} \rangle = Gal(M/K)$. Since p divides j, given β_0 in $U_{pi'+j}(T)$, there exists $\alpha_1 \in U_{pi'+j-(pj-1)}(T)$ so that $\beta_0 = \frac{\bar{\sigma}(\alpha_1)}{\alpha_1}\beta_1$ with $\beta_1 \in U_{pi'+j+1}(T)$. Repeating this, if we obtain $\beta_0 = \frac{\bar{\sigma}(\alpha_v)}{\alpha_v}\beta_v$ with $\beta_v \in U_{pi+j+v}(T)$, then $\beta_v = \frac{\bar{\sigma}(\alpha')}{\alpha'}\beta_{v+1}$ with $\beta_{r+1} \in U_{pi'+j+v}(T)$ provided $pi' + j - (pj - 1) + v$ is not divisible by p, i.e. provided $v + 1 < p$. Thus we may write $\beta_0 = \frac{\bar{\sigma}(\alpha)}{\alpha}\beta_{p-1}$ with $\beta_{p-1} \in U_{pi'+j+(p-1)}(T)$: that is, $\beta \equiv \frac{\bar{\sigma}(\alpha)}{\alpha} \pmod{t^{pi'+j+p-1}T}$.

Set $\beta' = \frac{\bar{\sigma}(\alpha)}{\alpha}$. Evidently $N(\beta') = 1$.

Now by (32.13), we can multiply β' by some $\delta \in U_{pi'+pj}(T)$ with norm ζ to get $\beta \in T$ of norm ζ. Then by (32.12) β yields a Kummer extension L/M of order p with L/K cyclic of order p^2 and with the desired ramification numbers. □

(32.15) COROLLARY. *Let $[\beta]$ be a class in $U_{pi'+j}(T)/U_{pi'+j+p-1}(T)$ such that $ord_T(\beta - 1) = pi' + j$, and let L/K be a corresponding Galois extension, cyclic of order p^2. If the associated order $\mathfrak{A}$ of L/K is a Hopf order then $\mathfrak{A} = E_v$ with $[\beta] = [v]$ for some $v \in R$. If $q = |R/\pi R|$, then there are at least $(q-1)(q^{p-2} - 1)$ extensions L/K with prescribed ramification whose associated orders are not Hopf.*

PROOF. If $\mathfrak{A}$ is a Hopf order, then $\mathfrak{A} = E_v$ by (32.7) and $\beta \equiv v^{-1} \pmod{t^{pi'+pj}T}$, hence $[\beta] = [v^{-1}]$.

For the count, let $q = |R/\pi R| = |T/tT|$, and let W be a set of representatives in R of the elements of T/tT. Any β with $ord_T(\beta - 1) = pi' + j$ may be expanded t-adically as

$$\beta \equiv 1 + a_0 t^{pi'+j} + a_1 t^{pi'+j+1} + \cdots + a_{p-2} t^{pi'+j+(p-2)} \pmod{t^{pi'+j+p-1}}$$

where $a_0, \ldots, a_{p-2} \in W$ and $a_0 \notin \pi R$. If $v \in R$ and $ord_T(v-1) = pi' + j$, then

$$v \equiv 1 + b_0 \pi^{i'+j/p} \pmod{t^{pi'+j+(p-1)}}$$

for unique $b_0 \notin \pi R$, since $\pi^{i'+(j/p)+1} \equiv 0 \pmod{t^{pi'+j+p-1}T}$. Thus there are $q-1$ possible classes $[v]$ with $v \in R$, and $(q-1)q^{p-2}$ possible classes $[\beta]$. Since $pi' + j + p - 1 \leq pi' + pj$, each class $[\beta]$ yields a different L/K by (32.6). □

For further work in this direction, see [Ch96], [Un98] and [By99].

CHAPTER 11

Formal Groups

There is a close connection between finite commutative, cocommutative Hopf algebras over the valuation ring R of a local field K and formal groups defined over R. In fact, the Oort Embedding Theorem asserts that any finite abelian group scheme is the kernel of an isogeny of commutative formal groups. If an R-Hopf algebra H is so described, then the Kummer theory of formal groups gives an explicit description as R-algebras of the principal homogeneous spaces for H.

This and the next chapter present these ideas.

§**33. Formal groups.** We begin with a review of formal groups. Throughout, R is the valuation ring of a finite extension K of $\mathbb{Q}_p$, with maximal ideal $\mathfrak{m}$.

(33.1) DEFINITION. An n-dimensional formal group law, or formal group, defined over R is an n-tuple of power series

$$F(x,y) = (F_1(x,y), F_2(x,y), \ldots, F_n(x,y))$$

where $x = (x_1, \ldots, x_n)$ and $y = (y_1, \ldots, y_n)$, satisfying, for all i,

$$(i) \qquad F_i(x,y) \equiv x_i + y_i \text{ modulo degree } 2$$

$$(ii) \qquad F_i(F(x,y),z) = F_i(x, F(y,z)) \text{ (associativity)}$$

$$(iii) \qquad F(x,0) = x,\ F(0,y) = y \text{ (identity)}.$$

We will assume also that F is commutative:

$$(iv) \qquad F_i(x,y) = F_i(y,x) \text{ (commutativity)}$$

Non-commutative formal groups of dimension > 1 exist, but we will not consider them here (for some examples, see [**CMSZ98**] or [**Ch98, page 6**].

The terminology "formal group law"reflects the fact that F can induce a new binary operation $+_F$ on certain sets, such as:

1) The ideal of power series $f \in R[[x_1, \ldots, x_n]]$ with $f(0) = 0$: $f +_F g = F(f,g)$

2) $\mathfrak{m}^n = \mathfrak{m} \times \ldots \times \mathfrak{m}$ (n copies), by

$$a +_F b = F(a,b).$$

Denote $\mathfrak{m}^n$ with the addition $+_F$ by $P(F,K)$ ("the K-points of F").

3) Same if $\mathfrak{m}_L$ is the maximal ideal of the valuation ring of some algebraic extension L of K: call $\mathfrak{m}_L^n$ with addition by $+_F$, $P(F,L)$. If $\overline{K}$ is an algebraic closure of K and $\overline{\mathfrak{m}} = \mathfrak{m}_{\overline{K}}$ is the maximal ideal of the valuation ring of $\overline{K}$, then $\overline{\mathfrak{m}}^n$ with addition by F is just denoted by $P(F)$, the points of F.

4) $Alg_{cont}(R[[x]], R)$, by convolution via F (c.f. (1.8): for ϕ, ψ in $Alg_{cont}(R[[x]], R)$,

$$\begin{aligned}(\phi * \psi)(f(x)) &= \mu(\phi \otimes \psi)(\Delta(f(x))) \\ &= \mu(\phi \otimes \psi)(f(F_1(x \otimes 1, 1 \otimes x), \ldots, F_n(x \otimes 1, 1 \otimes x))) \\ &= f(F_1(\phi(x), \psi(x)), \ldots, F_n(\phi(x), \psi(x))).\end{aligned}$$

Any element ϕ of $Alg_{cont}(R[[x]], R)$ is uniquely determined by $(\phi(x_1), \ldots, \phi(x_n)) \in \mathfrak{m}^n$, and the correspondence $\phi \mapsto (\phi(x_1), \ldots, \phi(x_n))$ defines an isomorphism of groups from $Alg_{cont}(R[[x]], R)$ to $P(F, K)$.

Condition (ii) implies that $+_F$ is associative, condition (iii) that $+_F$ is commutative, condition (iv) that 0 is the identity element under $+_F$. Thus the sets under $+_F$ are at least commutative monoids. To show that $+_F$ yields a group structure, one shows by the formal inverse function theorem ([**Fr68, Chapter I, Section 3, Proposition 1**] that there exists a unique n-tuple of formal power series $\lambda(x)$ so that $F(x, \lambda(x)) = 0$. Condition (i) implies that $\lambda(x)$ satisfies

$$\lambda(x) \equiv -x \pmod{\deg 2}.$$

(33.2) EXAMPLES. The two simplest examples of formal groups, and the most misleading (because they are polynomials), are the dimension one examples

$$\mathbb{G}_a(x, y) = x + y,$$

the additive formal group, and

$$\mathbb{G}_m(x, y) = x + y + xy,$$

the multiplicative formal group. If $\mathfrak{m}$ is the maximal ideal of R, then $P(\mathbb{G}_a, K) = \mathfrak{m}$ with the usual addition in R; $P(\mathbb{G}_m, K) \cong U_1(R) = 1 + \mathfrak{m}$, the multiplicative group of principal units of R.

The dimension n additive formal group is $\mathbb{G}_a(x, y) = x + y$.

(33.3) PROPOSITION. *If $F(x, y)$ is a dimension one polynomial formal group, then $F(x, y) = x + y + cxy$ for some $c \in R$.*

PROOF. Consider the identity

$$F(F(x, y), z) = F(x, F(y, z))$$

and observe that the degree of the right side in z can be the same as the degree of the left side in z only if that degree is 1; and similarly in x. The conditions $F(x, y) \equiv x + y \pmod{\deg 2}$, $F(x, 0) = 0 = F(0, y)$ then forces F to have the claimed form. □

(33.4). At the other extreme, one can define a universal one-dimensional formal group, as follows ([**Ha78, p. 7**]). Let

$$S = \mathbb{Z}[\{c_{i,j} | i, j \geq 1\}]$$

and let

$$F(x, y) = x + y + \sum_{i,j \geq 1} c_{i,j} x^i y^j.$$

Let

$$F(F(x,y),z) - F(x,F(y,z)) = \sum_{i,j,k} P_{i,j,k}(c) x^i j^j z^k$$

where $P_{i,j,k}(c)$ are polynomials in the $c_{i,j}$. Let J be the ideal of S generated by $c_{i,j} - c_{j,i}$ for all i, j and by $P_{i,j,k}(c)$ for all i, j, k. Then over S/J, F is a commutative dimension one formal group. If G is any dimension one formal group defined over a ring R, then specializing the coefficients $c_{i,j}$ to the coefficients of G yields a homomorphism from S/J to R under which F is mapped to G.

(33.5) DEFINITION. A *homomorphism* from an n-dimensional formal group F to an m-dimensional formal group G is an m-tuple $\phi(x) = (\phi_1(x), \dots, \phi_m(x))$ of power series in $x = (x_1, \dots, x_n)$ so that $\phi(0) = 0$ and

$$\phi_i(F(x,y)) = G_i(\phi(x), \phi(y))$$

for all i. The homomorphism ϕ is an isomorphism if $m = n$ and there is a homomorphism ψ from G to F so that $\psi(\phi(x)) = x$ and $\phi(\psi(x) = x$.

If $\phi : F \to G$ is a homomorphism, then ϕ induces a homomorphism

$$\phi : P(F) \to P(G)$$

since

$$\phi(a +_F b) = \phi(F(a,b)) = G(\phi(a), \phi(b)) = \phi(a) +_G \phi(b).$$

Isomorphisms over K provide a way to construct formal groups over R by using "logarithms", as follows.

Let $f(x) = x + a_2 x^2 + \dots$ be a formal power series with coefficients in K, and let $f^{-1}(x)$ be the composition inverse, so that $f^{-1}(f(x)) = x$. Then

$$F(x,y) = f^{-1}(f(x) + f(y))$$

is a dimension one formal group, isomorphic over K to $\mathbb{G}_a$. If R is the valuation ring of a local field K, and $f(x) \in K[[x]]$ is such that $F(x,y) = f^{-1}(f(x) + f(y))$ has coefficients in R, then F will be a formal group defined over R which is isomorphic to $\mathbb{G}_a$ over K but not necessarily over R. For example: $F = \mathbb{G}_m$ can be obtained in this way by letting $f(x) = log(1+x) = \sum_{n=1}^{\infty} \frac{(-1)^{n+1} x^n}{n}$, then $f^{-1}(x) = e^x - 1 = \sum_{n=1}^{\infty} \frac{x^n}{n!}$ and

$$\mathbb{G}_m(x,y) = e^{log(1+x)+log(1+y)} - 1 = (1+x)(1+y) - 1.$$

(33.6). It is a fact (see [**Fr68**]) that over a field of characteristic zero, every n-dimensional (commutative) formal group F is isomorphic (by an isomorphism called the logarithm map ℓ_F) to the n-dimensional additive formal group. In dimension one it is easily seen that if $\ell'_F(0) = 1$, that is, $\ell_F(x) \equiv x \pmod{\deg 2}$, then ℓ_F is unique, for if $f, g : F \to \mathbb{G}_a$ are two such logarithms, then their composition $g^{-1} \circ f = h$ is an endomorphism of $\mathbb{G}_a$, hence satisfies $h(x+y) = h(x) + h(y)$, and in characteristic zero the only formal power series satisfying that identity is $h(x) = x$. It follows that $\mathbb{G}_m$ is not isomorphic to $\mathbb{G}_a$ over the valuation ring R of a local field K because the logarithm map (over K) is not defined over R.

For F a (commutative) formal group, define the "multiplication by m" maps for m in $\mathbb{Z}$ by

$$\begin{aligned}
[1](x) &= x, \\
[2](x) &= F(x,x) \\
[m](x) &= F([m-1](x),x) \text{ for } m>1 \\
[-1](x) &= \lambda(x) \\
[-m](x) &= F([-(m-1)](x),\lambda(x)) \text{ for } m>1.
\end{aligned}$$

(33.7) PROPOSITION. *$[m](x)$ is an endomorphism of F for all m.*

PROOF. This is obvious for $[1](x)$. For $m>1$ we see this as follows: If

$$[m-1](F(x,y)) = F([m-1](x),[m-1](y)),$$

then by associativity and commutativity (33.1),

$$\begin{aligned}
[m](F(x,y)) &= F([m-1](F(x,y)),F(x,y)) \\
&= F(F([m-1](x),[m-1](y)),F(y,x)) \\
&= F([m-1](x),F([m-1](y),F(y,x))) \\
&= F([m-1](x),F(F([m-1](y),y),x)) \\
&= F([m-1](x),F(x,F([m-1](y),y))) \\
&= F(F([m-1](x),x),F([m-1](y),y)) \\
&= F([m](x),[m](y)).
\end{aligned}$$

Also, $[-1](F(x,y)) = F([-1](x),[-1](y))$. For by uniqueness of $[-1](x)$, it suffices to show that

$$F(F(x,y),F([-1](x),[-1](y))) = 0.$$

But

$$\begin{aligned}
F(F(x,y),F([-1](x),[-1](y))) &= F(F(F(x,y),[-1](x)),[-1](y)) \\
&= F(F(F(y,x),[-1](x)),[-1](y)) \\
&= F(F(y,F(x,[-1](x))),[-1](y)) \\
&= F(F(y,0),[-1](y)) \\
&= F(y,[-1](y)) = 0.
\end{aligned}$$

The argument for $m<-1$ is similar to that for $m>1$. □

(33.8) EXAMPLES. Let $F=\mathbb{G}_a$, $F(x,y)=x+y$. Then $[n](x)=nx$ for all n.

Let $F=\mathbb{G}_m$, $F(x,y)=x+y+xy$. Then $[n](x)=(1+x)^n-1$ for all $n>0$, as is quickly seen by induction.

More generally, for any $b\neq 0$, set $F_b(x,y)=x+y+bxy$. Then for $n>0$,

$$[n](x) = \frac{(1+bx)^n-1}{b}.$$

Over K, θ_b, defined by $\theta_b(x)=bx$, defines an isomorphism from F_b to $\mathbb{G}_m$:

$$\theta_b(F_b(x,y)) = \mathbb{G}_m(\theta(x),\theta(y)),$$

$$bx + by + b^2xy = bx + by + (bx)(by);$$

however, over R, θ_b is an isomorphism iff b is a unit of R.

§34. Formal groups and Hopf algebras. The interest in formal groups from the point of view of Hopf algebras starts with the next result.

(34.1) PROPOSITION. *Let F be a formal group of dimension n over R. Then F induces a "formal" Hopf algebra structure on $R[[x_1, \ldots, x_n]]$.*

SKETCH OF PROOF. To explain this we need to discuss the completed tensor product $\hat{\otimes}$ (see [**Fr68, p. 34**]). First, some notation.

Let $z = (z_1, \ldots, z_n)$.

Define $deg : R[[z]] \to \mathbb{Z}_{\geq 0} \cup \{\infty\}$ by $deg(0) = \infty$, $deg(z_i) = 1$, and if $J = (j_1, \ldots, j_n) \in \mathbb{Z}^n_{\geq 0}$ and z^J is the monomial $z_1^{j_1} \cdot \ldots \cdot z_n^{j_n}$, then z^J has degree $|J| = j_1 + \ldots + j_n$. For $f(z) = \sum a_J z^J \in R[[z]]$, set

$$deg(f) = min\{|J| : a_J \neq 0\}.$$

Now, let

$$\mu : R[[x]] \otimes R[[y]] \to R[[x, y]]$$

be the multiplication map; then μ is 1-1 but not onto (for example, $\sum_{i=1}^{\infty} x_1^i y_1^i$ is not in the image of μ). Define a topology on $R[[x]] \otimes R[[y]]$ by defining neighborhoods of 0 to be

$$\{\sum f_i \otimes g_i : deg(\mu(\sum f_i \otimes g_i)) \geq N\}$$

for $N > 0$. Complete $R[[x]] \otimes R[[y]]$ with respect to that topology. Equivalently, let

$$I = \langle x_1, \ldots, x_n \rangle \otimes R[[y]] + R[[x]] \otimes \langle y_1, \ldots, y_n \rangle$$

and complete $R[[x]] \otimes R[[y]]$ with respect to the basis $\{I^n\}_{n>0}$ of neighborhoods of 0 in $R[[x]] \otimes R[[y]]$.

Call the completion $R[[x]] \hat{\otimes} R[[y]]$. Then μ extends to an R-algebra isomorphism

$$\mu : R[[x]] \hat{\otimes} R[[y]] \to R[[x, y]].$$

Define a "formal" comultiplication

$$\Delta : R[[x]] \to R[[x]] \hat{\otimes} R[[x]]$$

by $\Delta(x_i) = F_i(x \hat{\otimes} 1, 1 \hat{\otimes} x)$, a counit map

$$\varepsilon : R[[x]] \to R$$

by $\varepsilon(x_i) = 0$, and an antipode

$$\lambda : R[[x]] \to R[[x]]$$

by $\lambda(x_i) = [-1]_i(x)$, extending each to $R[[x]]$ by requiring that each be an algebra homomorphism: thus for $f \in R[[x]]$,

$$\Delta(f)(x) = f(\Delta(x_1), \ldots, \Delta(x_n)).$$

Then using the formal group properties (33.1), one sees easily that $R[[x]]$ becomes a "formal" R-Hopf algebra, i.e. satisfies all the axioms for a Hopf algebra except that Δ maps to the completed tensor product $\hat{\otimes}$ of $R[[x]]$ with itself, rather than to the subalgebra $R[[x]] \otimes R[[x]]$. □

(34.2) DEFINITION. Let $f : F \to G$ be a homomorphism of formal groups. Let $H_f = R[[x]]/(f)$. Then f is an *isogeny* if H_f is a finite R-module.

(34.3) PROPOSITION. *If $f : F \to G$ is an isogeny of formal groups, then $H = R[[x]]/(f)$ is an R-Hopf algebra with operations induced from those on $R[[x]]$ by F.*

PROOF. We need to show that Δ, ε and λ, defined on $R[[x]]$ via F, induce well-defined operations on H_f. For Δ, we use the isomorphism $R[[x\hat{\otimes}1, 1\hat{\otimes}x]] \cong R[[y,z]]$ via $x\hat{\otimes}1 \mapsto y, 1\hat{\otimes}x \mapsto z$. If $R[[x]]/(f)$ is finite, then Δ induces a map

$$\begin{aligned} R[[x]] \to R[[x\hat{\otimes}1, 1\hat{\otimes}x]] \cong R[[y,z]] \cong R[[y]][[z]] &\to (R[[y]]/(f))[[z]] \\ \cong R[[y]]/(f) \otimes_R R[[z]] &\to R[[y]]/(f) \otimes R[[z]]/(f). \end{aligned}$$

The last isomorphism is valid since $R[[y]]/(f)$ is finitely generated and free over R. Now each component f_i of f satisfies

$$\begin{aligned} \Delta(f_i)(x) &= f_i(\Delta(x)) \\ &= f_i(F(x\hat{\otimes}1, 1\hat{\otimes}x)) \\ &= G_i(f(x\hat{\otimes}1), f(1\hat{\otimes}x)) \end{aligned}$$

which corresponds to $G_i(f(y), f(z))$ under the isomorphism $R[[x\hat{\otimes}1, 1\hat{\otimes}x]] \cong R[[y,z]]$. Since $G_i(0,0) = 0$, $\Delta(f_i)(x)$ in $R[x]$ maps to 0 in $R[[y]]/(f) \otimes R[[z]]/(f)$, and hence Δ on $R[[x]]$ induces a well-defined comultiplication on $H_f = R[[x]]/(f)$.

Similarly, one sees that $\varepsilon(f_i) = 0$ since for all i, f_i has no constant term. The antipode λ, defined by

$$\lambda(g)(x) = g(\lambda_1(x), \dots, \lambda_n(x))$$

for $g(x) \in R[[x]]$, satisfies

$$f_i(\lambda_1(x), \dots, \lambda_n(x)) = \lambda_{G,i}((f_1(x), \dots, f_n(x)),$$

which is in the ideal (f) since $\lambda_{G,i}$ has no constant term. □

PROPOSITION (34.4). *Let $\overline{K}$ be an algebraic closure of K with valuation ring $\overline{R}$ and maximal ideal $\overline{\mathfrak{m}}$. If $f : F \to G$ is an isogeny of formal groups of dimension n defined over R, then $\{x \in \overline{\mathfrak{m}}^n : f(x) = 0\}$ is finite.*

PROOF. We have $Alg_{cont}(\overline{R}[[x]], \overline{R}) \cong \overline{\mathfrak{m}}^n$ via the map ψ which takes an algebra homomorphism ϕ to the tuple $(\phi(x_1), \dots, \phi(x_n))$. Now $f : F \to G$ induces a homomorphism

$$\mathcal{F}_f : Alg_{cont}(\overline{R}[[x]], \overline{R}) \to Alg_{cont}(\overline{R}[[x]], \overline{R})$$

by $\mathcal{F}_f(\phi)(p(x)) = p(f(\phi(x)))$ for $\phi \in Alg_{cont}(\overline{R}[[x]], \overline{R})$ and $p(x) \in R[[x]]$. Thus

$$\begin{aligned} ker(\mathcal{F}_f) &= \{\phi \in Alg_{cont}(\overline{R}[[x]], \overline{R}) | \mathcal{F}_f(\phi) = \varepsilon\} \\ &= \{\phi | p(f(\phi(x))) = \mathcal{F}_f(\phi)(p(x)) = \varepsilon(p(x)) = p(0) \text{ for all } p \in \overline{R}[[x]]\} \\ &= \{\phi | f(\phi(x)) = 0\} \\ &= Alg_{cont}(\overline{R}[[x]]/(f), \overline{R}). \end{aligned}$$

Since $f\psi = \psi\mathcal{F}_f : Alg_{cont}(\overline{R}[[x]], \overline{R}) \to \overline{\mathfrak{m}}^n$ and ψ is an isomorphism, ψ induces an isomorphism

$$Alg_{cont}(\overline{R}[[x]]/(f), \overline{R}) \cong ker\{f : \overline{\mathfrak{m}}^n \to \overline{\mathfrak{m}}^n\}.$$

If we set $H_f = R[[x]]/(f)$, then

$$Alg_{\overline{R}}(\overline{R} \otimes H_f, \overline{R}) \cong ker\{f : \overline{\mathfrak{m}}^n \to \overline{\mathfrak{m}}^n\}.$$

Now $Alg_{\overline{R}}(\overline{R} \otimes H_f, \overline{R})$ maps 1-1 to $Alg_{\overline{K}}(\overline{K} \otimes H_f, \overline{K})$. Also $K \otimes H_f$ is a separable K-algebra, and so $\overline{K} \otimes H_f$ is isomorphic as $\overline{K}$-algebras to $\overline{K} \times \ldots \times \overline{K}$, the product of $[H_f : R]$ copies of $\overline{K}$. Hence $|Alg_{\overline{K}}(\overline{K} \otimes H_f, \overline{K})| = [H_f : R]$. □

If $f : F \to G$ is an isogeny of formal groups, and $H_f = R[[x]]/(f)$ with operations induced by F, then the finite abelian group scheme $Spec(H_f)$ is denoted $ker(f)$.

(34.5) DEFINITION. The formal group F is p-divisible if $[p]$ is an isogeny, i.e. $R[[x]]/([p])$ is a finitely generated R-module.

§**35. Height.** Let k be a field of characteristic p.

(35.1) PROPOSITION. *Let $f : F \to G$ be a non-zero homomorphism of dimension one formal groups over k. Then $f(x) = g(x^{p^h})$ for some $h \geq 0$ with $g'(0) \neq 0$.*

PROOF. (from [**Fr68**]) If $f'(x) \neq 0$ there is nothing to prove. If $f'(x) = 0$, differentiate the formula

$$f(F(x, y)) = G(f(x), f(y))$$

with respect to y to get

$$f'(F(x, y))F_y(x, y) = G_y(f(x), f(y))f'(y).$$

Set $y = 0$ and note that $F_y(x, 0)$ is an invertible power series in $k[[x]]$; thus since $f'(0) = 0$, then $f'(x) = 0$, in which case $f(x) = f_1(x^p)$. Then f_1 is a homomorphism from the formal group $F^{(p)}(x, y)$ to $G(x, y)$, where $F^{(p)}(x, y) = F(x^{1/p}, y^{1/p})^p$. Iterate this argument until $f_h'(x) \neq 0$, then set $g(x) = f_h(x)$. □

(35.2) DEFINITION. A homomorphism $f : F \to G$ has height h if $f(x) \neq 0$ and h is as in (35.1). f has infinite height if $f = 0$. A formal group F has height h if $[p]_F : F \to F$ has height h. F has height ∞ if $[p]_F = 0$.

(35.3) EXAMPLES. $\mathbb{G}_m$ has height 1. $\mathbb{G}_a$ has height ∞, and is the unique such example because of

(35.4) PROPOSITION. *F has height ∞ iff $F = \mathbb{G}_a$.*

The proof, in [**Fr68, p. 69, Corollary 2**], uses Lazard's Theorem and will be omitted here.

Now suppose k is the residue field of the valuation ring R of a local field $K \supset \mathbb{Q}_p$, with maximal ideal $\mathfrak{m} = \pi R$.

(35.5) DEFINITION. A dimension one formal group F defined over R has height h if its image modulo $\mathfrak{m}$ has height h as a formal group over k. A homomorphism $f : F \to G$ of formal groups over R has height h if its image modulo $\mathfrak{m}$ has height h over k.

Suppose $f : F \to G$ is a homomorphism (over R) of dimension one formal groups, and f has height $h < \infty$. Then

$$f(x) = a_1 x + a_2 x^2 + \ldots + a_{p^h} x^{p^h} + \ldots$$

where $a_1, \ldots, a_{p^h - 1} \in m$, while a_{p^h} is a unit of R. So we may apply the Weierstrass Preparation Theorem [**Fr68, p. 14**], [**Lg84, p. 215**] to factor $f(x)$ as

$$f(x) = p(x)q(x)$$

where $p(x)$ is a monic polynomial of degree p^h which, modulo m, is congruent to x^{p^h} and $q(x)$ is an invertible power series (i. e. $q(0) \neq 0$) .

(35.6) PROPOSITION. *If F, G are formal groups of dimension one over R and $f : F \to G$ is a homomorphism over R of height $h < \infty$, then f is an isogeny.*

PROOF. Let $q = p^h$. We show that $H_f = R[[x]]/(f(x))$ is a free R-module of rank q: more precisely, if t is the image of x in H_f, then $1, t, t^2, \ldots, t^{q-1}$ is an R-basis of H_f.

By the Weierstrass Preparation Theorem, $f(x) = p(x)q(x)$ with $q(x)$ a unit in $R[[x]]$. So $H_f = R[[x]]/(p(x))$. If

$$p(x) = x^q - \sum_{i=1}^{q-1} a_i x^i,$$

then in H_f,

$$t^q = \sum_{i=1}^{q-1} a_i t^i$$

where $a_i \in \mathfrak{m}$ for $i = 1, \ldots, q-1$. Thus

$$t^{kq} = (\sum_{i=1}^{q-1} a_i t^i)^k \equiv 0 \pmod{\mathfrak{m}^k},$$

so

$$t^{kq} = \sum_{i=1}^{q-1} a_{i,kq} t^i$$

with $a_{i,kq} \in \mathfrak{m}^k$. Then for $0 \le r < q$,

$$t^{kq+r} = \sum_{i=1}^{q-1} a_{i,kq} t^{i+r} = \sum_{i=1}^{q-1} a_{i,kq+r} t^i$$

where $a_{i,kq+r} \in \mathfrak{m}^k$ for each i.

Let $\sum_{i=0}^{\infty} b_i t^i$ be the image in H_f of an arbitrary power series in $R[[x]]$. Then

$$\begin{aligned}\sum_{i=0}^{\infty} b_i t^i &= \sum_{k=0}^{\infty}\sum_{r=0}^{q-1} b_{kq+r} t^{kq+r} \\ &= \sum_{k=0}^{\infty}\sum_{r=0}^{q-1} b_{kq+r} (\sum_{i=1}^{q-1} a_{i,kq+r} t^i) \\ &= \sum_{i=1}^{q-1}\sum_{k=0}^{\infty}\sum_{r=0}^{q-1} (a_{i,kq+r} b_{kq+r}) t^i.\end{aligned}$$

Let

$$c_{k,i} = \sum_{r=0}^{q-1} (a_{i,kq+r} b_{kq+r}),$$

then $c_{k,i} \in \mathfrak{m}^k$, and

$$\sum_{i=0}^{\infty} b_i t^i = \sum_{i=1}^{q-1} (\sum_{k=0}^{\infty} c_{k,i}) t^i,$$

where the coefficients of t^i converge in R for each i. Hence $1, t, \ldots, t^{q-1}$ are an R-basis of H_f. □

There exist dimension one formal groups over k of any height. They can be obtained from Lubin-Tate formal groups over R, which we construct in the next section.

The notion of height applies also to n-dimensional formal groups: an n-dimensional formal group over k has height r if $k[[x_1, \ldots, x_n]]$ is a finite free module of rank r over $k[[([p]_F)_1(x), \ldots, ([p]_F)_n(x)]]$. See [**Ha78, (18.3.8)**] for details.

§**36. Lubin-Tate formal groups.** Lubin-Tate formal groups were constructed to provide an elegant analogue [**Lu81**] for local fields to the Kronecker-Weber Theorem. They are dimension one formal groups defined over the valuation ring R of a local field $K \supset \mathbb{Q}_p$. Let π be a parameter for R and let q be the cardinality of the residue field $R/\pi R$.

Lubin-Tate formal groups F have the property that the bracket map $[\ ] : \mathbb{Z} \to End(F)$ extends to all of R, in such a way that

$$[\pi](x) \equiv \pi x \pmod{\deg 2}$$

and

$$[\pi](x) \equiv x^q \pmod{\pi}.$$

Lubin-Tate formal groups are needed in section 40, so in this section we describe their construction, following the elegant exposition in [**Se67**].

The construction is based on the following lemma from [**LT65**];

(36.1) PROPOSITION. *Let $\mathcal{F}$ denote the set of formal power series $f \in R[[x]]$ such that*

$$f(x) \equiv \pi x \pmod{deg\ 2}$$

and

$$f(x) \equiv x^q \pmod{\pi}.$$

Let $f, g \in \mathcal{F}$. Let $\phi_1(x_1, \dots, x_n) = \sum a_i x_i$ be a linear form, with $a_i \in R$. Then there exists a unique $\phi \in R[[x_1, \dots, x_n]]$ such that $\phi \equiv \phi_1 \pmod{deg\ 2}$ and $f(\phi(x_1, \dots, x_n)) = \phi(g(x_1), \dots, g(x_n))$.

This will be used to construct a formal group F admitting f as an endomorphism, and the uniqueness will be used to verify associativity and commutativity of F and other properties.

PROOF. Since $f(x) \equiv g(x) \equiv \pi x \pmod{\deg 2}$,

$$f(\phi_1(x_1, \dots, x_n)) = \phi_1(g(x_1), \dots, g(x_n)) \pmod{\deg 2}.$$

For $k \geq 1$, assume $\phi_k \in R[[x_1, \dots, x_n]]$ has been constructed so that

$$\phi_k \equiv \phi_1 \pmod{\deg 2}$$

and

$$f(\phi_k(x_1, \dots, x_n)) = \phi_k(g(x_1), \dots, g(x_n)) \pmod{\deg k+1}.$$

Then

$$f(\phi_k(x_1, \dots, x_n)) - \phi_k(g(x_1), \dots, g(x_n)) = \eta_{k+1}(x_1, \dots, x_n) \pmod{\deg k+2},$$

where η_{k+1} is homogeneous of degree $k+1$.

Let $\phi_{k+1} = \phi_k + \theta_{k+1}$ where θ_{k+1} is homogeneous of degree $k+1$. Then $f \circ \phi_{k+1} = f \circ (\phi_k + \theta_{k+1}) \equiv f \circ \phi_k + \pi\theta_{k+1} \pmod{\deg k+2}$ since $f(x) \equiv \pi x \pmod{\deg 2}$. On the other hand, $\phi_k(g(x_1), \dots, g(x_n)) + \theta_{k+1}(g(x_1), \dots, g(x_n)) = \phi_k(g(x_1), \dots, g(x_n)) + \pi^{k+1}\theta_{k+1}$ since $g(x) \equiv \pi x \pmod{\deg 2}$. Hence

$$f \circ \phi_{k+1} - \phi_{k+1} \circ (g \times \dots \times g) \equiv \eta_{k+1} + (\pi - \pi^{k+1})\theta_{k+1} \pmod{\deg k+2}.$$

This congruence uniquely determines θ_{k+1} and hence ϕ_{k+1} over K, by

$$\theta_{k+1} = \eta_{k+1}/(\pi^{k+1} - \pi).$$

To see that θ_{k+1} in fact has coefficients in R, we use the other property of elements of $\mathcal{F}$, namely that $f(x) \equiv x^q \pmod{\pi}$:

$$f \circ \phi_k - \phi_k \circ (g \times \dots \times g) \equiv (\phi_k(x))^q - \phi_k(x_1^q, \dots, x_n^q) \pmod{\pi}.$$

But for $\psi \in R[[x_1, \dots, x_n]]$, $\psi(x)^q \equiv \psi(x_1^q, \dots, x_n^q) \pmod{\pi}$ since $R/\pi R$ has q elements. Hence $\eta_{k+1} \in \pi R[x_1 \dots, x_n]$ and the construction yields $\phi \in R[[x]]$. □

Here is the existence of Lubin-Tate formal groups:

(36.2) THEOREM. *Let $f \in \mathcal{F}$. Then there is a unique formal group $F_f(x, y) \in R[[x, y]]$ with f as an endomorphism.*

PROOF. In (36.1) let $F(x,y)$ be the unique power series satisfying

$$F(x,y) \equiv x+y \pmod{\deg 2}$$

and

$$f(F(x,y)) = F(f(x),f(y)).$$

Then f will be in $End(F)$ if F is a formal group: that is,

$$\begin{aligned} F(F(x,y),z) &= F(x,F(y,z)) \\ F(x,y) &= F(y,x) \\ F(x,0) &= x. \end{aligned}$$

For associativity, notice that both $F(F(x,y),z)$ and $F(x,F(y,z))$ are solutions of the problem

$$H(x,y,z) \equiv x+y+z \pmod{\deg 2},$$
$$f(H(x,y,z)) = H(f(x),f(y),f(z)).$$

By uniqueness in (36.1), the two sides of the associativity equation are equal.

For commutativity, note that both $F(x,y)$ and $F(y,x)$ satisfy

$$H(x,y) \equiv x+y \pmod{\deg 2},$$

$$f(H(x,y)) = H(f(x),f(y)).$$

Again by uniqueness in (36.1), $F(x,y)=F(y,x)$.

Finally, for $F(x,0)=x$, note that both $F(x,0)$ and x satisfy

$$H(x) \equiv x \pmod{\deg 2},$$

$$f(H(x)) = H(f(x)).$$

Again by uniqueness, $F(x,0)=x$. Thus F is a formal group which admits f as an endomorphism. □

Useful properties of the Lubin-Tate formal group F are summarized in the following result:

(36.3) PROPOSITION. *Let $f,g \in \mathcal{F}$. Let F_f, F_g denote the formal groups of (36.2) which admit f, g, respectively, as endomorphisms. Then:-*

(1) *For any $a \in R$, there exists $[a]_{f,g}(x) \in R[[x]]$ satisfying*

$$\begin{aligned} g([a]_{f,g}(x)) &= [a]_{f,g}(f(x)), \\ [a]_{f,g}(x) &\equiv ax \pmod{\deg 2}; \end{aligned}$$

$[a]_{f,g}(x)$ is a homomorphism from F_f to F_g.

(2) *$[a]_{f,f} = [a]_f$ is an endomorphism of F_f;*

(3) *$[\pi]_f = f$*

(4) *$[\ ]: R \to End(F_f)$ is a homomorphism of rings;*

(5) *$F_f \cong F_g$.*

PROOF. Again we use the uniqueness in (36.1) often.

(1). Let $[a]_{f,g}(x) = H(x)$ satisfy

$$H(x) \equiv ax \pmod{\deg 2},$$

$$g(H(x)) = H(f(x)).$$

Then both $F_g([a]_{f,g}(x), [a]_{f,g}(y))$ and $[a]_{f,g}(F_f(x,y))$ satisfy

$$H(x,y) \equiv ax + ay \pmod{\deg 2},$$

$$g(H(x,y)) = H(f(x), f(y)).$$

So

$$F_g([a]_{f,g}(x), [a]_{f,g}(y)) = [a]_{f,g}(F_f(x,y))$$

and so $[a]_{f,g}(x)$ is a homomorphism from F_f to F_g. If $f = g$ then $[a]_f$ is an endomorphism of F which commutes with f, giving (2).

Applying this with $a = \pi$ shows that $f = [\pi]_f$, giving (3).

To show (4): recall that $End(F_f)$ is a ring under $+_F$ and composition. To show that [ ] is a homomorphism, we need to show that

$$[a+b]_f = [a]_f +_F [b]_f$$

or

$$[a+b]_f(x) = F([a]_f(x), [b]_f(x)).$$

and

$$[ab]_f = [a]_f \circ [b]_f$$

For addition, both sides commute with f and satisfy

$$H(x) \equiv (a+b)x \pmod{\deg 2},$$

so are equal by uniqueness; for multiplication, both sides commute with f and satisfy

$$H(x) \equiv (ab)x \pmod{\deg 2}.$$

The uniqueness of (36.1) again applies.

Finally, to show (5): $F_f \cong F_g$, observe that for $f, g, h \in \mathcal{F}$, again by (36.1) as applied to multiplication in (4),

$$[ab]_{f,h} = [a]_{f,g} \circ [b]_{g,h}.$$

Hence choosing a invertible in R, we find that $[a^{-1}]_{g,f}$ is a homomorphism from F_g to F_f which is the inverse of $[a]_{f,g}$ from F_f to F_g. Thus F_f and F_g are isomorphic. That completes all parts of the proof of (36.3). □

(36.4) PROPOSITION. *Let k be an algebraically closed field of characteristic p. Then for any $h > 0$ there is a formal group F defined over k of height h.*

PROOF. Let K be an unramified extension of $\mathbb{Q}_p$ of degree h, with valuation ring R. Then p is a parameter for R and $R/pR = k_0$ is a field of $q = p^h$ elements. Let $f(x) = px + x^q$. Then there is a Lubin-Tate formal group G for which $f(x) = [p]_G(x)$ by (36.2). Let F be the image of G modulo p, then F is defined over k_0 and $[p]_F = x^q$ with $q = p^h$, hence F has height h. □

Proposition (36.4) also follows from Lazard's theorem.

(36.5) REMARKS. The Lubin-Tate construction is not the only method for constructing formal groups of dimension one. In fact, there is a generic formal group of height h defined over $R[t_1, \ldots, t_{h-1}]$ which yields formal groups of height h over R by specializing the indeterminate $t_1, \ldots, t_{h-1}$ to elements of the maximal ideal of R [**LT66**], [**Lu79**]. If K is sufficiently ramified over $\mathbb{Q}_p$ one can, by appropriate specializations, get Hopf algebras H over R where the group associated to H, that is, $Alg_{\overline{K}}(\overline{K} \otimes H, \overline{K})$, is any finite abelian p-group. See [**CZ94**].

§**37. Oort embedding.** If $f : F \to G$ is an isogeny of commutative formal groups, then $ker(f) = Spec(H_f), H_f = R[[x]]/(f)$ is a finite abelian local R-Hopf algebra. This result has a converse.

(37.1) THE OORT EMBEDDING THEOREM. *Let R be the valuation ring of a local field K. Then any finite local abelian R- Hopf algebra may be identified as H_f for some isogeny $f : F \to F'$ of formal groups.*

This result was first proved in [**Oo67**], see [**Mz70, (2.4)**]. A proof is also given in [**MR70, (5.1)**] using the Hilbert scheme, which is rather beyond the scope of this book. See also [**Oo74, Section 3**].

For $G = Spec(H)$, H finite, here is a construction of a formal group F which maps onto G:

Let H be a commutative, cocommutative R-Hopf algebra which is free of rank $n+1$ as an R-module. The counit map $\varepsilon : H \to R$ is onto and splits, so yields an isomorphism of R-modules

$$H \cong R \cdot 1 \oplus ker(\varepsilon).$$

So $ker(\varepsilon)$ is free of rank n. Let $v_1, \ldots, v_n$ be an R- basis of $ker(\varepsilon)$, and set $v_0 = 1$.

Let

$$\Delta(v_i) = \sum_{j,k=0}^{n} c^i_{j,k} v_j \otimes v_k$$

for $c^i_{j.k} \in R$ and $i \leq n$. Then $c^i_{j,k} = c^i_{k,j}$ by cocommutativity. The counitary property $(1 \otimes \varepsilon)\Delta = id$ yields

$$v_i = \sum_{j,k=0}^{n} c^i_{j,k} v_j \otimes \varepsilon(v_k) = \sum_j c^i_{j,0} v_j,$$

hence $c^i_{j,0} = 0$ if $i \neq j$ and $c^i_{i,0} = 1$. Hence for $i > 0$,

$$\Delta(v_i) = v_i \otimes 1 + 1 \otimes v_i + \sum_{j,k=1}^{n} c^i_{j,k} v_j \otimes v_k.$$

Let $S = R[x_0, \ldots x_n]$. Define a comultiplication Δ and a counit ε on S by

$$\Delta(x_0) = x_0 \otimes x_0; \quad \varepsilon(x_0) = 1$$

and

$$\Delta(x_i) = x_i \otimes 1 + 1 \otimes x_i + \sum_{j,k=1}^{n} c^i_{j,k} x_j \otimes x_k$$

for $1 \leq i \leq n$; $\varepsilon(x_i) = 0$. Then S is a bialgebra, and the surjective algebra homomorphism from S to H defined by $x_i \mapsto v_i$ is a bialgebra homomorphism which is split as a coalgebra homomorphism by the coalgebra homomorphism $H \to S$ given by $v_i \mapsto x_i$. The bialgebra S is the universal commutative bialgebra on the coalgebra H [**Sw69, (3.2.4)**].

Now the comultiplication on S arises from the commutative, polynomial, degree 2, dimension n formal group law F defined by

$$F_i(x, y) = x_i + y_i + \sum_{j,k=1}^{n} c^i_{j,k} x_j y_k.$$

(where $c^0_{j,k} = 0$ for $j, k > 0$). Thus F makes $R[[x_0, \dots, x_n]]$ into a formal Hopf algebra (section 34) and the map $x_i \mapsto v_i$ induces a surjective Hopf algebra homomorphism $\phi : R[[x_0, \dots, x_n]] \to H$. Let $J = ker(\phi)$. Then

$$H = R[[x_0, \dots, x_n]]/J,$$

J a Hopf ideal of $R[[x_0, \dots, x_n]]$. Thus H is the quotient of the formal Hopf algebra $R[[x_0, \dots, x_n]]$ whose coalgebra structure is induced by a formal group law given by a polynomial of degree 2.

(37.2) EXAMPLE. Any free rank 2 R-Hopf algebra has the form $H = R[v]/(v^2 - v)$ for some b dividing 2 in R, with costructure given by $\Delta(v) = v \otimes 1 + 1 \otimes v + av \otimes v$, $\varepsilon(v) = 0$, $\lambda(v) = v$, where $ab = 2$ (see (1.5)). Then H is the quotient of the bialgebra $R[x]$ by the isogeny $\frac{1}{a}[2]_F = bx + x^2$ where

$$F(x, y) = x + y + axy,$$

$\varepsilon(x) = 0$ and Δ is induced by F.

(37.3) REMARKS.

1. The dimension of the formal group F defined in (37.1) is equal to the rank of H as an R-module. Thus even if $H \neq R$ is monogenic, F will never have dimension one. Oort and Mumford [**OM68, (5.2)**] prove that over a ring R with $p = 0$, if H is a finite local R-Hopf algebra with m parameters (i.e. the maximal ideal of H is generated by m elements), then there is a formal group F of dimension m mapping onto $Spec(H)$. However, they observe that in the proof of (37.1) given in [**Oo67**] the dimension of F is in general much bigger than the number of parameters of H.

2. The formal group law constructed in (37.1) is given by polynomials of degree ≤ 2. This observation in part motivated [**CGMSZ98**]; see Section 38.

3. The construction of (37.1) does not yield $G = Spec(H)$ as the kernel of an isogeny of formal groups. Finding such an isogeny for a given Hopf algebra seems difficult. A theorem of Lubin [**Lu67**] asserts that if G is a finite subgroup of a dimension one formal group, then G is the kernel of an isogeny f, but even in dimension one identifying f is not easy. This is illustrated by a construction by Lubin of an isogeny whose kernel is $Spec(H)$ for H a general (local) Tate-Oort Hopf algebra. For an exposition of Lubin's construction see [**Ch94, Theorem 3**]. See also (38.8) below.

4. Klapper [**Kl89, Proposition 2**] gives a proof based on [**Tt67**] that if F is an n-dimensional formal group of finite height and Γ is a any finite free subgroup of $ker[p^m]_F$, then $\Gamma = ker(h_1)$ for h_1 a homomorphism from F to some formal group

F_1 so that $[p^m]_F$ factors through h_1. As will be evident in the next section, degree 2 polynomial formal groups are rarely of finite height.

§38. Polynomial formal groups. The construction of Section 37 suggests the possible value of considering degree 2 polynomial formal groups and their isogenies. This was done in [**CS98**]. In this section we construct some examples of Hopf algebra orders in KC_p^n by modifying $[p]$ for certain n-dimensional degree 2 polynomal formal groups.

Recall that an element σ of a bialgebra H is grouplike if $\Delta(\sigma) = \sigma \otimes \sigma$. Similarly, we say that $\sigma \in R[[x]]$ is grouplike for the n-dimensional formal group F if

$$\sigma(F(x, y)) = \sigma(x) \cdot \sigma(y).$$

(38.1) Example. For the formal group F_b, let $\sigma = 1 + bx$. Then

$$\sigma(F_b(x, y)) = 1 + b(x + y + bxy) = (1 + bx)(1 + by) = \sigma(x)\sigma(y)$$

so σ is grouplike.

We begin with the dimension one case, in part to motivate the general case. As always, R is the valuation ring of a field K containing $\mathbb{Q}_p$.

(38.2). Recall $\mathbb{G}_m(x, y) = x + y + xy$, the multiplicative formal group. Using $\mathbb{G}_m$, $R[x]$ becomes an R-bialgebra with operations induced by $\varepsilon(x) = 0$ and

$$\begin{aligned}\Delta(x) &= \mathbb{G}_m(x \otimes 1 + 1 \otimes x) \\ &= x \otimes 1 + 1 \otimes x + x \otimes x.\end{aligned}$$

The antipode $\lambda(x)$ is defined by the equation

$$0 = \mathbb{G}_m(x, \lambda(x)) = x + \lambda(x) + x\lambda(x),$$

so

$$\lambda(x) = \frac{-x}{1 + x} = -x + x^2 - x^3 + \dots$$

Thus passing to the ring $R[[x]]$ of formal power series, Δ, ε and λ make $R[[x]]$ into a formal Hopf algebra. The element $\sigma = 1 + x$ is a grouplike element of the bialgebra $R[x]$, as one sees easily; since 1 together with the powers of σ form a K-basis of $K[x]$, and since grouplike elements are linearly independent (1.7), these are the only grouplike elements of $K[x]$.

The bracket endomorphisms $[m](x)$ are defined by

$$[m](x) = (1 + x)^m - 1$$

for all $m > 0$ (33.8). So $[p](x) = (1 + x)^p - 1$. If we set $H = R[x]/([p](x))$, then H is a local R-Hopf algebra of rank p, since $[p](x) \equiv x^p \pmod{p}$. Let σ be the image in H of $1 + x$, then $H = R[\sigma]$ with $\sigma^p = 1$ in K. Thus $H \cong RC_p$.

(38.3). Consider the formal group $F_b(x,y) = x+y+bxy$ for $b \in R$ (c.f (33.8)). Note that over K, if $b \neq 0$, then

$$bF_b(x,y) = \mathbb{G}_m(bx, by)$$

so the power series $f_b(x) = bx$ is an R-homomorphism $f_b : F_b \to \mathbb{G}_m$ (and an isomorphism over K, but not over R unless b is a unit of R). Then the map $[\ ]_b : \mathbb{Z} \to End(F_b)$ is given by

$$[m]_b(x) = \frac{(1+bx)^m - 1}{b} = f_b^{-1}([m]_{\mathbb{G}_m}(f_b(x))).$$

The map $f_{b^p} : F_{b^p} \to \mathbb{G}_m$ is a homomorphism over K, so the composite

$$\phi = f_{b^p}^{-1}[p]_{\mathbb{G}_m} f_b : F_b \to \mathbb{G}_m \to \mathbb{G}_m \to F_{b^p}$$

is a K-homomorphism which sends x to $\phi(x) = \frac{1}{b^p}((1+bx)^p - 1)$. If b^{p-1} divides p in R, then ϕ is defined over R, and

$$\phi(x) = \frac{p}{b^{p-1}}x + \cdots + \frac{pb^{p-2}}{b^{p-1}}x^{p-1} + x^p \ .$$

hence is an isogeny from F_b to F_{b^p}. Then $H = R[x]/(\phi(x))$ is an R-Hopf algebra of rank p, with comultiplication induced by F_b, that is,

$$\Delta(x) = F_b(x \otimes 1, 1 \otimes x) \ .$$

If σ is the image in H of the grouplike element $1+bx = 1+f_b(x)$, then $\sigma^p = 1$ and H is the Larson order

$$H = R[\frac{\sigma - 1}{b}] \ .$$

The Tate-Oort classification of Hopf algebras of rank p shows that by this construction we obtain all R-Hopf algebra orders in KC_p. To obtain Hopf orders in KC_{p^e} for $e > 1$, choose b so that b^{p^e-1} divides p in R. Dividing $[p^e]$ for $e > 1$ by b^{p^e-1} then yields the isogeny

$$\phi = f_{b^e}^{-1}[p^e]_{\mathbb{G}_m} f_b : F_b \to F_{b^{p^e}};$$

the image σ of $1+bx$ in $R[x]/(\phi(x))$ satisfies $\sigma^{p^e} = 1$, so $H = R[x]/(\phi(x))$ is a rank p^e Hopf order in KC_{p^e} of the form $H = R[\frac{\sigma-1}{b}]$, a one-parameter Larson order in KC_{p^e}.

To recapitulate this construction: given

$$f_b : F_b \to \mathbb{G}_m$$

we have

$$[p]_b(x) = f_b^{-1}[p]_{\mathbb{G}_m} f_b(x) \ ;$$

we divide $[p]_b(x)$ by the coefficient of x^p to get $\phi = f_{b^{-p}}[p]_{\mathbb{G}_m} f_b$; if ϕ is defined over R, then $\phi : F_b \to F_{b^p}$ is an isogeny and

$$R[x]/(\phi) \cong R[\frac{\sigma - 1}{b}]$$

is an R-Hopf algebra order in KC_p.

In [**CMSZ98**] and [**CS98**] this was generalized to dimensions 2 and n, respectively. We summarize that construction here.

(38.4). Suppose

$$\theta = \begin{pmatrix} u_{11} & & 0 \\ \vdots & \ddots & \\ u_{n1} & & u_{nn} \end{pmatrix} \in M_n(R) \cap GL_n(K)$$

is lower triangular. We set $F_\theta = \theta^{-1}(\mathbb{G}_m^n(\theta(\overline{x}), \theta(\overline{y}))$. Suppose F_θ is defined over R. Then $f_\theta : F_\theta \to \mathbb{G}_m^n$, $f_\theta(x) = \theta x$, is a homomorphism. We have $\theta[p]_\theta(x) = [p]_{\mathbb{G}_m}(\theta(x)) =$

$$\begin{pmatrix} (1 + u_{11}x_1)^p - 1 \\ (1 + u_{21}x_1 + u_{22}x_2)^p - 1 \\ \vdots \\ (1 + u_{n1}x_1 + \cdots + u_{nn}x_n)^p - 1 \end{pmatrix} = pv + \theta^{(p)}x^{(p)}$$

with $v \in R^n$, where for a matrix $A = (a_{ij})$ we set $A^{(p)} = (a_{ij}^p)$.

Consider then,

$$\gamma(x) = f_{\theta^{(p)}}^{-1}[p]_{\mathbb{G}_m} f_\theta(x) = (\theta^{(p)})^{-1}[p]_{\mathbb{G}_m}(\theta(x)) \ .$$

Over K, γ is a homomorphism from F_θ to $F_{\theta^{(p)}}$. Suppose

$$p(\theta^{(p)})^{-1} \in \pi M_n(R).$$

Then

$$\gamma(x) = (\theta^{(p)})^{-1}[p]_{\mathbb{G}_m}(\theta(x)) \equiv (\theta^{(p)})^{-1}pv + x^{(p)} \pmod{\pi R[x]^n}.$$

So γ is an isogeny and $H = R[x]/(\gamma)$ is an R-Hopf algebra, with operations induced from F_θ.

The degree one grouplike elements of $\mathbb{G}_m^n$ are $1 + x_i$, $i = 1, \ldots, n$. So the degree one grouplikes of F_θ are $\sigma_j = 1 + u_{j1}x_1 + \cdots + u_{jj}x_j$; hence

$$\begin{pmatrix} \sigma_1 - 1 \\ \vdots \\ \sigma_n - 1 \end{pmatrix} = \theta \begin{pmatrix} x_1 \\ \vdots \\ x_n \end{pmatrix} ,$$

and so H has generators

$$\begin{pmatrix} x_1 \\ \vdots \\ x_n \end{pmatrix} = \theta^{-1} \begin{pmatrix} \sigma_1 - 1 \\ \vdots \\ \sigma_n - 1 \end{pmatrix} . \tag{38.5}$$

Now $\sigma_i^p = 1$ in H. For over K, $K \otimes H \cong K[x]/[p]_{F_\theta}(x)$, since θ and $\theta^{(p)}$ are invertible in $M_n(K)$. But one sees by induction that if σ is grouplike in H then

$\sigma(x)^m = \sigma([m](x))$, hence $\sigma(x)^p = \sigma(0) = 1$ in H. Thus (38.5) describes the generators of H as an algebra in terms of generators of the group C_p^n. If

$$\theta^{-1} = \begin{pmatrix} w_{11} & & 0 \\ \vdots & \ddots & \\ w_{n1} & & w_{nn} \end{pmatrix}$$

then

$$H = R[\frac{\sigma_1 - 1}{w_{1,1}}, \dots, \frac{\sigma_1 - 1}{w_{n,1}} + \dots + \frac{\sigma_n - 1}{w_{n,n}}].$$

(38.6) EXAMPLE. Let $p = 3$, and let

$$\theta = \begin{pmatrix} \pi^m & 0 \\ u\pi^\ell & \pi^n \end{pmatrix} .$$

We first identify F_θ. Now

$$\mathbb{G}_m^2(x, y) = \begin{pmatrix} x_1 + y_1 + x_1 y_1 \\ x_2 + y_2 + x_2 y_2 \end{pmatrix} ,$$

and

$$\theta(\overline{x}) = \begin{pmatrix} \theta_1(\overline{x}) \\ \theta_2(\overline{x}) \end{pmatrix} = \begin{pmatrix} \pi^m x_1 \\ u\pi^\ell x_1 + \pi^n x_2 \end{pmatrix} .$$

So

$$F = \binom{F_1}{F_2} = \theta^{-1}(\mathbb{G}_m^2(\theta(\overline{x}), \theta(\overline{y}))$$

where

$$F_1 = x_1 + y_1 + \pi^m x_1 y_1$$

and

$$F_2 = x_2 + y_2 + (u^2\pi^{2\ell-n} - u\pi^{\ell+m-n}) x_1 y_1 + u\pi^\ell(x_1 y_2 + x_2 y_1) + \pi^n x_2 y_2 .$$

Hence F is defined over R iff $\ell, m, n \geq 0$ and $u^2\pi^{2\ell-n} - u\pi^{\ell+m-n} \in R$, that is,

$$u(\pi^{\ell-n})(u\pi^\ell - \pi^m) \in R. \tag{38.7}$$

If $\ell = m$, we need $\pi^{2\ell-n}(u-1) \in R$;
if $\ell < m$ we need $2\ell - n \geq 0$;
if $\ell > m$ we need $\ell + m - n \geq 0$.

Let us assume for the rest of the example that $\ell > m$. Then (38.7) is the condition $\ell + m \geq n$. (The other cases are similar.)

Now we identify the isogeny $\gamma(x) = (\theta^{(3)})^{-1}[3]_{\mathbb{G}_m}(\theta(x))$.

We have

$$\theta^{(3)} = \begin{pmatrix} \pi^{3m} & 0 \\ u^3\pi^{3\ell} & \pi^{3n} \end{pmatrix}$$

and

$$(\theta^{(3)})^{-1} = \begin{pmatrix} \pi^{-3m} & 0 \\ -u^3\pi^{3(\ell-m-n)} & \pi^{-3n} \end{pmatrix} .$$

Suppose $3(\theta^{(3)})^{-1} \in \pi M_2(R)$, that is,

$$\begin{aligned} 3m &< e \\ 3n &< e \text{ and} \\ 3m + 3n &< e + 3\ell \end{aligned}$$

where $e = ord_K(3)$. Then

$$\gamma(x) = (\theta^{(3)})^{-1}[3]_{\mathbb{G}_m}\theta(x)$$

$$= \begin{pmatrix} \pi^{-3m} & 0 \\ -u^3\pi^{3(\ell-m-n)} & \pi^{-3n} \end{pmatrix} \begin{pmatrix} (1+\pi^m x_1)^3 - 1 \\ (1+u\pi^\ell x_1 + \pi^n x_2)^3 - 1 \end{pmatrix} = \begin{pmatrix} \gamma_1 \\ \gamma_2 \end{pmatrix}$$

is an isogeny from F_θ to $F_{\theta^{(3)}}$, where

$$\gamma_1 = \pi^{-3m}[(1+\pi^m x_1)^3 - 1] = 3\pi^{-2m}x_1 + 3\pi^{-m}x_1^2 + x_1^3$$

and

$$\begin{aligned} \gamma_2 = {} & 3(u\pi^{\ell-3n} - u^3\pi^{3\ell-2m-3n})x_1 + 3\pi^{-2n}x_2 + (3u^2\pi^{2\ell-3n} - 3u^3\pi^{3\ell-m-3n})x_1^2 \\ & + 6u\pi^{\ell-2n}x_1x_2 + 3\pi^{-n}x_2^2 + 3u^2\pi^{2\ell-2n}x_1^2x_2 + 3u\pi^{\ell-n}x_1x_2^2 + x_2^3. \end{aligned}$$

Assuming $\ell > m$, F is defined over R and γ is an isogeny if the following inequalities hold:

$$\begin{aligned} \ell &> m \\ \ell + m &\geq n \quad \text{(from (38.7))} \\ e &\geq 3m \\ e &\geq 3n, \text{ and} \\ e + 3\ell &\geq 3m + 3n, \text{ which follows from the other inequalities.} \end{aligned}$$

Then

$$H = R[x_1, x_2]/(\gamma_1, \gamma_2) = R[t_1, t_2]$$

where

$$\begin{pmatrix} t_1 \\ t_2 \end{pmatrix} = \begin{pmatrix} \pi^{-m} & 0 \\ -u\pi^{\ell-m-n} & \pi^{-n} \end{pmatrix} \begin{pmatrix} \sigma_1 - 1 \\ \sigma_2 - 1 \end{pmatrix},$$

hence

$$\begin{aligned} t_1 &= \frac{\sigma_1 - 1}{\pi^m} \\ t_2 &= -u\pi^{\ell-n}\frac{(\sigma_1 - 1)}{\pi^m} + \frac{\sigma_2 - 1}{\pi^n}. \end{aligned}$$

If $\ell - n \geq 0$ then H is the Larson order $R[\frac{\sigma_1-1}{\pi^m}, \frac{\sigma_2-1}{\pi^n}]$. We obtain non-Larson examples, for example, for any ℓ, m, n with $0 < m \leq \ell < n \leq \ell + m \leq \lfloor \frac{e}{3} \rfloor$.

(38.8) REMARKS. Greither [**Gr92, Part I**] classified Hopf orders in KG where G is elementary abelian of order p^2 and also where G is cyclic of order p^2. The question of which Hopf orders in the elementary abelian case can be obtained via a polynomial isogeny of the type described in (38.4) (with $n = 2$) was answered in [**CS98, Theorem 4.3**]. Greither identified the Hopf orders in KG, $G = C_p^2$, with a quotient of groups of principal units $U_{m'+\frac{n}{p}}/U_{m'+n}$. Those arising from isogenies of the form in (38.5) correspond to $U_{m'+\frac{n}{2}}/U_{m'+n}$. Underwood [**Un99**] showed that for G cyclic of order p^2, Hopf orders in KG corresponding to the same quotient of units groups $U_{m'+\frac{n}{2}}/U_{m'+n}$ (see Chapter 9) can be obtained via a polynomial isogeny of degree 2 polynomial formal groups. None of the Hopf orders obtained by Underwood are realizable, i. e. are the associated orders of valuation rings of Galois extensions of K with Galois group G (see (32.10)). It remains open how to represent realizable Hopf orders in KG as Hopf algebras of kernels of isogenies of formal groups.

(38.9). One strategy to classify finite abelian R-Hopf algebras is to classify k-Hopf algebras for k the residue field of R, and identify all their lifts to R. The first results in this direction were in [**OM68**], namely, that the local-local group scheme α_p over $\mathbb{F}_p$ of order p whose Hopf algebra is constructed in (15.2) does not lift to $\mathbb{Z}_p$, but that every abelian k-group scheme lifts to some ramified extension of the Witt ring W(k). In fact, [**Rb92**] shows that for $p > 3$ the ramification index $e_{L/K}$ can have any value between 1 and $p - 1$.

In view of (39.9), below, it is of particular interest to classify and lift monogenic local-local k-Hopf algebras H, where *monogenic* means that as an algebra, H is the quotient of a polynomial ring in one variable. A. Koch [**Kc99**] has obtained results in this area. Using Dieudonné module theory, Koch has classified the monogenic local-local abelian k-Hopf algebras for $k = \mathbb{F}_q$ any finite field, and has determined which k-Hopf algebras lift to the Witt ring $W(k)$ (= the unique unramified extension of $\mathbb{Z}_p$ with residue field k). If a given k-Hopf algebra H lifts to $W(k)$, he counts the number of lifts, and in particular, determines which H lift uniquely. If $k = \mathbb{F}_p$ then every lift is unique, but for $q = p^m$ with $m > 1$, there exist k-Hopf algebras with more than one lift to $W(k)$. His result yield (complicated) formulas for the number of lifts of a given k-Hopf algebra, and also for the number of monogenic local-local abelian $W(k)$-Hopf algebras of rank p^n. See [**Kc95**], [**Kc98a**] and [**Kc98b**] for related work.

For ramified extensions of $W(k)$ the situation is yet more complicated: Chapter 6 and [**CZ94**] hint at the complexity.

CHAPTER 12

Principal Homogeneous Spaces and Formal Groups

This chapter develops the Kummer theory of formal groups and applies it to a class of examples of Byott which have unsatisfactory classical Galois module structure but satisfactory non-classical structure.

§**39. Kummer extensions.** In this section we show: if the dual of a finite abelian R-Hopf algebra H can be presented as H_f for some isogeny $f : F \to G$ of formal groups, then there is an explicit classification and description as algebras of all the H_f-Galois extensions, or H_f-Galois objects. The idea, a generalization of classical Kummer theory, arose in an informal manuscript of Martin Taylor [**Ta86**]; the theory was developed in [**CM94**] for dimension one formal groups, and in [**Mo94**], [**Mo96**] for dimension n formal groups.

The construction generalizes the construction of pure extensions.

A pure extension of a field K is a field $L = K[z]$ where $z^q \in K$. Pure extensions play a special role in algebraic number theory. If $z^q = 1$ then L is a cyclotomic extension of K; if K already contains a primitive qth root of unity, and $z^q \in K$ then $L = K[z]$ is a Kummer extension of K. A polynomial $f(x)$ with coefficients in K is solvable by radicals iff the roots of f are contained in a field L obtained from K by a succession of pure extensions. The Kronecker-Weber theorem is that every abelian extension of $\mathbb{Q}$ is contained in a cyclotomic extension of $\mathbb{Q}$. Kummer extensions are used in class field theory to construct class fields of ideal groups [**Ja73, Chapter V**].

Any pure extension $L = K[z], z^q = a \neq 0 \in K$, of degree q, is an $H = KC_q^*$-Hopf Galois extension of K, or equivalently, a KC_q-Galois object, since the C_q-grading on L defines in a natural way a KC_q-comodule algebra structure on L,

$$\alpha : L \to L \otimes KC_q,$$

by $\alpha(z) = z \otimes \sigma$ if z has grade σ. Similarly, the Kummer extensions looked at in this chapter are most naturally looked at as Galois objects, or principal homogeneous spaces.

To illustrate the general construction we first consider the formal group $\mathbb{G}_m(x, y) = x + y + xy$. Recall that

$$[n]_{\mathbb{G}_m}(x) = (1 + x)^n - 1;$$

Let

$$K_a = K[x]/([n]_{\mathbb{G}_m}(x) - a)$$

then setting $1+x=z$, the extension $K_a \cong K[z]/(z^n-(1+a))$ is a pure extension. If $a=0$ then the roots of $[n]$ in $\overline{K}$,

$$\mu_n = \{\alpha \in \overline{K} | [n]_{\mathbb{G}_m}(\alpha) = 0\},$$

are translates of the nth roots of unity in $\overline{K}$, and the smallest subfield of $\overline{K}$ containing μ_n is a cyclotomic extension of K. Let $H_n = K[x]/([n]_{\mathbb{G}_m}(x))$, then $H_n \cong KC_n$ via the map sending σ to $x+1$; if K contains a primitive nth root of unity, then K_a is a Kummer extension of K (i.e. is Galois with group C_n). With or without roots of unity in K, K_a is an H_n-Galois object.

When the formal group $\mathbb{G}_m$ is replaced by an arbitrary p-divisible formal group F of dimension n defined over K one obtains cyclotomic and pure extensions of K for the formal group F. An element α of $\overline{\mathfrak{m}}^n$ so that $[n]_F(\alpha)=0$, an n-torsion point, can be viewed as a generalization of an nth root of unity, and so $K[\ker[n]_F]$ is an F-cyclotomic extension. The K-algebra $K[x]/([n]_F(x)-a)$ is an F-pure extension.

The Lubin-Tate version of local class field theory embeds any abelian extension of a local field K into an F-cyclotomic extension of K for F a Lubin-Tate formal group, thereby providing an elegant local analogue of the Kronecker-Weber Theorem – every abelian extension of $\mathbb{Q}_p$ is contained in a cyclotomic extension of $\mathbb{Q}_p$. See [**Se67**].

The theory of pure extensions works for any isogeny, not just for $[n]_F$, as we now show.

Suppose R is the valuation ring of a finite extension K of $\mathbb{Q}_p$, with maximal ideal $\mathfrak{m}$. Let F, G be formal groups of dimension n defined over R and f an isogeny from F to G, also defined over R. Then, setting $X=(X_1,\dots,X_n)$, $x=(x_1,\dots,x_n)$, the algebra

$$H = R[[X]]/(f) = R[[x]],$$

with $f_i(x)=0$ for $i=1,\dots,n$, is a finite R-Hopf algebra representing $\ker(f)$. The comultiplication is induced by F.

(39.1) Proposition. *For any* $c=(c_1,\dots,c_n)\in\mathfrak{m}^n$, *let*

$$S = R[[T]]/(f(T)-c) = R[[t]],\ t=(t_1,\dots,t_n)$$

with $f_i(t)=c_i$ *for* $i=1,\dots,n$. *Then* S *is an* H*-Galois object.*

Proof. Define $\alpha: S\to S\otimes H$ by

$$\alpha(t_i) = F_i(t_1\otimes 1,\dots,t_n\otimes 1, 1\otimes x_1,\dots,1\otimes x_n) = F_i(t\otimes 1, 1\otimes x)$$

for $i=1,\dots,u$. To show that α yields a well-defined algebra homomorphism, we need to show that $\alpha(f_i)=\alpha(c_i)$ for $i=1,\dots,n$. This follows since f is an isogeny: we have

$$\alpha(c_i) = c_i\alpha(1) = c_i(1\otimes 1)$$

while

$$\begin{aligned}
&\alpha(f_i(t_1,\dots,t_n)) \\
&\quad = f_i(\alpha(t_1),\dots,\alpha(t_n)) \\
&\quad = f_i(F_1(t\otimes 1, 1\otimes x),\dots,F_n(t\otimes 1,1\otimes x)) \\
&\quad = G_i(f_1(t\otimes 1),\dots,f_n(t\otimes 1), f_1(1\otimes x),\dots,f_n(1\otimes x)) \\
&\quad = G_i(f_1(t)\otimes 1,\dots,f_n(t)\otimes 1, 1\otimes f_1(x),\dots,1\otimes f_n(x)) \\
&\quad = G_i(c_1\otimes 1,\dots,c_n\otimes 1,0,\dots,0) \\
&\quad = c_i\otimes 1 \quad .
\end{aligned}$$

Now we verify that α makes S into a right H-comodule: that is, $(1\otimes\Delta)\alpha = (\alpha\otimes 1)\alpha$ and $(1\otimes\varepsilon)\alpha = id$. The first follows by associativity of F:

$$\begin{aligned}
(\alpha\otimes 1)\alpha(t_i) &= (\alpha\otimes 1)(F_i(t\otimes 1,1\otimes x)) \\
&= F_i(\alpha(t)\otimes 1, 1\otimes 1\otimes x) \\
&= F_i(F(t\otimes 1,1\otimes x)\otimes 1, 1\otimes 1\otimes x)) \\
&= F_i(F(t\otimes 1\otimes 1, 1\otimes x\otimes 1), 1\otimes 1\otimes x) \\
&= F_i(t\otimes 1\otimes 1, F(1\otimes x\otimes 1, 1\otimes 1\otimes x)) \\
&= F_i(t\otimes 1\otimes 1, 1\otimes\Delta(x)) \\
&= (1\otimes\Delta)(F_i(t\otimes 1,1\otimes x)) = (1\otimes\Delta)\alpha(t_1) \quad .
\end{aligned}$$

The second, $(1\otimes\varepsilon)\alpha = id$:

$$\begin{aligned}
&(1\otimes\varepsilon)\alpha(t_i) \\
&\quad = (1\otimes\varepsilon)(F_i(t\otimes 1,1\otimes x)) \\
&\quad = F_i(t\otimes 1,0) = t_i \quad .
\end{aligned}$$

Thus S is a right H-comodule.

Now we show that S is an H-Galois object, that is, that the R-module homomorphism

$$\begin{aligned}
\gamma : S\otimes S &\to S\otimes H \ , \\
\gamma(s\otimes t) &= (s\otimes 1)\alpha(t)
\end{aligned}$$

is bijective. Since both domain and range of γ are free of rank $[H:R]^2$, it suffices to show that γ is surjective. For that, it suffices to show that for any $h\in H$, $1\otimes h$ is in the image of γ.

Let $h = m(x)\in H$. We claim $\gamma(m(F([-1](t)\otimes 1, 1\otimes t))) = 1\otimes m(x)$. To see this, we have

$$\gamma(m(F([-1](t)\otimes 1,1\otimes t))) = m(\gamma(F([-1](t)\otimes 1,1\otimes t)));$$

but

$$\begin{aligned}\gamma(F_i([-1](t)\otimes 1, 1\otimes t)) &= F_i(\gamma([-1](t)\otimes 1), \gamma(1\otimes t))\\ &= F_i(([-1](t)\otimes 1)\alpha(1), \alpha(t))\\ &= F_i([-1](t)\otimes 1, F(t\otimes 1, 1\otimes x))\\ &= F_i(F([-1](t)\otimes 1, t\otimes 1), 1\otimes x)\\ &= F_i(F([-1](t), t)\otimes 1, 1\otimes x)\\ &= F_i(0, 1\otimes x) = 1\otimes x_i.\end{aligned}$$

So

$$\gamma(m(F([-1](t)\otimes 1, 1\otimes t) = m(1\otimes x) = 1\otimes m(x),$$

as we wished to show. Thus S is an H Galois object. □

(39.2) EXAMPLES. Suppose F, G are formal groups of dimension one, and f_0 is a homomorphism from F to G of height $h < \infty$. Then, modulo the maximal ideal $\mathfrak{m}$ of R, $f_0(x) = g(x^q)$ with $g'(0) \neq 0$, where $q = p^h$. Thus

$$f_0(x) = a_1 x + a_2 x^2 + \ldots + a_q x^q + \ldots$$

where $a_1, \ldots, a_{q-1} \in \mathfrak{m}$ and a_q is a unit of R. Let $c \in \mathfrak{m}$. Then $f(x) = f_0(x) - c$ is congruent to $f_0(x)$ modulo $\mathfrak{m}$. So by the Weierstrass Preparation Theorem,

$$f(x) = p(x)q(x)$$

where $q(x)$ is an invertible power series in $R[[x]]$ and $p(x)$ is a monic polynomial of degree q which is congruent to x^q modulo $\mathfrak{m}$.

Suppose $c \in \mathfrak{m} \setminus \mathfrak{m}^2$, so is a parameter for R. Then, comparing constant terms, we see that $p(0) \in \mathfrak{m} \setminus \mathfrak{m}^2$, so $p(x)$ is an Eisenstein polynomial. Thus $p(x)$ is irreducible over K and $S_c = R[[x]]/(f(x)) \cong R[[x]]/(p(x)) \cong R[x]/(p(x))$ (see (35.5)), so L_c is a field extension of K of degree q, and S_c is the valuation ring of L_c and is an H_f-Galois object.

Then L_c is a Hopf Galois object for $K \otimes H_f$.

(39.3) PROPOSITION. *L_c is normal over K (i.e. a classical Galois extension of K) iff $ker(f_0) \subset L_c$.*

PROOF. Write $L_c = K[\gamma]$, where $\gamma \in \overline{K}$ is a root of $f(x)$, i.e. $f(\gamma) = 0$, or $f_0(\gamma) = c$. Using the Weierstrass Preparation Theorem, write $f_0(x) = p_0(x)q_0(x)$ with $q_0(x)$ invertible and $p_0(x)$ a monic polynomial of degree q. Then $ker(f_0) = ker(p_0)$. Thus the cardinality of $ker(f_0)$ is q, provided that $f_0(x)$ has distinct roots in $\overline{K}$. Assuming that for the moment, define a map

$$\phi : ker(f_0) \to Aut_K(L_c)$$

by

$$\phi(\alpha)(g(\gamma)) = g(\gamma +_F \alpha)$$

for $g(x)$ any polynomial with coefficients in K and α in $ker(f_0)$.

To see that $\phi(\alpha)$ is well-defined on L_c, note that

$$f_0(\gamma +_F \alpha) = f_0(F(\gamma, \alpha)) = G(f_0(\gamma), f_0(\alpha)) = G(c, 0) = c.$$

Hence $\gamma +_F \alpha$ is a root of $f(x)$ for any $\alpha \in ker(f_0)$. Evidently ϕ is a homomorphism (where $ker(f_0)$ is a group via $+_F$) and is 1-1, since if $\gamma +_F \alpha = \gamma$, then adding $[-1]_F(\gamma)$ to both sides yields $\alpha = 0$. So ϕ will be onto, hence an isomorphism, if f_0, equivalently p_0, has q distinct roots.

To see this, differentiate

$$f_0(F(x,y)) = G(f_0(x), f_0(y))$$

with respect to y as in (35.1): one obtains

$$f_0'(x)F_y(x,0) = G_y(f_0(x),0)f_0'(0),$$

where $G_y(x,0)$ and $F_y(x,0)$ are invertible power series. Thus if $f_0'(0) = 0$, then $f_0'(x) = 0$, hence $f_0(x) = 0$. Thus $f_0'(0) \neq 0$. Setting $x = \alpha$, a root of $f_0(x)$ (or $p_0(x)) \in \overline{\mathfrak{m}}$, we find that since $G_y(0,0)f'(0) \neq 0$, $f_0'(\alpha) \neq 0$. Thus α is not a multiple root of $f_0(x)$.

Hence $\phi : ker(f_0) \to Aut_K(L_c)$ is an isomorphism, and L_c is a Galois extension of K with Galois group isomorphic to $ker(f_0)$. □

Let $PH(H_f)$ denote the set of H_f-comodule algebra isomorphism classes of H_f-Galois objects, or H_f-principal homogeneous spaces (2.10). Let $[S]$ be the class of the H_f-Galois object S. Then $PH(H_f)$ has a group structure defined by $[S][T] = [S \cdot T]$ where

$$\begin{aligned} S \cdot T = \Big\{ \sum_i s_i \otimes t_i \in S \otimes T \,\Big| \\ \sum_{i,(t_i)} s_i \otimes t_{i,(0)} \otimes t_{i,(1)} = \sum_{i,(s_i)} s_{i,(0)} \otimes t_i \otimes s_{i,(1)} \Big\} \\ = \Big\{ \sum_i s_i \otimes t_i) | (1 \otimes \alpha_T)(\sum s_i \otimes t_i) = (1 \otimes \tau)(\alpha_s \otimes 1)(\sum_i s_i \otimes t_i) \Big\} \end{aligned}$$

with comodule structure $1 \otimes \alpha_T : S \cdot T \to S \cdot T \otimes H$ (c.f. [**Bt76**], [**Ch86**]). The identity element is $[H_f]$, the class of the trivial H_f-Galois object (see (2.10a)).

(39.4) PROPOSITION. *Let $\mathfrak{m}_G^n$ be $\mathfrak{m}^n$ with the group structure induced by the formal group G. Then*

$$S : \mathfrak{m}_G^n \to PH(H_f)$$

is a homomorphism of groups.

PROOF. Let

$$S_c = R[[t]] \text{ with } f(t) = c\ ,$$
$$S_{c'} = R[[w]] \text{ with } f(w) = c'\ ,$$

and $T = S_{G(c,c')} = R[[z]]$ with $f(z) = G(c,c')$. We need to show that

$$S_c \cdot S_{c'} = S_{G(c,c')}\ .$$

To do this we show that $S_{G(c,c')}$ maps into $S_c \cdot S_{c'}$ by an R-algebra, H_f-comodule homomorphism.

Let $\beta : T \to S_c \otimes S_{c'}$ by $\beta(z_i) = F_i(t \otimes 1, 1 \otimes w)$. Then β is a well-defined R-algebra homomorphism:

$$\begin{aligned}\beta(f_i(z)) &= f_i(\beta(z))\\ &= f_i(F(t \otimes 1, 1 \otimes w))\\ &= G_i(f(t) \otimes 1, 1 \otimes f(w))\\ &= G_i(c \otimes 1, 1 \otimes c')\\ &= G_i(c, c')(1 \otimes 1)\\ &= \beta(G_i(c, c')) \ .\end{aligned}$$

We show that the image of β is contained in $S_c \cdot S_{c'}$. To do this, we check that $(1 \otimes \tau)(\alpha_{s_c} \otimes 1)\beta = (1 \otimes \alpha_{s_{c'}})\beta$:

$$\begin{aligned}(1 \otimes \alpha_{s_{c'}})\beta(z_i) &= (1 \otimes \alpha_{s_{c'}})(F_i(t \otimes 1, 1 \otimes w)\\ &= F_i(t \otimes 1 \otimes 1, F(1 \otimes w \otimes 1, 1 \otimes 1 \otimes x)) \in S_c \otimes S_{c'} \otimes H_f\end{aligned}$$

while

$$\begin{aligned}&(1 \otimes \tau)(\alpha_{S_c} \otimes 1)(\beta(z_i)\\ &\quad = (1 \otimes \tau)(\alpha_{S_{c'}} \otimes 1)(F_i(t \otimes 1, 1 \otimes w))\\ &\quad = (1 \otimes \tau)F_i(F(t \otimes 1, 1 \otimes x) \otimes 1, 1 \otimes w)\\ &\quad = F_i(F(t \otimes 1 \otimes 1, 1 \otimes 1 \otimes x), 1 \otimes w \otimes 1) \ .\end{aligned}$$

By associativity and commutativity of F, these are equal. Thus β is an R-algebra homomorphism from

$$T = S_{G(c,c')} \text{ to } S_c \cdot S_{c'} \ .$$

We check that β is an H-comodule homomorphism, that is,

$$(1 \otimes \alpha_{S_{c'}})\beta = (\beta \otimes 1)\alpha_T.$$

We just showed that

$$(1 \otimes \alpha_{S_{c'}})\beta(z_i) = F_i(t \otimes 1 \otimes 1, F(1 \otimes w \otimes 1, 1 \otimes 1 \otimes x)) \ .$$

On the other hand,

$$\begin{aligned}&(\beta \otimes 1)\alpha_T(z_i)\\ &\quad = (\beta \otimes 1)(F_i(z \otimes 1, 1 \otimes x))\\ &\quad = F_i(\beta(z) \otimes 1, 1 \otimes x)\\ &\quad = F_i(F(t \otimes 1, 1 \otimes w) \otimes 1, 1 \otimes x) \ ;\end{aligned}$$

thus $(1 \otimes \alpha_{S_{c'}})\beta = (\beta \otimes 1)\alpha_T$ by associativity of F.

Since β is an R-algebra, H-comodule homomorphism from one H-Galois object to another, β is an isomorphism, by (14.4).

Hence the map S is a homomorphism from $\mathfrak{m}_G^n$ to $PH(H_f)$. □

Now we show that the map $c \mapsto S_c$ induces a 1-1 homomorphism from $\mathfrak{m}_G^n / f(\mathfrak{m}_F^n)$ to $PH(H_f)$. Recall that

$$S_c = R[[t]] \text{ with } f(t) = c\ .$$

(39.5) PROPOSITION. *S_c is isomorphic to H_f as H_f-Galois objects iff $c = f(d)$ for some $d \in \mathfrak{m}^n$.*

PROOF. Suppose $c = f(d)$ for $d \in \mathfrak{m}^n$. Define $\phi : S_c \to H_f$ by

$$\phi(t) = F(x, d)$$

and

$$\psi : H_f \to S_c$$

by

$$\psi(x) = F(t, [-1]_F(d))\ .$$

Then ϕ and ψ are inverses, for

$$\begin{aligned}(\phi \cdot \psi)(x) = \phi(\psi(x)) &= \phi(F(t, [-1]_F(d)) \\ &= F(F(x, d), [-1]_F(d)) \\ &= F(x, F(d, [-1]_F(d)) \\ &= F(x, 0) = x\end{aligned}$$

and the opposite composition is similar.

We show that ϕ is a homomorphism of H_f-Galois objects, that is, ϕ is an R-algebra, H_f-comodule homomorphism.

To show ϕ is a well-defined R-algebra homomorphism we need to show that

$$\phi(f_i(t) - c_i) = 0$$

for $i = 1, \ldots, n$. But $\phi(c_i) = c_i$, while since $\phi(q(t)) = q(\phi(t)) = q(F(x, d))$ for any $q(T) \in R[[T]]$,

$$\begin{aligned}\phi(f_i(t)) &= f_i(F(x, d)) \\ &= G_i(f(x), f(d)) \\ &= G_i(f(x), c) \\ &= G_i(0, c) = c_i\end{aligned}$$

since $f(x) = 0$ in H_f. The argument that ψ is well- defined is similar.

Now we need to show that ϕ is an H_f-comodule homomorphism, that is,

$$\begin{array}{ccc} S_c & \xrightarrow{\phi} & H_f \\ \alpha \downarrow & & \downarrow \Delta \\ S_c \otimes H_f & \xrightarrow{\phi \otimes 1} & H_f \otimes H_f \end{array}$$

commutes, or

$$(\phi \otimes 1)\alpha(t) = \Delta\phi(t)\ .$$

Now

$$\begin{aligned}(\phi \otimes 1)\alpha(t) &= (\phi \otimes 1)(F(t \otimes 1, 1 \otimes x) \\ &= F(\phi(t) \otimes 1, 1 \otimes x) \\ &= F(F(x, d) \otimes 1, 1 \otimes x) \\ &= F(F(x \otimes 1, d \otimes 1), 1 \otimes x).\end{aligned}$$

while

$$\begin{aligned}\Delta\phi(t) &= \Delta(F(x, d)) \\ &= F(\Delta(x), \Delta(d)) \\ &= F(F(x \otimes 1, 1 \otimes x), d \otimes 1) \\ &= F(x \otimes 1, F(1 \otimes x, d \otimes 1)) \\ &= F(x \otimes 1, F(d \otimes 1, 1 \otimes x)) \\ &= F(F(x \otimes 1, d \otimes 1), 1 \otimes x) .\end{aligned}$$

Thus $S_c \cong H_f$ as H_f-Galois objects if $f(d) = c$ for some $d \subset \mathfrak{m}^n$.

For the opposite implication, suppose

$$\varphi : S_c \to H_f$$

is an R-algebra, H-comodule isomorphism. Then $\varphi(t_i) = q_i(x) \in H_f$ for some polynomial $q_i(X) \in R[X]$. We have

$$0 = \varphi(f_i(t) - c_i) = f_i(q(x)) - c_i$$

so

$$\begin{aligned}0 &= \varepsilon(f_i(q(x)) - c_i \\ &= f_i(q(0)) - c_i \\ &= f_i(d) - c_i\end{aligned}$$

for $d = q(0) \in R^n$.

To see that $d \in \mathfrak{m}^n$, observe that if $k = R/\mathfrak{m}$ then $j : k \otimes_R H_f \cong k \otimes_R S_c$ by $j(1 \otimes x_i) = 1 \otimes t_i$ is an isomorphism of k-algebras since $c \in \mathfrak{m}^n$. Thus $(1 \otimes \varphi)j : k \otimes H_f \to k \otimes H_f$ is a k-algebra isomorphism and $(1 \otimes \varphi)j(1 \otimes x_i) = 1 \otimes q_i(x)$. Now H_f is a local ring since it is the quotient of $R[[x_1, \ldots, x_n]]$ by an ideal $(f_1, \ldots f_n) \subset (x_1, \ldots, x_n) = \ker \varepsilon$. Thus $k \otimes H_f$ is a local ring with maximal ideal $\ker(\varepsilon)$, and $(1 \otimes \varphi)j$ maps $\ker(\varepsilon)$ to $\ker(\varepsilon)$. Since $1 \otimes x_i \in \ker(\varepsilon)$, we have $(1 \otimes \varphi)j(1 \otimes x_i) \in \ker(\varepsilon)$, for all i, and so

$$\begin{aligned}0 = \varepsilon(1 \otimes \varphi)j(1 \otimes x_i)) &= \varepsilon(1 \otimes q_i(x)) \\ &= 1 \otimes \varepsilon(q_i(x)) \\ &= 1 \otimes q_i(0) \\ &= 1 \otimes d_i \quad .\end{aligned}$$

Hence $d_i \in \mathfrak{m}$.

That completes the proof. □

If H_f represents the kernel of an isogeny $f : F \to G$ of formal groups, then the map $S : c \mapsto S_c$ induces an isomorphism from $\mathfrak{m}_G^n / f(\mathfrak{m}_F^n)$ onto $PH(H_f)$. One may see this by identifying $PH(H_f)$ as a first cohomology group H^1, and applying results of Mazur and Roberts [**MR70**], namely

i) $\mathfrak{m}_G^n / F(\mathfrak{m}_F^n) \cong H^1$ and

ii) $\mathfrak{m}_G^n / f(\mathfrak{m}_F^n)$ is finite (cf. also [**Mo96**], section 4).

Since S induces a 1-1 homomorphism between finite sets of the same cardinality, S is onto. Thus the map S provides an explicit realization of an isomorphism

$$\mathfrak{m}_G^n / f(\mathfrak{m}_F^n) \to H^1 \to PH(H_f)$$

described cohomologically by Mazur and Roberts.

(39.6) EXAMPLE. Let $F(x, y) = x + y + bxy$ with b in R. Then

$$[q](x) = \frac{(1 + bx)^q - 1}{b} \text{ for all } q \geq 0 \; .$$

For any c in K, $f(x) = \frac{1}{c}x$ is an isogeny from F_b to F_{bc}: for

$$\begin{aligned} f(F_b(x, y)) &= \frac{1}{c}(x + y + bxy) \\ &= f(x) + f(y) + bcf(x)f(y) \\ &= F_{bc}(f(x), f(y)) \; . \end{aligned}$$

Thus $g(x) = \frac{1}{b^{q-1}}[q](x)$ is an isogeny from F_b to F_{b^q}, provided that $g(x)$ is defined over R.

Suppose $q = p^d$, then

$$g(x) = \frac{1}{b^q}\left(\sum_{r=1}^{q-1} \binom{p^d}{r} b^r x^r\right) + x^q$$

is defined over R iff b^{q-r} divides $\binom{p^d}{r}$ for $r = 1, \ldots, q - 1$, iff $b^{p^d - p^{d-1}}$ divides $\binom{p^d}{p^{d-1}}$, iff $b^{p^{d-1}(p-1)}$ divides p. In that case, $H_g = R[x]/(g(x))$ is an R-Hopf algebra. Then $K \otimes H_g \cong R(1 + bx)$ with $(1 + bx)^q = 1$, so $K \otimes H_g \cong K[G]$, G cyclic of order q, and if $G = \langle\sigma\rangle$, $\sigma = 1 + bx$, then

$$H_g = R\left[\frac{\sigma - 1}{b}\right] \; .$$

The Galois objects for H_g are $S_c = R[t]$ where $g(t) = c$ for some $c \in \mathfrak{m}$. The equation $g(t) = c$ is

$$\frac{1}{b^q}((1 + bt)^q - 1) = c$$

or

$$(1 + bt)^q = 1 + b^q c \; .$$

Setting $z = 1 + bt$, we have $z^q = 1 + b^q c$ and $S_c \cong R[\frac{z-1}{b}]$ with $z^q = 1 + b^q c$.

For the isogeny g we have

$$PH(H_g) \cong \mathfrak{m}_{F_{b^q}}/g(\mathfrak{m}_{F_b}) \ .$$

Given $d \in R$, let $U_d(R) = (1 + dR) \cap U(R)$.

(39.7) PROPOSITION ([**CMS98**], PROP. 1.9).

$$PH(H_g) \cong U_{b^q}(R)/U_b(R)^q \ .$$

PROOF. The isomorphism is induced by the map

$$\mathfrak{m}_{F_{b^q}} \xrightarrow{\psi} U_{b^q}(R)$$

given by $\psi(c) = 1 + b^q c$. It is easy to check that ψ is a homomorphism, and induces a 1-1 homomorphism

$$\overline{\psi} : \mathfrak{m}_{F_{b^q}}/g(\mathfrak{m}_{F_b}) \to U_{b^q}(R)/U_b(R)^q \ .$$

To show onto, let $u = 1 + b^q s \in U_{b^q}(R)$. If $s \in \mathfrak{m}$ then $u \in Im(\overline{\psi})$. If not, find $t \in R$ so that $-s \equiv t^q \pmod{\mathfrak{m}}$ and let $v = 1 + bt$. Then

$$\begin{aligned} uv^q &= (1 + b^q s)(1 + b^q t^q + py) \\ &= 1 + b^q(s + t^q) + b^q y' \\ &= 1 + b^q z \\ &= \psi(z) \end{aligned}$$

for some z, y, y' in $\mathfrak{m}$. So $\overline{\psi}$ is onto. □

For $q = p$ we obtain

$$PH(H_b) \cong U_{b^p}(R)/U_b(R)^p,$$

a result in [**CM94**].

(39.8) REMARKS. (1) In contrast to the case $n = 1$, if H represents the kernel of an isogeny f on a formal group of dimension > 1 it is not clear how to find principal homogeneous spaces for H which are integrally closed. The strategy in the dimension one case of simply letting c be a parameter, thereby generating S generated by a root of an Eisenstein polynomial, doesn't work. The difficulty is compounded by the obvious fact that if H is the kernel of an isogeny of n-dimensional formal groups, then H is naturally presented as an R-algebra generated as algebra by n elements, and by the following observation of Byott:

(39.9) PROPOSITION. *Let L/K be totally ramified with valuation rings $S \supset R$. If S is an H-Hopf Galois extension of R for some finite cocommutative R-Hopf algebra H, then H^* is monogenic.*

PROOF. Since S is an H^*-comodule algebra and an H^*-principal homogeneous space, the map

$$\gamma : S \otimes_R S \to S \otimes_R H^*$$

by $\gamma(s \otimes t) = \sum_{(t)} st_{(0)} \otimes t_{(1)}$ is an S-algebra isomorphism (see (2.19)). Since L/K is totally ramified, $S = R[w] \cong R[x]/(f(x))$ where w is a parameter for S and $f(x)$ is an Eisenstein polynomial of degree $q = [L : K]$ (22.1). Hence $S \otimes H^* \cong S[x]/(f(x))$. If $\mathfrak{m}, \mathfrak{m}_S$ are the maximal ideals of R, S, respectively, then $R/\mathfrak{m} \cong S/\mathfrak{m}_S = k$ and

$$k \otimes_R H^* \cong S/\mathfrak{m}_S[w] \cong k[x]/(f(x)) \cong k[x]/(x^q).$$

Hence $k \otimes_R H^*$ is monogenic. By Nakayama's Lemma, H^* is monogenic. $\square$

A Hopf algebra $H_f = R[[x_1, \dots, x_n]]/(f)$ with f an isogeny on an n-dimensional formal group, $n > 1$, may still be monogenic. An example is $H = RC_p^2$, representing the kernel of the isogeny $[p]_{\mathbb{G}_m^2}$ on the 2-dimensional multiplicative formal group. If K is sufficiently ramified over $\mathbb{Q}_p$, then one can obtain H as the kernel of an isogeny of generic dimension one formal groups, using methods of [**CZ94**] (c.f. also [**Gr95**]), and it is easy to see, using (24.9), that H has principal homogeneous spaces which are integrally closed.

(2) The group $PH(H)$ is often denoted by $Gal(H^*)$ or $H^1(X, G)$ (with $X = Spec(R), G = Spec(H)$) in the literature. For alternative descriptions of the product on $PH(H)$, see [**Ha65**], [**Ch86**], [**TB92**], [**By97b**] and [**Ca98b, Chapter 10**]. The group $PH(H)$ has been computed in a variety of cases other than those cited in (21.4) or described in section 39: see, for example, [**CS69**], [**Or69**], [**Sh69**], [**Wa71**], [**Ch71**], [**Sm72**], [**Ch77**], [**Kr83**], [**Ma84**], [**KM84**], [**Wl87**], [**KM88**], [**Ta90**], [**TB92, III, Sec. 4**], [**Gr92b**], [**Un97**] (using [**SS94**], [**SS96**]) and [**Ca98b**],

§40. **Which is the correct Hopf Galois structure.** In Chapter 2 we found that most Galois extensions L/K have more than one Hopf Galois structure on them. This offers the possibility that the Galois module structure of the valuation ring S of L may be more attractive with respect to a non-classical Hopf Galois structure on L/K than with respect to the classical Galois structure. For example, the associated order for some non-classical structure may be Hopf even though the associated order in KG is not.

In Chapters 8 and 10 we showed that there are rather stringent conditions on L/K in order that its associated order in KG be Hopf. However, in the last section we constructed Kummer extensions for Hopf algebras that are kernels of isogenies of formal groups. These Kummer extensions have a natural non-classical Hopf Galois structure for which the associated order is Hopf.

In this section we present a class of examples found by Byott [**By99b**], Kummer extensions for an isogeny of Lubin-Tate formal groups, which are also classically Galois but for which Byott's necessary congruence condition on ramification numbers from Chapter 8 fails. These examples confirm the value of expanding Galois module theory to look at non-classical Hopf Galois structures on Galois extensions L/K.

Let K be a local field, a finite extension of $\mathbb{Q}_p$, with valuation ring R and parameter π. Let q be the cardinality of the residue field $R/\pi R$. Let

$$h(x) = \pi x + x^q.$$

Let $G(x,y)$ be the Lubin-Tate formal group (section 36) which admits $h(x)$ as an endomorphism; then $h(x) = [\pi](x)$ and

$$[\ ] : R \to End(G)$$

is an isomorphism.

The maximal ideal $\overline{\mathfrak{m}}$ of the val uation ring $\overline{R}$ of a fixed algebraic closure $\overline{K}$ of K is a group with respect to the operation $\alpha +_G \beta = G(\alpha, \beta)$: we denote this group by $\overline{\mathfrak{m}}_G$.

Let

$$G_n = \{x \in \overline{\mathfrak{m}}_G | [\pi^n](x) = 0\},$$

the group of π^n-torsion points of $\overline{\mathfrak{m}}_G$. Let $K_n = K(G_n)$, the field obtained from K by adjoining all π^n-torsion points of $\overline{\mathfrak{m}}_G$.

Since G_n is the set of roots of $[\pi^n](x)$, G_n has cardinality q^n.

If ω_1 is a non-zero root of $[\pi](x) = x(x^{q-1} - \pi)$, then ω_1 is a root of the Eisenstein polynomial $x^{q-1} - \pi$ over K. Hence $K_1 = K[\omega_1]$ is a totally ramified field extension of K with valuation ring $R[\omega_1]$ and ω_1 is a parameter for K_1. Given K_i and ω_i, let ω_{i+1} be a root of $[\pi](x) - \omega_i$, an Eisenstein polynomial over K_i, then $K_i[\omega_{i+1}] = K_{i+1}$ is a totally ramified field extension of K_i of degree q with valuation ring $R_{i+1} = R_i[\omega_{i+1}]$, and ω_{i+1} is a parameter for K_{i+1}. Thus the sequence of totally ramified field extensions $K \subset K_1 \subset K_2, \dots, \subset K_n$, $K_i = K(G_i)$, has a sequence of parameters $\omega_1, \dots, \omega_n$ so that $[\pi](\omega_1) = 0$ and $[\pi](\omega_j) = \omega_{j-1}$ for $j = 2, \dots, n$.

Let $\omega = \omega_n$.

Now $R \cong End(G)$ acts on $\overline{\mathfrak{m}}_G$ and for $\alpha \in R$, $[\alpha](x) \equiv \alpha x \pmod{x^2}$; this action makes $\overline{\mathfrak{m}}_G$ into an R-module (note that

$$[\alpha](x +_G y) = [\alpha](x) +_G [\alpha](y)$$

since $[\alpha]$ is an endomorphism of G; also

$$([\alpha] + [\beta])(x) = [\alpha](x) +_G [\beta](x)$$

because of the way addition is defined in $End(G)$.) Then G_n is an R-submodule of $\overline{\mathfrak{m}}_G$. The kernel of the map from R to G_n given by $\alpha \mapsto [\alpha](\omega)$ is $\pi^n R$, hence

$$R/\pi^n R \cong R\omega \subset G.$$

Since $R/\pi^n R$ has cardinality q^n = the cardinality of G_n, therefore G_n is a cyclic $R/\pi^n R$-module, generated by ω.

For determining ramification numbers, it is useful to note that for $\phi \in G_n$, $\phi \notin G_{n-1}$ iff

$$\phi = [\alpha](\omega) \equiv \alpha\omega \pmod{\omega^2}$$

with α a unit of R; in that case, ϕ is also a parameter for K_n.

Now we consider a class of Galois extensions of K_r.

(40.1) PROPOSITION (BYOTT). *Let* $m > r > 0$ *and let* $c \in \mathfrak{m}_r = R_r\omega_r$. *Let*

$$s = \omega_r +_G [\pi^{m-r}](c) \in R_r.$$

Let $d \in \overline{\mathfrak{m}}_G$ with $[\pi^r](d) = c$, let $z = \omega_{m+r} +_G d$ and let $N_c = K_r[z]$. Then $[\pi^m](z) = s$, $[N_c : K_r] = q^m$, and N_c is totally ramified and Galois over K_r with Galois group $\cong G_m$ where G_m acts via right translation: $z \mapsto z +_G \phi$ for $\phi \in G_m$

PROOF. To show that N_c is a field, observe that s is a parameter for K_r: for since $c \in \mathfrak{m}_r$ and $m > r$, $[\pi^{m-r}](c) \in \mathfrak{m}_r^2$, hence $s \equiv \omega_r \pmod{\omega_r^2 R_r}$. Thus z satisfies the Eisenstein polynomial $[\pi^m](x) - s$ over R_r of degree q^m, and so N_c/K_r is a totally ramified field extension of degree q^m, and the valuation ring of N_c, $\mathcal{O}_c = R_r[z]$.

We show that N_c is normal over K_r with Galois group isomorphic to G_m. For $\phi \in G_m, [\pi^m](z +_G \phi) = [\pi^m](z) +_G [\pi^m](\phi) = s$, so $z +_G \phi$ is a root of $[\pi^m](x) - s$. To see that $z +_G \phi$ is in N_c, since $z = \omega_{m+r} +_G d$ we have

$$\omega_m = [\pi^r](\omega_{m+r}) = [\pi^r](z) -_G [\pi^r](d) = [\pi^r](z) -_G c$$

which is in N_c. Hence $K_m = K[\omega_m] = K[G_m] \subset N_c$. Thus any $\phi \in G_m$ is in N_c. Thus the homomorphism ρ_ϕ from N_c to $\overline{K}$ induced by $\rho_\phi(z) = z +_G \phi$ yields a K_r-automorphism of N_c.

Since $\rho : G_m \to Aut_{K_r}(N_c)$, $\phi \mapsto \rho_\phi$, is 1-1 and both G_m and the set of K_r-algebra isomorphisms from N_c to $\overline{K}$ have cardinality q^m, ρ is onto, N_c is normal over K_r and ρ is an isomorphism of G_m onto the Galois group Γ of N_c/K_r. □

(40.2) THEOREM. *The associated order of $\mathcal{O}_c$ in $K_r[G_m]$ is not Hopf.*

PROOF. By (30.8), it suffices to show that not all ramification numbers of the Galois group Γ are congruent to -1 modulo q^m. In fact, we show that t_1 is not congruent to -1 modulo q^m.

Now z is a parameter for N_c, so to compute ramification groups for Γ it is enough to look at $\rho_\phi(z) - z$ for $\phi \in G_m$. We have

$$\rho_\phi(z) - z = (z +_G \phi) - z$$

and

$$z +_G \phi = G(z, \phi) \equiv z + \phi \pmod{z\phi},$$

hence

$$(z +_G \phi) - z \equiv \phi \pmod{z\phi}.$$

Therefore, if $\phi \in G_m \setminus G_{m-1}$, since N_c/K_m is totally ramified of degree q^r and ϕ is a parameter of K_m,

$$\begin{aligned} ord_{N_c}(\rho(\phi)(z) - z) &= ord_{N_c}(\phi) \\ &= [N_c : K_m] ord_{K_m}(\omega_m) = q^r, \end{aligned}$$

while for $\phi \in G_{m-1}$, $ord_{N_c}(\phi) \geq [N_c : K_{m-1}] = q^{r+1}$. Thus the ramification groups of Γ satisfy

$$\Gamma_1 = \ldots = \Gamma_{q^r-1} = \Gamma,$$

while Γ_{q^r} has index q, and so

$$t_1(\Gamma) = \ldots = t_{q-1}(\Gamma) = q^r - 1.$$

Since $r < m, q^r - 1 \not\equiv -1 \pmod{q^m}$. Thus by Theorem (30.8), the associated order of $\mathcal{O}_c$ in $K_r[G_m]$ is not Hopf. □

On the other hand, N_c is a Kummer extension with respect to the Lubin-Tate formal group G, since it is obtained by adjoining a root of $[\pi^m](x) - s$ for some $s \in \mathfrak{m}_r$. Hence $\mathcal{O}_c$ is a principal homogeneous space for the Hopf algebra $H = R_r[x]/[\pi^m](x)$, and hence is a Hopf Galois extension of R_r with Hopf algebra H^*. It follows that $\mathcal{O}_c$ is a free H^*-module, and hence that H^* is the associated order of $\mathcal{O}_c$ in the dual of the K_r-Hopf algebra $K_r[x]/[\pi^m](x)$.

These examples are examples of normal field extensions where the Hopf Galois structure on the extension given by some other Hopf algebra is more satisfactory than that given by the Galois group.

Byott in [**By99d**] considers Galois extensions L/K of local fields with Galois group $C_p \times C_p$ and obtains a precise description of when the Galois group gives the best Galois module structure, when some other Hopf Galois structure on L/K is better, and when it doesn't matter. These results contrast with results of [**Ch95**], where for $G = C^{p^2}$ the Hopf Galois structures on the extension by any of the p Hopf algebras acting on L/K are equally good: if one Hopf algebra has an associated order which is Hopf, they all do.

More information on the Galois module structure of field extensions constructed via Lubin-Tate formal groups may be found in [**Ta85**], [**Ta87**], [**Sa93**], [**By97a**], [**By97c**] and [**By99a**].

Bibliography

Numbers in parentheses are section(s) in which the reference is cited.

[Ag99] A. Agboola, *On primitive and realisable classes*, (to appear) (1999), (21).

[AS98] N. Andruskiewitsch and H.-J. Schneider, *Lifting of quantum linear spaces and pointed Hopf algebras of order p^3*, J. Algebra **209** (1998), 658-691, (15).

[Bt76] M. Beattie, *A direct sum decomposition for the Brauer group of dimodule algebras*, J. Algebra **43** (1976), 686–693, (39).

[BDG99] M. Beattie, S. Dascalescu and L. Grünenfelder, *On the number of types of finite dimensional Hopf algebras*, Invent. Math. **136** (1999), 1-7, (15).

[Be78] A. Bergé, *Arithmetique d'une extension Galoisienne à groupe d'inertie cyclique*, Ann. Inst. Fourier, Grenoble **28** (1978), 17–44, (12).

[Bg85] G. Bergman, *Everybody knows what a Hopf algebra is*, A. M. S. Contemporary Math **43** (1985), 25–48, (2).

[BF72] F. Bertrandias, M.-J. Ferton, *Sur l'anneau des entiers d'une extension cyclique de degré premier d'un corps local*, Paris, C.R. Acad. Sci. **274** (1972), 1330–1333, (12).

[BBF72] F. Bertrandias, J.-P. Bertrandias, M.-J. Ferton, *Sur l'aneau des entiers d'une extension cyclique de degré premier d'un corps local*, C.R. Acad. Sci. Paris **274** (1972), 1388–1391, (12).

[By91] N. P. Byott, *Some self-dual local rings of integers not free over their associated orders*, Math. Proc. Cambridge Philos. Soc. **110** (1991), 5–10; *Corrigendum*, Math. Proc. Cambridge Philos. Soc. **116** (1994), 569.

[By93] N. Byott, *Cleft extensions of Hopf algebras* I, J. Algebra **157** (1993), 405–429; II, Proc. London Math. Soc. **67** (1993), 277–304, (31).

[By95a] N. P. Byott, *Local Galois module structure and Hopf orders.*, Proceedings of the 2nd Gauss Symposium. Conference A: Mathematics and Theoretical Physics (Munich, 1993), Sympos. Gaussiana, de Gruyter, Berlin, 1995, pp. 249–255.

[By95b] N. P. Byott, *Hopf orders and a generalization of a theorem of L. R. McCulloh*, J. Algebra **177** (1995), 409–433, (0, 14).

[By95c] N. P. Byott, *Tame and Galois extensions with respect to Hopf orders*, Math. Z. **220** (1995), 495–522, (27, 29).

[By96a] N.P. Byott, *Uniqueness of Hopf Galois structure of separable field extensions*, Comm. Algebra **24** (1996), 3217–3228, 3705, (0, 7, 8).

[BL96b] N. P. Byott; G. Lettl, *Relative Galois module structure of integers of abelian fields*, J. Theor. Nombres Bordeaux **8** (1996), 125–141.

[By97a] N. P. Byott, *Associated orders of certain extensions arising from Lubin-Tate formal groups*, J. Theor. Nombres, Bordeaux **9** (1997), 449–462, (40).

[By97b] N. P. Byott, *Two constructions with Galois objects*, Comm. Algebra **25** (1997), 3513–3520, (39).

[By97c] N.P. Byott, *Galois structure of ideals in wildly ramified abelian p-extensions of a p-adic field, and some applications*, J. Theor. Nombres Bordeaux **9** (1997), 201–219, (0, 27, 30, 40).

[By98] N. P. Byott, *Tame realisable classes over Hopf orders*, J. Algebra **201** (1998), 284–316, (20).

[By99a] N. P. Byott, *Integral Galois module structure of some Lubin-Tate extensions*, J. Number Theory **77** (1999), 252–273, (40).

[By99b] N. P. Byott, *Galois module structure and Kummer theory for Lubin-Tate formal groups*, (to appear), Proc. Conf. on Algebraic Number Theory and Diophantine Analysis, Graz, 1998, (0, 40).

[By99d] N. P. Byott, *Integral Hopf Galois structures on extensions of degree p^2*, (to appear) (1999), (32, 40).

[By00] N. P. Byott, *Picard invariants of Galois algebras over dual Larson orders*, Proc. London Math. Soc. **80** (2000), 1-30, (21).

[Ca98a] S. Caenepeel, *Kummer theory for monogenic Larson orders*, Rings, Hopf Algebras and Brauer Groups (Antwerp/Brussels, 1996), Lecture Notes in Pure and Applied Mathematics vol. 197, Marcel Dekker, New York, 1998, pp. 85–102, (21).

[Ca98b] S. Caenepeel, *Brauer Groups, Hopf Algebras and Galois Theory*, K-Monographs in Mathematics, vol 4, Kluwer, Dordrecht, 1998, (21, 39, 40).

[CC99] S. Carnahan, L. N. Childs, *Counting Hopf Galois structures on non-abelian Galois extensions*, J. Algebra **218** (1999), 81–92, (0, 10).

[CHR65] S. Chase, D. Harrison, A. Rosenberg, *Galois theory and Galois cohomology of commutative rings*, Memoirs Amer. Math. Soc. **52** (1965), 15–33, (2).

[CS69] S. Chase, M. Sweedler, *Hopf Algebras and Galois Theory*, Lecture Notes in Math. Vol. 97, Springer-Verlag, New York, 1969, (2, 6).

[Cs76] S. U. Chase, *Infinitesimal group scheme actions on finite field extensions*, Amer. J. Math. **98** (1976), 441-480, (6).

[Ch71] L. N. Childs, *Abelian Galois extensions of rings containing roots of unity*, Illinois J. Math. **15** (1971), 273–280, (21, 39).

[CM74] L. N. Childs, A. Magid, *The Picard invariant of a principal homogeneous space*, J. Pure Appl. Algebra **4** (1974), 273–286, (21).

[Ch77] L. Childs, *The group of unramified Kummer extensions of prime degree*, Proc. London Math. Soc. **35** (1977), 407–422, (21, 39).

[CH86] L.N. Childs, S. Hurley, *Tameness and local normal bases for objects of finite Hopf algebras*, Trans. Amer. Math. Soc. **298** (1986), 763–778, (0, 13, 14).

[Ch86] L.N. Childs, *Products of Galois objects and the Picard invariant map*, Math. J. Okayama Univ. **28** (1986), 29–36, (39).

[Ch87] L.N. Childs, *Taming wild extensions with Hopf algebras*, Trans. Amer. Math. Soc. **304** (1987), 111–140, (0, 12, 24).

[Ch88] L. N. Childs, *On Hopf Galois extensions of fields and number rings,*, Perspectives in Ring Theory (F. van Oystaeyen, L. LeBruyn,, eds.), Kluwer, 1988, pp. 117–128, (0, 26).

[Ch89] L. N. Childs, *On the Hopf Galois theory for separable field extensions*, Comm. Algebra **17** (1989), 809–825, (7, 8).

[CZ94] L. N. Childs, K. Zimmermann, *Congruence-torsion subgroups of dimension one formal groups*, J. Algebra **170** (1994), 929–955, (36, 38, 39).

[Ch94] L. N. Childs, *Galois extensions over local number rings*, Rings, Extensions and Homology (A. Magid, ed.), Marcel Dekker, Inc., New York, 1994, pp. 41–66, (37).

[CM94] L. N. Childs, D. J. Moss, *Hopf algebras and local Galois module theory*, Advances in Hopf Algebras (J. Bergen, S. Montgomery, eds.), Marcel Dekker, Inc., New York, 1994, pp. 1–24, (0, 12, 39).

[Ch95] L. Childs, *Hopf Galois structures on degree p^2 cyclic extensions of local fields*, New York J. Math **2** (1995), 86–102, (0, 32, 40).

[CGMSZ98] L. N. Childs, C. Greither, D. Moss, J. Sauerberg, K. Zimmermann, *Hopf algebras, polynomial formal groups and Raynaud orders*, Memoirs Amer. Math. Soc., vol. 136, 1998, (0, 31, 33, 37).

[Ch98] L. N. Childs,, *Introduction to polynomial formal groups and Hopf algebras*, Memoirs Amer. Math. Soc. **136** (1998), 1–10, (33).

[CMS98] L. N. Childs, D. Moss J. Sauerberg, *Dimension one polynomial formal groups*, Memoirs Amer. Math. Soc. **136** (1998), 11–20, (21, 39).

[CMSZ98] L. N. Childs, D. Moss J. Sauerberg K. Zimmermann, *Dimension two polynomial formal groups and Hopf algebras*, Memoirs Amer. Math. Soc. **136** (1998), 21–54, (33).

[CS98] L. N. Childs, J. Sauerberg, *Degree 2 formal groups and Hopf algebras*, Memoirs Amer. Math. Soc. **136** (1998), 55–90, (15, 21, 38).

[Cn85] T. Chinburg, *Exact sequences and Galois module structure*, Annals of Math. (2) **121** (1985), 351–376, (12).

[CEPT96] T. Chinburg, B. Erez, G. Pappas, M. J. Taylor, *Tame actions of group schemes: integrals and slices*, Duke Math. J. **82** (1996), 270–308, (13).

[CP94] M. Cox Paul, *The Image of the Picard Invariant Map for Hopf Algebra Extensions*, Ph. D. thesis, SUNY at Albany, 1994, (21).

[CR81] C. W. Curtis, I. Reiner, *Methods of Representation Theory*, Wiley, 1981, (0, 2, 13, 14).

[Da82] E. Dade, *The equivalence of various generalizations of group rings and modules*, Math.Z. **181** (1982), 335–344, (2).

[De84] Desrochers, M., *Self-duality and torsion Galois modules in number fields*, J. Algebra **90** (1984), 230–246, (12).

[EM94] G. G. Elder, M. L. Madan, *Galois module structure of the integers in wildly ramified cyclic extensions*, J. Number Theory **47** (1994), 138–174, (12).

[El95] G. G. Elder, *Galois module structure of ideals in wildly ramified cyclic extensions of degree p^2*, Ann. Inst. Fourier (Grenoble) **45** (1995), 625–647; *Corrigendum* **48** (1998), 609-610, (12).

[Fe74] M.-J. Ferton, *Sur l'anneau des entiers d'extensions cyclique d'un corps local*, Bull. Soc. Math. France, Memoire **37** (1974), 69–74, (12).

[FV93] I.B. Fesenko, S.V. Vostokov, *Local Fields and Their Extensions*, Translations of Mathematical Monographs, vol. 121, Amer. Math. Soc., 1993, (24).

[F62] A. Fröhlich, *The module structure of Kummer extensions over Dedekind domains*, J. reine angew. Math. **209** (1962), 39–53, (25).

[F67] A. Fröhlich, *Local Fields*, Algebraic Number Theory (J.W.S. Cassels and A. Frohlich, eds.), Thompson, Washington, D.C., 1967, pp. 1–41, (12).

[Fr68] A. Fröhlich, *Formal Groups*, Lecture Notes in Math., vol. 74, Springer Verlag, New York, 1968, (33, 34, 35).

[Fr83] A. Fröhlich, *Galois Module Structure of Algebraic Integers*, Springer Verlag, Berlin, 1983, (12).

[AS98] S. Gelaki, *Pointed Hopf algebras and Kaplansky's 10th conjecture*, J. Algebra **209** (1998), 637-657, (15).

[Go82] D. Gorenstein, *Finite Simple Groups: An Introduction to Their Classification*, Plenum, New York/London, 1982, (10).

[GP87] C. Greither B. Pareigis, *Hopf Galois theory for separable field extensions*, J. Algebra **106** (1987), 239–258, (0, 6, 7).

[Gr89] C. Greither, *Unramified Kummer extensions of prime power degree*, Manuscripta Math. **64** (1989), 261–290, (25).

[Gr92] C. Greither, *Extensions of finite group schemes, and Hopf Galois theory over a discrete valuation ring*, Math. Z. **220** (1992), 37–67, (0, 2, 24, 31, 38).

[GR92b] C. Greither, *Cyclic Galois Extensions of Commutative Rings*, vol. 1534, Springer LNM, 1992, (25).

[Gr95] C. Greither, *Constructing monogenic Hopf algebras over p-adic rings of integers.*, J. Algebra **174** (1995), 794–800, (39).

[GC98] C. Greither, L. Childs, *p-elementary group schemes–constructions, and Raynaud's theory*, Memoirs Amer. Math. Soc. **136** (1998), 91–117, (0, 15, 21, 31).

[GRRS99] C. Greither, D. Replogle, K. Rubin, A. Srivastav, *Swan modules and Hilbert-Speiser number fields*, J. Number Theory **79** (1999), 164–173, (0).

[Ha65] D. K. Harrison, *Abelian extensions of commutative rings*, Memoirs Amer. Math. Soc. **52** (1965), 1 14, (39).

[Has36] H. Hasse, *Die Gruppe der p-primaren Zahlen fur einen Primteiler p von p*, Crelle **174** (1936), 174–183, (25).

[Ha78] M. Hazewinkel, *Formal Groups and Applications*, Academic Press, New York, 1978, (33, 35).

[Hi97] D. Hilbert, *Die Theorie der algebraischen Zahlen*, Gesammelte Abhandlungen, bd. I, Chelsea, New York, 1965, pp. 63–363, (0).

[Hu84] S. Hurley, *Tame and Galois Hopf objects with normal bases*, Ph. D. thesis, Univ. at Albany, (1984).

[Hu87] S. Hurley, *Galois objects with normal bases for free Hopf algebras of prime degree*, J. Algebra **109** (1987), 292–318, (21).

[Ja87] J. C. Jantzen, *Representations of Algebraic Groups*, Academic Press, Boston, 1987, (3).

[Ja73] G. J. Janusz, *Algebraic Number Fields*, Academic Press, New York and London, 1973, (39).

[Kr83] I. Kersten, *Eine neue Kummertheorie fur zyklische Galoiserweiterungen vom Grad* p^2, Algebra Berichte **45** (1983), (39).

[KM88] I. Kersten and J. Michalichek, *Kummer theory without roots of unity*, J. Pure Applied Algebra **50** (1988), 21-72, (39).

[KM84] I. Kersten and J. Michaelicek, *Applications of Kummer theory without roots of unity*, Methods of Ring Theory (F. van Oystaeyen, D. Reidel, eds.), 1984, pp. 201–205, (39).

[Kl89] A. Klapper, *Selmer group estimates arising from the existence of canonical subgroups*, Comp. Math. **72** (1989), 121–137, (37).

[KO74] M.-A. Knus, M. Ojanguren, *Theorie de la Descent et Algebres d'Azumaya*, Springer LNM No. 389, 1974, (2).

[Kc95] A. Koch, *Cyclic Dieudonne modules, Witt subgroups, and their lifts to characteristic zero*, Ph. D. thesis, Univ. at Albany, 1995, (38).

[Kc96] A. Koch, *Monogenic Hopf algebras*, 10 pages, submitted (1996), (38).

[Kc98a] A. Koch, *Lifting Witt subgroups to characteristic zero*, New York J. Math. **4** (1998), 127–136, (38).

[Kc98b] A. Koch, *Witt subgroups and cyclic Dieudonné modules killed by p*, preprint (1998), (38).

[Ko96] T. Kohl, *Classification of abelian Hopf algebra forms acting on radical extensions*, Ph. D. thesis, University at Albany, 1996.

[Ko98] T. Kohl, *Classification of the Hopf algebra structures on prime power radical extensions*, J. Algebra **207** (1998), 525–546, (0, 9, 11).

[KC76] H.F. Kreimer and P. Cook, *Galois theories and normal bases*, J. Algebra **43** (1976), 115–121.

[Kr82] H.F. Kreimer, *Quadratic Hopf algebras and Galois extensions*, Contemp. Math **13** (1982), 353–361, (1, 15).

[KT81] H.F. Kreimer, M. Takeuchi, *Hopf algebras and Galois extensions of an algebra*, Indiana Univ. Math. J. **30** (1981), 675–692.

[Lm68] T. Y. Lam, *Induction theorems for Grothendieck groups and Whitehead groups of finite groups*, Ann. Sci. Ecole Norm. Sup. **4** (1968), 91–148, (13).

[LL93] T. Y. Lam, D. B. Leep, *Combinatorial structure on the automorphism group of* S_6, Expositiones Math. **11** (1993), 289–308, (10).

[Lg84] S. Lang, *Algebra*, Addison-Wesley, 1984, (0, 35).

[LS69] R. G. Larson, M. E. Sweedler, *An associative orthogonal bilinear form for Hopf algebras*, Amer. J. Math. **91** (1969), 75–93, (3).

[La72] R. G. Larson, *Orders in Hopf algebras*, J. Algebra **22** (1972), 201–210.

[La76] R.G. Larson, *Hopf algebras defined by group valuations*, J. Algebra **38** (1976), 414–452, (0, 16, 31).

[Lz55] M. Lazard, *Sur les groupes de Lie formels a un parametre*, Bull. Soc. Math. France **83** (1955), 251–274.

[Lz63] M. Lazard, *Les zeros des fonctions analytiques d'une variable sur un corps value complet*, Publ. Math. IHES **14** (1963), 223–251.

[Le59] H.W. Leopoldt, *Uber die Hauptordnung der ganzen Elemente eines abelschen Zahlkorpers*, J. reine angew. Math. **201** (1959), 119–149, (0, 12).

[Lt98] G. Lettl, *Relative Galois module structure of integers of local abelian fields*, Acta Arithmetica **85** (1998), 235–248, (0, 12).

[Lu64] J. Lubin, *One-parameter formal Lie groups over p-adic integer rings*, Annals of Math. **80** (1964), 464–484.

[LT65] J. Lubin, J. Tate, *Formal complex multiplication in local fields*, Ann. of Math. (2) **81** (1965), 380–387, (36).

[LT66] J. Lubin and J. Tate, *Formal moduli for one-parameter formal Lie groups.*, Bull. Soc. Math. France **94** (1966), 49–59, (36).

[Lu67] J. Lubin, *Finite subgroups and isogenies of one-parameter formal Lie groups.*, Ann. of Math. (2) **85** (1967), 296–302, (37).

[Lu79] J. Lubin, *Canonical subgroups of formal groups*, Trans. Amer. Math. Soc. **251** (1979), 103–127, (36).

[Lu81] J. Lubin, *The local Kronecker-Weber theorem*, Trans. Amer. Math. Soc. **267** (1981), 133–138, (36).

[Ma84] D. Maurer, *Stickelberger's criterion, Galois algebras and tame ramification in algebra number fields*, J. Pure Appl. Algebra **33** (1984), 281–293, (39).

[Mz70] B. Mazur, *Local flat duality*, Amer. J. Math. **92** (1970), 343-361, (0, 37).

[MR70] B. Mazur, L. Roberts, *Local Euler characteristics*, Inv. Math **9** (1970), 201–234, (37, 39).

[MBD16] G. Miller, H. Blichtfeldt, L.E. Dickson, *Theory and Applications of Finite Groups*, Dover, New York, 1961 (1916 original), (9).

[Mt93] S. Montgomery, *Hopf Algebras and Their Actions on Rings*, CBMS Regional Conference Series in Mathematics, vol. 82, Amer. Math. Soc., Providence, R. I., 1993, (0, 1, 4).

[Mo94] D. J. Moss, *Kummer Theory of Formal Groups*, Ph. D. thesis, University at Albany, 1994, (0, 39).

[Mo96] D. Moss, *Hopf Galois Kummer theory of formal groups*, Amer. J. Math. **118** (1996), 301–318, (0, 39).

[No31] E. Noether, *Normalbasis bei Körpern ohne höhere Verzweigung,*, J. reine angew. Math. **167** (1931), 147–152, (0, 12).

[Oo67] F. Oort, *Embeddings of finite group schemes into abelian schemes*, mimeographed notes, Bowdoin College (1967), (37).

[OM68] F. Oort, D. Mumford, *Deformations and liftings of finite commutative group schemes*, Inv. Math. **5** (1968), 317–334, (37, 38).

[Oo74] F. Oort, *Dieudonne modules of finite local group schemes*, Indag. Math. **36** (1974), 284–292, (37).

[Or69] M. Orzech, *A cohomological description of abelian Galois extensions*, Trans. Amer. Math. Soc. **137** (1969), 481–499, (39).

[Par71] B. Pareigis, *When Hopf algebras are Frobenius algebras*, J. Algebra **18** (1971), 588–596, (3).

[Pa73] B. Pareigis, *On K-Theory of Hopf Algebras of Finite Type*, Algebra Berichte **3** (1973), (13).

[Pa90] B. Pareigis, *Forms of Hopf algebras and Galois theory*, Banach Center Publications **26** (1990), 75–93, (8).

[Pa98] B. Pareigis, *Fourier transforms over finite quantum groups*, Seminarberichte FB Mathematik FernUniversität Hagen **63** (1998), 561–570, (3).

[PS80] B. Parshall, L. Scott, *An imprimitivity theorem for algebraic groups*, Proc. Nederl. Ak. Wet. **A83** (1980), 39-47, (3).

[Ra74] M. Raynaud, *Schémas en groupes de type* $(p, \ldots, p)$, Bull. Soc. Math. France **102** (1974), 241–280, (3, 15).

[Rb92] J. Roubaud, *Schémas en Groupes Finis sur un Anneau de Valuation Discrète et Systèmes de Honda Associés*, Université de Paris-Sud, Paris, 1992, (38).

[Ro73] L. Roberts, *The flat cohomology of group schemes of rank p*, Amer. J. Math. **95** (1973), 688–702, (21).

[Ro82] D. J. S. Robinson, *A Course in the Theory of Groups*, Springer-Verlag, New York, 1982, (8).

[Sa93] J. J. Sauerberg, *Kummer Theory for Lubin-Tate Formal Groups*, Ph.D. thesis, Brown Univ., 1993, (40).

[Sch77] H.-J. Schneider, *Cartan matrix of liftable finite group schemes*, Comm. Algebra **5** (1977), 795–819, (0, 13).

[Sch90] H.-J. Schneider, *Principal homogeneous spaces for arbitrary Hopf algebras*, Israel J. Math. **72** (1990), 167–195, (2).

[Sch94] H.-J. Schneider, *Hopf Galois extensions, crossed products and Clifford theory*, Advances in Hopf Algebras (J. Bergen, S. Montgomery, eds.), Marcel Dekker, Inc., New York, 1994, pp. 267–298, (2).

[SS94] T. Sekiguchi, N. Suwa, *Theories de Kummer-Artin-Schreier-Witt*, C.R. Acad. Sci. Paris **319, I** (1994), 105–110, (39).

[SS96] T. Sekiguchi, N. Suwa, *On the unified Kummer-Artin-Schreier-Witt theory*, preprint (1996), (39).

[Se67] J. P. Serre, *Local class field theory*, Algebraic Number Theory (J. W. S. Cassels, A. Fröhlich, ed.), Thompson, Washington, D. C.,, 1967, pp. 129–162, (36, 39).

[CL] J.-P. Serre, Corps Locaux,, *Local Fields*, Hermann Paris, 1962, Springer, 1979, (22, 27, 28, 30, 32).

[Sh69] S. S. Shatz, *Principal homogeneous spaces for finite group schemes*, Proc. Amer. Math. Soc. **22** (1969), 678–680, (39).

[Sh86] S. S. Shatz, *Group schemes, formal groups and p-divisible groups*, Arithmetic Geometry (G. Cornell, J. H. Silverman, ed.), Springer-Verlag, New York, 1986, pp. 29–78, (15).

[Sm72] C. Small, *The group of quadratic extensions*, J. Pure Appl. Algebra **2** (1972), 83–105, 395, (39).

[Sm97] Harold H. Smith, III, *Constructing Hopf orders in Elementary Abelian Group Rings*, Ph.D. thesis, SUNY at Albany, 1997, (0, 31).

[ST90] A. Srivastav, M.J. Taylor, *Elliptic curves with complex multiplication and Galois module structure*, Invent. Math. **99** (1990), 165–184, (26).

[Sw60] R.G. Swan, *Induced representations and projective modules*, Annls. of Math. **71** (1960), 552–578, (0, 13).

[Sw69] M. Sweedler, *Hopf Algebras*, W.A. Benjamin, New York, 1969, (0, 1, 3, 4, 37).

[Tt67] J. Tate, *p-divisible groups*, Proc. of a Conference on Local Fields, NUFFIC Summer School in Dreibergen, Springer, Berlin, 1967, pp. 158 183, (37).

[TO70] J. Tate and F. Oort, *Group schemes of prime order*, Ann. Scient. Ec. Norm. Sup. **3** (1970), 1–21, (0, 1, 15).

[Ta85] M.J. Taylor, *Formal groups and the Galois module structure of local rings of integers*, J. reine angew. Math. **358** (1985), 97–103, (40).

[Ta86] M.J. Taylor, *A note on Galois modules and group schemes*, informal manuscript (1986), (39).

[Ta87] M.J. Taylor, *Hopf structure and the Kummer theory of formal groups*, J. reine angew. Math. **375/376** (1987), 1–11, (26, 40).

[Ta88] M.J. Taylor, *Mordell-Weil groups and the Galois module structure of rings of integers*, Illinois J. Math. **32** (1988), 428–452, (26).

[Ta90] M.J. Taylor, *Resolvendes et espaces homogènes principaux de schemas en groupe*, Sem. de Theorie des Nombres, Bordeaux **2** (1990), 255–271, (39).

[TB92] M.J. Taylor, N.P. Byott, *Hopf orders and Galois module structure*, DMV Seminar 18, Birkhauser Verlag, Basel, 1992, pp. 154–210, (39).

[Ta92] M. Taylor, *The Galois module structure of certain arithmetic principal homogeneous spaces*, J. Algebra **153** (1992), 203–214.

[Ta95] M. J. Taylor, *On the Galois module structure of rings of integers of wild abelian extensions*, J. London Math. Soc. (2) **52** (1995), 73–87, (12).

[Ts97] M.-Y. Tse, *Hopf Algebra Actions on Elementary Abelian Extensions of Degree p^2*, Ph. D. thesis, SUNY at Albany, 1997, (8).

[Ul81] K.-H. Ulbrich, *Vollgraduierte Algebren*, Abh. Math. Sem. Univ. Hamburg **51** (1981), 136–148.

[Un92] R. G. Underwood, *Hopf Algebra Orders Over a Complete Discrete Valuation Ring, Their Duals, and Extensions of R-Groups*, Ph. D. thesis, Univ. at Albany, 1992.

[Un94] R. G. Underwood, *R-Hopf algebra orders in KC_{p^2}*, J. Algebra **169** (1994), 418–440, (31).

[Un96] R. G. Underwood, *The valuative condition and R-Hopf algebra orders in KC_{p^3}*, Amer. J. Math. **118** (1996), 701–743, (0, 21, 31).

[Un97] R. G. Underwood, *The group of Galois extensions over orders in KC_{p^2}*, Trans. Amer. Math. Soc. **349** (1997), 1503-1514, (39).

[Un98] R. G. Underwood, *The structure and realizability of R-Hopf algebra orders in KC_{p^3}*, Comm. Algebra **26** (1998), 3447–3462, (32).

[Un99] R. G. Underwood, *Isogenies of polynomial formal groups*, J. Algebra **212** (1999), 428–459, (38).

[Vo78] S. V. Vostokov, *Ideals of an abelian p-extension of a local field as Galois modules*, J. Soviet Math. **9** (1978), 299–317, (28).

[Wa71] W. C. Waterhouse, *Principal homogeneous spaces and group scheme extensions*, Trans. Amer. Math. Soc. **153** (1971), 181–189, (39).

[Wa79] W. C. Waterhouse, *Introduction to Affine Group Schemes*, Springer-Verlag, New York, 1979, (2, 5, 13, 15).

[Wa87] W. C. Waterhouse, *A unified Kummer-Artin-Schreier sequence*, Math. Ann. **277** (1987), 447–451.

[Wa88] W. C. Waterhouse, *Tame Objects for finite commutative Hopf algebras*, Proc. Amer. Math. Soc. vol 103 (1988), 354–356.

[Wa92] W. C. Waterhouse, *Normal basis implies Galois for coconnected Hopf algebras*, preprint (1992), (14).

[We90] David Weinraub, *Cofinite Induction and Noether's Theorem for Hopf Orders in Group Algebras*, Ph. D. thesis, University at Albany, 1990.

[We94] D. Weinraub, *Noether's theorem for Hopf orders in group algebras*, Trans. Amer. Math. Soc. **342** (1994), 563–574, (13).

[We95] D. Weinraub, *Cofinite induction and Artin's theorem for Hopf algebras*, J. Algebra **175** (1995), 799–810, (13).

[Wl87] T. Wyler, *Torsors under abelian p-groups*, J. Pure Appl. Algebra **45** (1987), 273–286, (39).

[Wy69] B.F. Wyman, *Wildly ramified gamma extensions*, Amer. J. Math. **91** (1969), 135–152, (24).

[Zu94] Y. Zhu, *Hopf algebras of prime dimension*, Int. Math. Research Notices **1** (1994), 53–59, (15).

Index